State of the World 2000

Other Norton/Worldwatch Books

State of the World 1984 through *1999* (an annual report on
   progress toward a sustainable society)
   Lester R. Brown et al.

Vital Signs 1992 through *1999* (an annual report on the
   environmental trends that are shaping our future)
   Lester R. Brown et al.

Saving the Planet
   Lester R. Brown
   Christopher Flavin
   Sandra Postel

How Much is Enough?
   Alan Thein Durning

Last Oasis
   Sandra Postel

Full House
   Lester R. Brown
   Hal Kane

Power Surge
   Christopher Flavin
   Nicholas Lenssen

Who Will Feed China?
   Lester R. Brown

Tough Choices
   Lester R. Brown

Fighting for Survival
   Michael Renner

The Natural Wealth of Nations
   David Malin Roodman

Life Out of Bounds
   Chris Bright

Beyond Malthus
   Lester R. Brown
   Gary Gardner
   Brian Halweil

Pillar of Sand
   Sandra Postel

State of the World
2000

A Worldwatch Institute Report on
Progress Toward a Sustainable Society

Project Director	Lester R. Brown
Associate Project Directors	Christopher Flavin, Hilary French
Senior Fellow	Sandra Postel
Editor	Linda Starke

Contributing Researchers

Janet N. Abramovitz   Brian Halweil

Lester R. Brown   Ashley Tod Mattoon

Chris Bright   Anne Platt McGinn

Seth Dunn   Molly O'Meara

Christopher Flavin   Sandra Postel

Hilary French   Michael Renner

Gary Gardner

W · W · NORTON & COMPANY

NEW YORK   LONDON

The STATE OF THE WORLD and WORLDWATCH INSTITUTE trademarks are registered in the U.S. Patent and Trademark Office.

The views expressed are those of the authors and do not necessarily represent those of the Worldwatch Institute; of its directors, officers, or staff; or of its funders.

The text of this book is composed in Galliard, with the display set in Franklin Gothic and Gill Sans. Book design by Elizabeth Doherty; composition by Worldwatch Institute; manufacturing by the Haddon Craftsmen, Inc.

First Edition

ISBN 0-393-04848-9
ISBN 0-393-31998-9 (pbk)

W. W. Norton & Company, Inc., 500 Fifth Avenue, New York, N.Y. 10110
www.wwnorton.com

W. W. Norton & Company Ltd., 10 Coptic Street, London WC1A IPU

1 2 3 4 5 6 7 8 9 0

⊛ This book is printed on recycled paper.

Acknowledgments

Only the combined efforts of a dedicated team of researchers, writers, editors, communications specialists, and support staff—all backed by a generous group of funders—could produce our annual review of planetary health. Worldwatch is blessed with just such a group of individuals.

To begin, we are indebted to the foundation community, long an important mainstay of our work. For providing specific funding for *State of the World*, we are especially grateful to the John D. and Catherine T. MacArthur Foundation and the U.N. Population Fund. Our deep appreciation also goes to the following foundations for providing both unrestricted and program support this past year: the Geraldine R. Dodge, Ford, William and Flora Hewlett, W. Alton Jones, Charles Stewart Mott, Curtis and Edith Munson, David and Lucile Packard, Summit, Turner, Wallace Genetic, Weeden, Wege, and Winslow Foundations; Rockefeller Financial Services; and the Wallace Global Fund.

We are indebted, as well, to the Institute's individual donors, starting with our Council of Sponsors—Tom and Cathy Crain, Vicki and Roger Sant, Robert Wallace, and Eckart Wintzen—who generously contribute $50,000 each year to support the Institute. We also recognize the contributions of the many Friends of Worldwatch whose ongoing support and broadening ranks demonstrate that educating the world on environmental issues strikes an increasingly responsive chord. The solid expression of confidence in our work from these two groups of individual donors is a great boost to our morale as well as a source of flexibility in our research program.

Many chapter authors were assisted in their work by an able and enthusiastic group of colleagues. Staff researcher Lisa Mastny tracked down and managed a complicated flow of information for Chapter 10. The 1999 interns—Gerard Alleng, Shivani Chaudhry, and Sarah Porter—were instrumental in feeding essential information to researchers. Gerard ably compiled and analyzed statistical data and other information for Chapter 8; Shivani conscientiously assembled obscure information for Chapters 7 and 9; and Sarah lent her ample energies to data collection for Chapters 2, 5, and 6.

The Institute's library staff is also indispensable in this information gathering process. Research librarian Lori Brown and library assistant Anne Smith tracked down books and articles from around the world and organized the daily inflow of information into a manageable form for researchers. When Anne left in August for Berkeley, office assistant Jonathan Guzman stepped in at our busy chapter draft stage and kept the information flowing without a pause.

Once the research and writing are finished, each chapter benefits from an extensive review process that helps ensure that our findings are accurately and clearly presented. At this year's internal review meeting, authors, researchers, and interns were joined by *World Watch* magazine staffers Ed Ayres and Curtis Runyan, Worldwatch alum John Young, and researchers Payal Sampat and David Roodman for a day-long session of friendly bloodletting. We are grateful for their insights.

Reviews also come in from outside experts, who are especially generous with their time and expertise. We are grateful to the following for their comments: Linda Brinbaum, Kate Clancy, Marc Cohen, Pat Costner, Clif Curtis, Rafael Flores, Thomas Goreau, Gail Gorham, Maryanne Grieg-Gran, Jim Han, Ray Hayes, Allen Hershkowitz, Susan Kinsella, John Klugness, Yoichi Kuroda, Tim Lang, Orie Loucks, Norman Myers, Bruce Nordman, Karen Perry, Barry Popkin, Jeff Privette, Sonya Rabeneck, Jon Ranson, Roger Rowell, Arnold Schecter, Frances Seymour, Tina Skaar, Ken Skog, Maureen Smith, Marguerite Sykes, John Smith, Ted Smith, Michael Stanley Jones, Chad Tolman, Erwin Villiger, Ted Wagner, and John Young.

Chapters mature still further under the guidance of editor Linda Starke, who skillfully crafts the disparate writing styles of multiple authors into a coherent whole. Under Art Director Elizabeth Doherty's talented hand, the edited chapters are laid out and figures and tables are designed, so that the edited chapters begin to look like a book. This year, Liz also undertook the daunting task of redesigning the whole publication, and we are delighted with the results. Ritch Pope rounds out the process through his work in constructing the index.

State of the World is influential only if the book and its message reach those who need it, a challenge that is ably handled by our communications staff. Mary Caron organizes the *State of the World* press briefing, and ensures that the media is plugged in to our message. Amy Warehime helped administer the Institute's sales and marketing efforts until October 1999, when Liz Hopper came aboard and stepped into this role. Dick Bell oversees these operations as well as the work on our Web site—our newest communications tool, whose development is now the immediate responsibility of Christine Stearn. The efforts of the communications team typically result in a flurry of mail, telephone, fax, and e-mail orders, which are processed efficiently by Millicent Johnson, Joseph Gravely, and Sharon Lapier.

All these efforts would be impossible without strong administrative backing. Reah Janise Kauffman greases the managerial wheels through a broad range of responsibilities, including fundraising, managing relationships with our network of foreign publishers, and assisting the president. She is ably assisted in these diverse activities by Mary Redfern. Barbara Fallin oversees the office operations and handles the Institute's finances, ensuring that the staff is paid, healthy, and happy. Suzanne Clift backs up the work of the vice presidents for research and others on the research staff so that travel and other logistical matters unfold without a hitch.

A special word of thanks is due our overseas colleagues for their extraordinary efforts to disseminate our vision of a sustainable future. *State of the World* is available in some 30 languages due to the often heroic efforts of a host of publishers, nongovernmental groups, and individuals who are dedicated to translating and promoting the Institute's research. For a list of these publishers, see <www.worldwatch.org/foreign/index.html>. In particular, we would like to acknowledge the work of Magnar Norderhaug, who for many years has arranged for the publication of *State of the World* in the Nordic countries and an intensive week of press launchings in each nation; Hamid Taravaty and Farsaneh Bahar, who publish *State of the World* and many of our other books in Persian in Iran; Soki Oda, who manages to get all of the Institute's books and the magazine published in Japanese; Gianfranco Bologna, head of the World Wildlife Fund–Italy, who teams with Edizione Ambiente to produce *State of the World* and *Vital Signs*; and Eduardo Athayde of the Atlantic Forest Open University in Bahia, Brazil, who this year launched the Brazilian edition of *State of the World* and *World Watch* magazine.

The superhuman efforts of these individuals to get this book out quickly are matched by our friends and colleagues at W.W. Norton & Company. Amy Cherry, Andrew Marasia, and Nomi Victor provide unflagging help with the production and promotion of *State of the World*, as well as all the other Worldwatch books. And one reason our books are used as texts in more than 1,000 U.S. colleges and universities is the excellent work done by Norton's college department.

We close this year's acknowledgments by noting with delight the addition of the newest members of the Worldwatch family: Benjamin Huy Hoàng Phạm Roodman and Ileanna Là Tavia Guzman. It's been awhile since we have had the pleasure of having babies around to remind us of who will inherit the Earth.

Lester R. Brown, Christopher Flavin,
and Hilary French

Contents

Acknowledgments   vii

List of Tables and Figures   xiii

Foreword   xvii

1   **Challenges of the New Century**   3
Lester R. Brown
    Environmental Trends Shaping the
      New Century
    Replacing Economics with Ecology
    Crossing the Sustainability Threshold
    Crossing the Decline Threshold
    Two Keys to Regaining Control of
      Our Destiny

2   **Anticipating Environmental**   22
"Surprise" *Chris Bright*
    Tropical Rainforests: The Inferno
      Beneath the Canopy
    Coral: Death in the Warming Seas
    The Atmosphere: An Invisible
      Confluence of Poisons
    An Agenda for the Unexpected

3   **Redesigning Irrigated**   39
Agriculture *Sandra Postel*
    Mounting Water Deficits
    The New Water Wars
    The Productivity Frontier

    Expanding Irrigation to Poor Farmers
    The Policy Challenge

4   **Nourishing the Underfed**   59
and Overfed *Gary Gardner*
and Brian Halweil
    A Malnourished World
    The Roots of Hunger
    The Creation of Overeating
    Diet and Health
    Societal Costs of Poor Diet
    Nutrition First

5   **Phasing Out Persistent**   79
Organic Pollutants *Anne Platt*
McGinn
    The World of POPs
    Routes of Exposure and
      Environmental Fate
    Health Consequences
    The Policy Response to POPs
    Retooling Regulations, Business, and
      Agriculture

6   **Recovering the Paper**   101
Landscape *Janet N. Abramovitz*
and Ashley T. Mattoon
    The Paper Landscape
    Uncovering the Costs of Paper
    Reducing the Burden of Production

Trimming the Costs of Consumption
Designing a Sustainable Paper
   Economy

7 Harnessing Information   *121*
Technologies for the Environment
Molly O'Meara

  An Expanding Global Network
  Squandering or Saving Natural
    Resources?
  Monitoring and Modeling
  Networking for Sustainable
    Development
  Information Tools for a Healthy
    Planet

8 Sizing Up Micropower   *142*
Seth Dunn and Christopher Flavin

  Miniaturized Machines
  Cool Power
  Is Smaller Better?
  Remaking Market Rules
  How Far, How Fast?

9 Creating Jobs, Preserving   *162*
the Environment *Michael Renner*

  The World of Work
  Boosting Resource Productivity
  Environment Policy: Job Killer or
    Creator?
  Restructuring Energy, Creating Jobs
  Durability and Remanufacturing
  Shifting Taxes
  Rethinking Work

10 Coping with Ecological   *184*
Globalization *Hilary French*

  Trading on Nature
  The WTO Meets the Environment
  Greening the International Financial
    Architecture
  Innovations in Global Environmental
    Governance

Notes   203

Index   263

Worldwatch Database Disk

The data from all graphs and tables contained in this book, as well as from those in all other Worldwatch publications of the past two years, are available on 3½-inch disks for use with PC or Macintosh computers. This includes data from the State of the World *and* Vital Signs *series of books, other Norton/Worldwatch books, Worldwatch Papers, and* World Watch *magazine. The data (in spreadsheet format) are provided as Microsoft Excel 5.0/95 workbook (*.xls) files. Users must have spreadsheet software installed on their computer that can read Excel workbooks for Windows. Information on how to order the Worldwatch Database Disk can be found at the back of this book.*

Visit our Web site at www.worldwatch.org

List of Tables and Figures

Tables

1 Challenges of the New Century

1–1 Current and Projected Populations in Selected Developing
        Countries, 1999 and 2050   15
1–2 Trends in Global Energy Use, by Source, 1990–98   17

2 Anticipating Environmental "Surprise"

2–1 Three Types of Environmental Surprise   24
2–2 Some Major Causes of Discontinuities and Synergisms   25

3 Redesigning Irrigated Agriculture

3–1 Irrigated Area in the Top 20 Countries and the World, 1995   40
3–2 Water Deficits in Key Countries and Regions, Mid-1990s   44
3–3 Populations in Selected Hot Spots of Water Dispute, 1999, with
        Projections to 2025   47
3–4 Menu of Options for Improving Irrigation Water Productivity   49
3–5 Efficiencies of Selected Irrigation Methods, Texas High Plains   51
3–6 Low-Cost Irrigation Methods for Small Farmers   53

4 Nourishing the Underfed and Overfed

4–1 Types and Effects of Malnutrition, and Number Affected Globally, 2000   60
4–2 Share of Children Who Are Underweight and Adults Who Are Overweight,
        Selected Countries, Mid-1990s   61
4–3 Underweight Children in Developing Countries, 1980 and 2000   62
4–4 Health Problems That Could Be Avoided Through Dietary Change   69
4–5 Costs of a Range of Nutrition Interventions   75

5 Phasing Out Persistent Organic Pollutants

5–1 Production and Use of the "Dirty Dozen" POPs   82

5–2   Known or Estimated Dioxin and Furan Air Emissions, Selected Countries,
        Total and Per Capita, 1995   *84*
5–3   Regulatory Status of the "Dirty Dozen" POPs   *91*
5–4   Status, Strengths, and Weaknesses of Selected International Conventions
        Complementary to the Global POPs Treaty   *94*
5–5   Alternatives to the "Dirty Dozen" POPs, by Use or Source   *98*

6 Recovering the Paper Landscape

6–1   Top 10 Producers of Paper and Paperboard and Share of World
        Production, 1997   *102*
6–2   Top 10 Consumers of Paper and Paperboard and Share of World
        Consumption and Population, 1997   *102*

7 Harnessing Information Technologies for the Environment

7–1   Key Events in Communications History   *124*
7–2   Internet Hosts by Region, January 1999   *125*
7–3   Mapping the Environment: Global Digital Data Sets on the Internet   *132*
7–4   Selected Global Environmental Networks   *139*

8 Sizing Up Micropower

8–1   Power Plant Downsizing   *145*
8–2   Combustion-Based Micropower Options   *146*
8–3   Non-Combustion-Based Micropower Options   *148*
8–4   Seven Hidden Benefits of Micropower   *152*
8–5   Eight Barriers to Micropower   *156*
8–6   Ten Micropower Market Accelerators   *158*
8–7   Small-Scale Power Applications, Developing Countries   *159*

9 Creating Jobs, Preserving the Environment

9–1   Unemployment Rates by Region and Selected Countries, 1987 and 1997   *166*
9–2   Value-Added, Employment, Energy Use, and Toxics Releases, Selected
        U.S. Manufacturing Industries, Mid-1990s   *169*
9–3   Total Labor Force, Industrial and Developing Countries, by Economic
        Sector, 1960 and 1990   *170*
9–4   Job Impact Findings, Selected Studies on Climate Policy   *175*
9–5   Likely Employment Effects of Durable, Repairable, and Upgradable Products   *178*
9–6   Average Workweek for Manufacturing Employees, Selected Industrial
        Countries, 1950–98   *181*

Figures

1 Challenges of the New Century

1–1	Average Temperature at Earth's Surface, 1950–98	*6*

2 Anticipating Environmental "Surprise"

2–1	Number of Reef Provinces with Moderate to Severe Bleaching, 1979–98	*30*
2–2	World Sulfur and Nitrogen Emissions from Fossil Fuels, 1970–94	*33*

4 Nourishing the Underfed and Overfed

4–1	World Vegetable Oil Consumption Per Person, 1964–99	*66*

5 Phasing Out Persistent Organic Pollutants

5–1	U.S. Synthetic Organic Chemical Production, 1966–94	*81*
5–2	Pesticide Exports from the Top 30 Countries, Selected Years	*87*

6 Recovering the Paper Landscape

6–1	World, Industrial, and Developing Country Paper Consumption, 1961–97, with Projections to 2010	*103*
6–2	World Paper Production, Major Grades, 1961–97	*104*
6–3	World Fiber Supply for Paper, 1961–97	*106*

7 Harnessing Information Technologies for the Environment

7–1	Growth of Information Technologies	*123*
7–2	Density of Telephone Penetration, by Region	*125*

8 Sizing Up Micropower

8–1	Example of a Centralized Power System	*154*
8–2	Example of a Distributed Power System	*155*

9 Creating Jobs, Preserving the Environment

9–1	Selected Factor Productivities in U.S. Manufacturing, 1950–96	*168*
9–2	U.S. Goods and Services-Related Jobs, 1950–99	*170*
9–3	U.S. Coal Mining, Output and Jobs, 1958–98	*173*

10 Coping with Ecological Globalization

10–1	World Trade in Forest Products, 1970–98	*186*
10–2	World Exports of Agricultural Products, 1970–97	*187*
10–3	World Fish Exports, 1970–97	*188*
10–4	Private Capital Flows to Developing Countries, 1970–98	*195*
10–5	International Environmental Treaties, 1921–98	*201*

Foreword

When the first edition of *State of the World* was published in 1984, the year 2000 seemed a long way off. Record rates of population growth, soaring oil prices, debilitating levels of international debt, and extensive damage to forests from the new phenomenon of acid rain were all causing concern. By the end of the century, we then hoped, the world would be well on the way to creating a sustainable global economy.

Far from it. As we complete this seventeenth *State of the World* report, we are about to enter a new century having solved few of these problems, and facing even more profound challenges to the future of the global economy. The bright promise of a new millennium is now clouded by unprecedented threats to humanity's future.

Just weeks before the new millennium began, world population reached 6 billion. But in a world where 1.2 billion people are hungry, 1.2 billion lack access to clean water, and nearly 1 billion adults are illiterate, passing this demographic landmark was not a cause for celebration. India, which reached the 1 billion mark in August, joining China as the only other member of the 1-billion club, is one of many countries still failing to take its population problems—or related social concerns—seriously. Some 53 percent of Indian children are undernourished and underweight, yet the government has used a disconcertingly large share of its resources to become a nuclear power. As a result, India now has the distinction of having a nuclear arsenal to protect the largest concentration of impoverished people on Earth.

In this *State of the World* report, we look at food through two different lenses: water and nutrition. In addition to the traditional problems associated with irrigation, such as waterlogging, salting, and silting, some of which led to the demise of earlier civilizations, we now face a new threat: aquifer depletion. Only within the last half-century have we acquired the ability to use powerful diesel and electric pumps to empty aquifers in a matter of decades.

We estimate the overpumping of aquifers in key regions where data are available at 160 billion tons. Using the rule of thumb that 1,000 tons of water are needed to produce one ton of grain, this suggests that if overpumping stopped, world grain production would fall by at least 160 million tons—enough to feed 480 million of the world's 6 billion people.

As we look at the global food situation at the dawn of a new century, the symptoms of overnutrition and undernutrition have converged: both cause increased susceptibility to illness, reduced productivity, and reduced life expectancy. Our tabulations show that, for the first time, the number of people in the world who are overnourished and overweight rivals the number who are undernourished and underweight—some 1.2 billion. Americans seem particularly sensitive to their weight problems: 400,000 liposuction (fat removal) surgical procedures were performed in the United States last year.

One of the most disturbing trends as we begin a new century is the way in which infectious disease—the scourge of earlier centuries—is devastating societies in sub-Saharan Africa. A reported 23 million Africans are beginning the new century with a death sentence imposed by HIV. For the first time in the modern era, life expectancy—a sentinel indicator of development—is declining for a major region, threatening the economic future of the 800 million people in sub-Saharan Africa. The HIV epidemic is reducing life expectancy in some countries by 20 years or more. In Zimbabwe, before the onslaught of AIDS, life expectancy was 65 years; by 2010, it is projected to fall to 39 years. With more than one fifth of the adult population infected by the virus, Africa is losing many of the agronomists, engineers, and teachers needed to sustain development.

In Chapter 2 of this year's *State of the World*, we look at synergies among environmental and social trends, and the growing potential for "surprises" as their complexity exceeds our analytical capacity. We do not know when the surprises will arrive or what form they will take. We only know that the potential number and magnitude is much greater than ever before. It could take the form of unexpected effects on

human health or an acceleration of biological extinction. We know, for example, that the bodies of everyone reading this book contain measurable amounts of some 500 chemicals that did not exist a century ago. What the health effects of these may be, much less of some of them acting synergistically, no one knows.

Some of the biggest surprises may come from climate change. In the past year alone, evidence that Earth's climate is getting warmer has mounted dramatically. Some 3,000 square kilometers of the west Antarctica ice sheet broke up within a one-year period. One iceberg that broke free, some 65 kilometers in length, is still floating in the Southern Ocean. Another sign of climate change was the emergence of one of our ancestors—in Canada's western Yukon Territory—from a glacier last year as the ice that had encased him for more than 2,000 years melted. This followed the discovery of another "ice man" protruding from an Alpine glacier several years earlier, this one some 5,000 years old.

But people alive today are being affected even more profoundly by climate change. Rising temperatures and increased precipitation have already allowed infectious diseases such as cholera and malaria to begin spreading far beyond their normal ranges in the tropics. Other possible signs of climate change in the last year include U.S. government estimates that as much as two thirds of the world's coral reefs are showing signs of bleaching from rising temperatures—a condition that could eventually lead to the decimation of many of those reefs.

Not all is bleak as the new century begins, however. Even as climate change accelerates, so does investment in the technologies needed to combat it. Wind power, for example, is a $3-billion industry that is beginning to show its potential as a cornerstone of a new solar economy that may replace fossil fuels. In the United States, wind power has spread beyond the early stronghold in California, with several large wind farms coming online in the Great Plains. China has opened its first wind farm in Inner Mongolia. Meanwhile, the use of solar cells, particularly in roofing tiles, to generate electricity is gaining momentum in Japan and Europe. And the use of coal, in contrast, did not increase at all during the 1990s, suggesting that its role may soon turn downward as investors shift to more environmentally benign energy sources, such as wind and solar cells.

As for *State of the World* itself, we are starting the new millennium with an updated design of both the cover and the interior of the book's U.S. edition, thanks to the creative efforts of our Art Director, Elizabeth Doherty. We have replaced the former cover image of Earth showing political boundaries with a composite photograph taken from outer space. It shows the planet as a whole, a view more appropriate for a volume that focuses attention on global environmental trends and challenges.

One of our goals for *State of the World* in the years ahead is adding to the more than 30 languages it has been published in so far. This past year, we added Estonian and two Portuguese editions (in both Portugal and Brazil), and we resumed publication of the Russian edition—the first since the breakup of the Soviet Union.

We were pleased that NHK, Japan's national television network, purchased the television rights for *State of the World 1999* and is now working with CNN and a European consortium of networks on a six-part series. It is scheduled for airing in the latter half of 2000, and is expected to be one of the leading television projects marking the new millennium.

We are also expanding our publishing effort. In addition to our two annuals (*State of the World* and *Vital Signs*), the Worldwatch/Norton environmental books, the Worldwatch Papers, and *World Watch* magazine, we have begun a series of frequent four-page Worldwatch News Briefs on developments that either are in the news or we think should be. These are distributed electronically to some 1,650 editors and reporters worldwide—and are available at <www.worldwatch.org/alerts/index.html>. We now release something at the Institute every other week, whether it's a new issue of the magazine, a News Brief, a Worldwatch Paper, or a book.

In the year ahead, we also plan to develop our Web site further, making it a premier source of environmental information. Among other things, we plan to install a search engine, making it possible to search the entire library of Worldwatch publications.

We hope you will find this first *State of the World* of the new millennium useful in thinking through your own priorities and activities. If you have any comments on this edition or suggestions for future editions, please let us know by letter, fax (202-296-7365), or e-mail (worldwatch@worldwatch.org). And please visit our Web site (www.worldwatch.org) to check on the latest releases.

Lester R. Brown
Christopher Flavin
Hilary French

Worldwatch Institute
1776 Massachusetts Ave., NW
Washington, DC 20036

December 1999

State of the World 2000

Chapter 1

Challenges of the New Century

Lester R. Brown

As we look back at the many spectacular achievements of the century just ended, the landing on the Moon in July 1969 by American astronauts Neil Armstrong and Buzz Aldrin stands out. At the beginning of the century, few could imagine humans flying, much less breaking out of Earth's field of gravity to journey to the Moon. And few could imagine how quickly the world would go from air travel to space exploration.

Indeed, when the century began, the Wright brothers were still working in their bicycle shop in Dayton, Ohio, trying to design a craft that would fly. Just 66 years elapsed from their first precarious flight in 1903 on the beach at Kitty Hawk, North Carolina, to the landing on the Moon. Although their first flight was only 120 feet, it opened a new era, setting the stage for a century of breathtaking advances in technology.[1]

In 1945, engineers at the University of Pennsylvania's Moore School of Electrical Engineering successfully designed what many consider to be the first electronic computer, the ENIAC (Electronic Numerical Integrator and Computer). This advance was to have an even more pervasive effect than the Wright brothers' invention, as it set the stage for the evolution of the information economy. Computer technology progressed even more rapidly, going from the era of large mainframes to personal computers in just a few decades.[2]

A new industry evolved. New firms were created. IBM, Hewlett-Packard, Dell, Apple, Microsoft, Intel, and America On-Line became household names. Fortunes were made overnight. When the listed stock value of Microsoft overtook that of General Motors in 1998, it marked the beginning of a new era—a shift from a period dominated by heavy industry to one dominated by information.[3]

The stage was set for the evolution of the Internet, a novel concept that has tied the

---

Units of measure throughout this book are metric unless common usage dictates otherwise.

world together as never before. Although still in its early stages as the new century begins, the Internet is already affecting virtually every facet of our lives—changing communication, commerce, work, education, and entertainment. It is creating a new culture, one that is evolving in cyberspace.

Some 23 million Africans are beginning a new century with a death sentence imposed by HIV.

In the United States, the information technology industry, including computer and communications hardware, software, and the provision of related services, was a major source of economic growth during the 1990s. Creating millions of new, higher paying jobs, it has helped fuel the longest peacetime economic expansion in history. It has also induced a certain economic euphoria, one that helped drive the Dow Jones Industrial Average of stock prices to a long string of successive highs, raising it from less than 3,000 in early 1990 to over 11,000 in 1999.[4]

Caught up in this economic excitement, we seem to have lost sight of the deterioration of environmental systems and resources. The contrast between our bright hopes for the future of the information economy and the deterioration of Earth's ecosystem leaves us with a schizophrenic outlook.

Although the contrast between our civilization and that of our hunter-gatherer ancestors could scarcely be greater, we do have one thing in common—we, too, depend entirely on Earth's natural systems and resources to sustain us. Unfortunately, the expanding global economy that is driving the Dow Jones to new highs is, as currently structured, outgrowing those ecosystems. Evidence of this can be seen in

shrinking forests, eroding soils, falling water tables, collapsing fisheries, rising temperatures, dying coral reefs, melting glaciers, and disappearing plant and animal species.

As pressures mount with each passing year, more local ecosystems collapse. Soil erosion has forced Kazakhstan, for instance, to abandon half its cropland since 1980. The Atlantic swordfish fishery is on the verge of collapsing. The Aral Sea, producing over 40 million kilograms of fish a year as recently as 1960, is now dead. The Philippines and Côte d'Ivoire have lost their thriving forest product export industries because their once luxuriant stands of tropical hardwoods are largely gone. The rich oyster beds of the Chesapeake Bay that yielded more than 70 million kilograms a year in the early twentieth century produced less than 2 million kilograms in 1998. As the global economy expands, local ecosystems are collapsing at an accelerating pace.[5]

Even as the Dow Jones climbed to new highs during the 1990s, ecologists were noting that ever growing human demands would eventually lead to local breakdowns, a situation where deterioration would replace progress. No one knew what form this would take, whether it would be water shortages, food shortages, disease, internal ethnic conflict, or external political conflict.

The first region where decline is replacing progress is sub-Saharan Africa. In this region of 800 million people, life expectancy—a sentinel indicator of progress—is falling precipitously as governments overwhelmed by rapid population growth have failed to curb the spread of the virus that leads to AIDS. In several countries, more than 20 percent of adults are infected with HIV. Barring a medical miracle, these countries will lose one fifth or more of their adult population during this decade. In the absence of a low-cost cure, some 23 million Africans are beginning a new century with a death sentence imposed by the virus. With the failure of governments in the region to control the spread of HIV, it is becoming an epidemic of epic proportions. It is also a tragedy of epic proportions.[6]

Unfortunately, other trends also have the potential of reducing life expectancy in the years ahead, of turning back the clock of economic progress. In India, for instance, water pumped from underground far exceeds aquifer recharge. The resulting fall in water tables will eventually lead to a steep cutback in irrigation water supplies, threatening to reduce food production. Unless New Delhi can quickly devise an effective strategy to deal with spreading water scarcity, India—like Africa—may soon face a decline in life expectancy.[7]

Environmental Trends Shaping the New Century

As the twenty-first century begins, several well-established environmental trends are shaping the future of civilization. This section discusses seven of these: population growth, rising temperature, falling water tables, shrinking cropland per person, collapsing fisheries, shrinking forests, and the loss of plant and animal species.

The projected growth in population over the next half-century may more directly affect economic progress than any other single trend, exacerbating nearly all other environmental and social problems. Between 1950 and 2000, world population increased from 2.5 billion to 6.1 billion, a gain of 3.6 billion. And even though birth rates have fallen in most of the world, recent projections show that population is projected to grow to 8.9 billion by 2050, a gain of 2.8 billion. Whereas past growth occurred in both industrial and developing countries, virtually all future growth will occur in the developing world, where countries are already overpopulated, according to many ecological measures. Where population is projected to double or even triple during this century, countries face even more growth in the future than in the past.[8]

Our numbers continue to expand, but Earth's natural systems do not. The amount of fresh water produced by the hydrological cycle is essentially the same today as it was in 1950 and as it is likely to be in 2050. So, too, is the sustainable yield of oceanic fisheries, of forests, and of rangelands. As population grows, the shrinking per capita supply of each of these natural resources threatens not only the quality of life but, in some situations, even life itself.

A second trend that is affecting the entire world is the rise in temperature that results from increasing atmospheric concentrations of carbon dioxide (CO_2). When the Industrial Revolution began more than two centuries ago, the CO_2 concentration was estimated at 280 parts per million (ppm). By 1959, when detailed measurements began, using modern instruments, the CO_2 level was 316 ppm, a rise of 13 percent over two centuries. By 1998, it had reached 367 ppm, climbing 17 percent in just 39 years. This increase has become one of Earth's most predictable environmental trends.[9]

Global average temperature has also risen, especially during the last three decades—the period when CO_2 levels have been rising most rapidly. The average global temperature for 1969–71 was 13.99 degrees Celsius. By 1996–98, it was 14.43 degrees, a gain of 0.44 Celsius (0.8 degrees Fahrenheit). (See Figure 1–1.)[10]

If CO_2 concentrations double pre-industrial levels during this century, as projected,

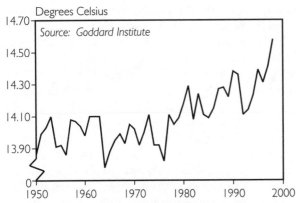

Figure 1–1. Average Temperature at Earth's Surface, 1950–98

global temperature is likely to rise by at least 1 degree Celsius and perhaps as much as 4 degrees (2–7 degrees Fahrenheit). Meanwhile, sea level is projected to rise from a minimum of 17 centimeters to as much as 1 meter by 2100.[11]

This will alter every ecosystem on Earth. Already, coral reefs are being affected in nearly all the world's oceans, including the rich concentrations of reefs in the vast eastern Pacific and in the Indian Ocean, stretching from the east coast of Africa to the Indian subcontinent. For example, record sea surface temperatures over the last two years may have wiped out 70 percent of the coral in the Indian Ocean. (See Chapter 2.) Coral reefs, complex ecosystems that are sometimes referred to as the rainforests of the sea, not only serve as breeding grounds for many species of marine life, they also protect coastlines from storms and storm surges.[12]

The modest temperature rise in recent decades is melting ice caps and glaciers. Ice cover is shrinking in the Arctic, the Antarctic, Alaska, Greenland, the Alps, the Andes, and the Quinghai-Tibetan Plateau. A team of U.S. and British scientists reported in mid-1999 that the two ice shelves on either

side of the Antarctic peninsula are in full retreat. Over roughly a half-century through 1997, they lost 7,000 square kilometers. But then within a year or so they lost 3,000 square kilometers. The scientists attribute the accelerated ice melting to a regional rise in average temperature of some 2.5 degrees Celsius (4.5 degrees Fahrenheit) since 1940.[13]

In the fall of 1991, hikers in the southwestern Alps near the border of Austria and Italy discovered an intact human body, a male, protruding from a glacier. Believed to have been trapped in a storm some 5,000 years ago and quickly covered with snow and ice, his body was remarkably well preserved. And in the late summer of 1999, another body was found protruding from a melting glacier in the Yukon territory of western Canada. Our ancestors are emerging from the ice with a message for us: Earth is getting warmer.[14]

One of the least visible trends that is shaping our future is falling water tables. Although irrigation problems, such as waterlogging, salting, and silting, go back several thousand years, aquifer depletion is a new one, confined largely to the last half-century, when powerful diesel and electric pumps made it possible to extract underground water at rates that exceed the natural recharge from rainfall and melting snow. According to Sandra Postel of the Global Water Policy Project, overpumping of aquifers in China, India, North Africa, Saudi Arabia, and the United States exceeds 160 billion tons of water per year. Since it takes roughly 1,000 tons of water to produce 1 ton of grain, this is the equivalent of 160 million tons of grain, or half the U.S. grain harvest. In consumption terms, the food supply of 480 million of the world's 6 billion people is being produced with the unsustainable use of water.[15]

The largest single deficits are in India and China. As India's population has tripled since 1950, water demand has climbed to where it may now be double the sustainable yield of the country's aquifers. As a result, water tables are falling in much of the country and wells are running dry in thousands of villages. The International Water Management Institute, the world's premier water research body, estimates that aquifer depletion and the resulting cutbacks in irrigation water could drop India's grain harvest by up to one fourth. In a country that is adding 18 million people a year and where more than half of all children are malnourished and underweight, a shrinking harvest could increase hunger-related deaths, adding to the 6 million worldwide who die each year from hunger and malnutrition.[16]

In China, the quadrupling of the economy since 1980 has raised water use far beyond the sustainable yield of aquifer recharge. The result is that water tables are falling virtually everywhere the land is flat. Under the north China plain, which produces 40 percent of the country's grain harvest, the water table is falling by 1.6 meters (5 feet) a year. As aquifer depletion and the diversion of water to cities shrink irrigation water supplies, China may be forced to import grain on a scale that could destabilize world grain markets.[17]

Also making it more difficult to feed the projected growth in population adequately over the next few decades is the worldwide shrinkage in cropland per person. Since the mid-twentieth century, grainland area per person has fallen in half, from 0.24 hectares to 0.12 hectares. If the world grain area remains more or less constant over the next half-century (assuming that cropland expansion in such areas as Brazil's cerrado will offset the worldwide losses of cropland to urbanization, industrialization, and land degradation), the area per person will shrink to 0.08 hectares by 2050.[18]

Our ancestors are emerging from the ice with a message for us: Earth is getting warmer.

Among the more populous countries where this trend threatens future food security are Ethiopia, Nigeria, and Pakistan—all countries with weak family planning programs. As a fixed area of arable land is divided among ever more people, it eventually shrinks to the point where people can no longer feed themselves. Unfortunately, in the poorer nations of sub-Saharan Africa and the Indian subcontinent, subsistence farmers may not have access to imports. For them, land scarcity translates into hunger.

Pakistan's population, for example, is projected to grow from 146 million today to 345 million in 2050, shrinking the grainland area per person in this crowded nation to a minuscule 0.04 hectares by 2050—less than half of what it is today, and an area scarcely the size of a tennis court. A family of six will then have to produce all its food on roughly one fifth of a hectare, or half an acre—the equivalent of a small suburban building lot in the United States. Similar prospects lie ahead for Nigeria, where numbers are projected to double to 244 million over the next half-century, and for Ethiopia, where more than half the children are undernourished and where population is projected to nearly triple. In these and dozens of other developing countries, grainland area per person will shrink dramatically.[19]

If world grainland productivity, which climbed by 170 percent over the last half-century, were to rise rapidly over the next half-century, the shrinkage in cropland area per person might not pose a serious threat. Unfortunately, the rise is slowing. From 1950 to 1990, world grain yield per hectare increased at more than 2 percent a year, well ahead of world population growth. But from 1990 to 1999 it grew at scarcely 1 percent a year. While biotechnology may reduce insecticide use through insect-resistant varieties, it offers little potential for raising yields.[20]

Humanity also depends heavily on the oceans for food, particularly animal protein. From 1950 until 1997, the oceanic fish catch expanded from 19 million tons to more than 90 million tons. This fivefold growth since mid-century has pushed the catch of most oceanic fisheries to their limits or beyond. If, as most marine biologists believe, the oceans cannot sustain an annual catch of more than 95 million tons, the catch per person will decline steadily in the decades ahead as world population continues to grow. This also means that all future growth in demand for food will have to be satisfied from land-based sources.[21]

These three parallel trends—falling water tables, shrinking cropland area per person, and the leveling off of the oceanic fish catch—all suggest that it will be far more difficult to keep up with the growth in world demand for food over the next half-century if the world remains on the U.N. medium population trajectory of adding nearly 3 billion people and if incomes continue to rise.[22]

Forests, too, are being overwhelmed by human demands. Over the past half-century, the world's forested area has shrunk substantially, with much of the loss occurring in developing countries. And the forested area per person worldwide is projected to shrink from 0.56 hectares today to 0.38

hectares in 2050. This figure reflects both population growth and the conversion of some forestland to cropland. In many situations, the rising worldwide demand for forest products—lumber, paper, and fuelwood—is already overwhelming the sustainable yield of forests.[23]

In some ways, the trend that will most affect the human prospect is an irreversible one—the accelerating extinction of plant and animal species. The share of birds, mammals, and fish vulnerable or in immediate danger of extinction is now measured in double digits: 11 percent of the world's 8,615 bird species, 25 percent of the world's 4,355 mammal species, and an estimated 34 percent of all fish species. The leading cause of species loss is habitat destruction, but habitat alterations from rising temperatures or pollution can also decimate both plant and animal species. As human population grows, the number of species with which we share the planet shrinks. As more and more species disappear, local ecosystems begin to collapse; at some point, we will face wholesale ecosystem collapse.[24]

Replacing Economics with Ecology

As noted earlier, global economic trends during the 1990s were remarkably bullish, but environmental trends were disastrous. The contrast could scarcely be greater. An economic system that worked well in times past when the demands of a smaller economy were well within the capacities of Earth's ecosystems is no longer working well. If the trends outlined in the last section cannot be reversed, we face a future where continuing environmental deterioration almost certainly will lead to economic decline. The challenge is to redesign the economic system so that it will not destroy its environmental support systems, so that economic progress can continue.

The time has come for what science historian Thomas Kuhn describes as a paradigm shift. In his classic work *The Structure of Scientific Revolutions,* Kuhn observes that as the scientific understanding of reality in a field advances, reaching a point where existing theory no longer adequately explains reality, then theory has to change. It has to be updated, replacing the old paradigm with a new one. Perhaps history's best known example of this is the shift from the Ptolemaic view of the world, in which the sun revolved around Earth, to the Copernican view, which argued that Earth revolved about the sun. Once the Copernican model was accepted, relationships not only within the solar system but between the solar system and the rest of the universe suddenly made sense to those who studied the heavens, leading to an era of steady advances in astronomy.[25]

We are now facing such a situation with the global economy. The market is a remarkably efficient device for allocating resources and for balancing supply and demand, but it does not respect the sustainable yield thresholds of natural systems. In a world where demands of the economy are pressing against the limits of natural systems, relying exclusively on economic indicators to guide investment decisions is a recipe for disaster. Historically, for example, if the supply of fish was inadequate, the price would rise, encouraging investment in additional fishing trawlers. This market system worked well. But today, with the fish catch already exceeding the sustainable yield of many fisheries, investing in more trawlers in response to higher seafood prices will simply accelerate the collapse of

fisheries. A similar situation exists with forests, rangelands, and aquifers.

The gap between economists and ecologists in their perception of the world as the new century begins could not be wider. Economists look at grain markets and see the lowest grain prices in 20 years—a sure sign that production capacity is outrunning effective demand, that supply constraints are not likely to be a problem for the foreseeable future. But ecologists see water tables falling in key food-producing countries. Knowing that 480 million of the world's 6 billion people are being fed with grain produced by overpumping aquifers, they are worried about the effect of eventual aquifer depletion on food production.[26]

Economists see a world economy that has grown by leaps and bounds over the last half-century, but ecologists see growth based on the burning of vast quantities of cheap fossil fuels, which is destabilizing the climate. They are keenly aware that someone buying a gallon of gasoline pays the cost of pumping the oil, of refining it into gasoline, and of distributing the gasoline to the service station, but not the cost to society of future climate disruptions. Again, while economists see booming economic indicators, ecologists see an economy that is altering the climate with consequences that no one can foresee.

Today ecologists look at the deteriorating ecosystem and see a need to restructure the economy, the need for a paradigm shift. For example, stabilizing Earth's climate now depends on reducing carbon emissions by shifting from fossil fuels to a solar/hydrogen energy economy. Solar is here defined broadly, including not only direct sunlight but also indirect forms of solar energy—wind power, hydropower, and biological sources, such as wood. Fortunately, the technologies for tapping this enormous source of energy already exist. We can now see electricity generated from wind being used to electrolyze water and to produce hydrogen. Hydrogen then becomes the basic fuel for the new economy, relying initially on the distribution and storage facilities of the natural gas industry. Put simply, the principles of ecological sustainability now require a shift from a carbon-based to a hydrogen-based energy economy.

There is a similar need for restructuring the world food economy. Some 40 percent of the world's food is produced on irrigated land, with much of the water used for irrigation being heavily subsidized. Encouraging water use with subsidies at a time when water tables are falling sends the wrong signal, one that encourages the inefficient use of water. As world water use has tripled over the last half-century, often pressing against the limits of local supply, water has become scarcer than land. With water emerging as the principal constraint on efforts to expand food production, restructuring the world food economy to make it more water-efficient is a necessary, though not sufficient, precondition to feeding an expanding world population adequately. Among other things, this means shifting to more water-efficient crops and more grain-efficient sources of animal protein, such as poultry.[27]

As the global economy outgrows the various natural capacities of Earth, as just described, it imposes new demands on the political system that is responsible for managing the interaction between the two. Managing this increasingly stressed relationship between the global market economy, which is expanding by a trillion dollars per year, and Earth's ecosystems, whose capacities are essentially fixed, becomes ever more demanding. The demands on political institutions to reverse deterioration will

intensify. At issue is whether our political institutions are capable of incorporating ecological principles into economic decisionmaking.[28]

Crossing the Sustainability Threshold

Most environment ministers understand the need to restructure the global economy so that progress can continue, but unfortunately not enough of their constituents understand this. The ministers must also contend with interests that are vested in the existing economic system, interests that are more than willing to bribe political leaders either directly or in the form of campaign contributions and to mount disinformation campaigns to confuse the public about the need for change. Eventually, if enough people in a country are convinced of the need for change, they can override these vested interests, crossing a threshold of social change.

A threshold—a concept widely used in ecology in reference to the sustainable yield of natural systems—is a point that, when crossed, can bring rapid and sometimes unpredictable change. In the social world, the thresholds of sudden change are no less real, though they may be more difficult to identify and anticipate. Among the more dramatic recent threshold crossings are the ones that led to the political revolution in Eastern Europe and to the dramatic decline in cigarette smoking in the United States.

The change in political systems in Eastern Europe came with no apparent warning. It almost seemed that people woke up one morning and understood that the era of the one-party political system and the centrally planned economy was over. Even those in power at the time realized this. No analysts writing in political science journals forecast this essentially bloodless political revolution, one that led to a fundamental change in the form of governance. Although we do not understand the process well, we do know that at some point a critical mass had been reached—that a time arrived when so many people were convinced of the need for change that the process achieved an irresistible momentum.

A similar story can be told about smoking. In the early 1960s, smoking in the United States was an increasingly popular habit, one that was being aggressively promoted by the cigarette manufacturers. Then in 1964, the U.S. Surgeon General released a report on the relationship between smoking and health, the first in a series that appeared almost every year for the rest of the century. These reports, and media coverage of the thousands of research projects they spawned, fundamentally altered the way people think not only about their own smoking but also about inhaling secondhand smoke from the cigarettes of others.[29]

This shift in thinking was so strong that in November 1998 the tobacco industry, which for decades had argued under oath that there was no proof of a link between smoking and health, agreed to reimburse state governments for the past Medicare costs of treating the victims of cigarette smoking. This settlement with 46 state governments, plus separate agreements reached earlier with 4 other states, totaled $251 billion, nearly $1,000 for every person in the United States. (In September 1999, in a stunning display of the new official attitude toward cigarettes, the U.S. Department of Justice announced that it was filing suit to reclaim smoking-related federal health care expenditures.)[30]

This revolution in attitudes toward smoking is spreading fast in other countries.

Following the industry agreement with state governments in the United States, governments of several developing countries—including Bolivia, Guatemala, Nicaragua, Panama, and Venezuela—filed suits in U.S. courts for similar reimbursements because their people, who had been smoking the cigarettes manufactured by the same companies, were suffering from the same smoking-induced illnesses.[31]

In effect, the tobacco industry's payments to state governments are a retroactive tax on the sale of cigarettes over decades.

The agreement in the United States represented an implicit acceptance by the tobacco industry of responsibility for the indirect effects of using their products. In effect, the payments to state governments are a retroactive tax on the sale of cigarettes over the past several decades. This is a massive precedent for the idea of a carbon tax on fossil fuels. As with the health-care-related costs of smoking, an analysis of the indirect effects of burning fossil fuels, including air pollution, acid rain, and climate disruptions, would be needed to determine the amount of the carbon tax.

Whether a similar revolution in the environmental field will follow that in smoking remains to be seen. Some 34 years passed between the first Surgeon General's report and the landmark agreement between the tobacco industry and state governments. In Eastern Europe, it was fully four decades from the imposition of socialism until its demise. Thirty-eight years have passed since biologist Rachel Carson published *Silent Spring*, issuing the wake-up call that gave rise to the modern environmental movement.

Signs that the world is approaching a key

environmental threshold are perhaps as strong within the corporate community as in any sector. The shifts have been particularly dramatic in the oil industry, led by Royal Dutch Shell and British Petroleum. And in February 1999, Mike Bowlin, the chief executive officer of ARCO, startled an energy conference in Houston by saying: "We've embarked on the beginning of the Last Days of the Age of Oil." He went on to discuss the need to convert our carbon-based energy economy into a hydrogen-based energy economy.[32]

Two months later, Shell Oil and DaimlerChrysler announced they were leading a consortium of corporations whose goal is to make Iceland the world's first hydrogen-based economy. Iceland—with an abundance of geothermal energy, widely used for heating buildings, and cheap hydropower—is an ideal place to begin. Cheap electricity from hydropower makes it economically feasible to split the water molecule by electrolysis, producing hydrogen that can be used in new, highly efficient fuel cell engines that are under development. DaimlerChrysler, a leader in the development of these engines, which are expected to replace the traditional internal combustion engine, plans to market its new fuel cell-powered automobiles in Iceland within the next few years. (See Chapter 8.) Shell has also opened its first hydrogen station—the future equivalent of today's gasoline station—in Hamburg, Germany.[33]

In the United States, the threshold for responsible forest management appears to have been crossed. In effect, the principles of ecology are replacing basic economics in the management of national forests. After several decades of building roads with taxpayers' money to help logging companies clearcut publicly owned forests, the Forest Service announced in early 1998 that it was imposing a moratorium on road building. For decades the goal of the forest management system, which had built some 600,000 kilometers of roads to facilitate clearcutting, had been to maximize the timber harvest in the short run.[34]

The new chief of the Forest Service, Michael Dombeck, responding to a major shift in public opinion, introduced a new management system—one designed to maintain the integrity of the ecosystem and to be governed by ecology, not by economics. Henceforth, the 78 million hectares of national forests—more than the area planted to grain in the United States—will be managed with several goals in mind. The system will recognize the need, for example, to manage the forest so as to eliminate the excessive flooding, soil erosion, and silting of rivers, and the destruction of fisheries associated with the now banned practice of clearcutting. Under the new policy, the timber harvest from national forests, which reached an all-time high of 12 billion board feet per year during the 1980s, has been reduced to 3 billion board feet.[35]

The United States is not the only country to institute a radical change in forest management. In mid-August 1998, after several weeks of near-record flooding in the Yangtze river basin, Beijing acknowledged for the first time that the flooding was not merely an act of nature but was exacerbated by the deforestation of the upper reaches of the watershed. Premier Zhu Rongji personally ordered a halt not only to the tree-cutting in the upper reaches of the Yangtze basin, but also to the conversion of some state timbering firms into tree-planting firms. The official view in Beijing now is that trees standing are worth three times as much as those cut, simply because of the water storage and flood control capacity of forests.[36]

A chastened tobacco industry, oil companies investing in hydrogen, reformed forest management in the United States and China—these are just some of the signs that the world may be approaching the kind of paradigm shift that Thomas Kuhn wrote about. Across a spectrum of activities, places, and institutions, attitudes toward the environment have changed markedly in just the last few years. Among giant corporations that could once be counted on to mount a monolithic opposition to serious environmental reform, a growing number of high-profile CEOs have begun to sound more like environmentalists than representatives of the bastions of global capitalism.

If the evidence of a global environmental awakening were limited to only government initiatives or a few corporate initiatives, it might be dubious. But with the evidence of growing momentum now coming on both fronts, the prospect that we are approaching the threshold of a major transformation becomes more convincing. The question is, Will it happen soon enough? Will it happen before the deterioration of natural support systems reaches a point of no return?

Crossing the Decline Threshold

As noted earlier, collapsing fisheries, shrinking forests, and falling water tables illustrate how human demands are exceeding the sustainable yield of natural systems. Exactly when these sustainable yield thresholds are exceeded is not always evident. When expanding demand for a resource first surpasses the sustainable yield, the additional demand can be satisfied, and it typically is, by consuming the resource base. At first the shrinkage is scarcely detectable, but with each passing year the excess of demand over

sustainable yield typically increases and is satisfied by consuming ever more of the fish stocks, the forests, or the aquifers.

As a result, fisheries that appear stressed and show signs of a shrinking catch can suddenly collapse. Similarly with an overpumped aquifer. At first the year-to-year fall of the water level is barely perceptible. But over time, as the gap between the rising demand for water and the sustainable yield of the aquifer widens, each successive annual drop in the water table is greater than the year before. Almost overnight a falling water table can become a depleted aquifer, reducing the rate of pumping to the rate of recharge.

The risk in a world adding nearly 80 million people annually is that so many sustainable yield thresholds will be crossed in such a short period of time that the consequences will become unmanageable. Historically, when early civilizations lived largely in isolation, the consequences of threshold crossings were strictly local. Today, in the age of global economic integration, a threshold crossing in one major country can put additional pressure on resources in other countries. When Beijing banned logging in the upper reaches of the Yangtze River basin in 1998, for example, the increased demand for forest products from neighboring countries in Southeast Asia intensified the pressure on the region's remaining forests.[37]

A similar situation exists with water. As countries press against the limits of their water supplies, they often satisfy growing urban demand by diverting irrigation water from the countryside to the cities. They then offset the reduction in food production by increasing imports. Since it takes 1,000 tons of water to produce 1 ton of grain, this is a highly efficient way of importing water.[38]

This helps explain why the semiarid

North Africa and Middle East region, stretching from Morocco eastward through Iran, has been the world's fastest-growing grain market in recent years. Both rapid population growth and oil-driven gains in affluence are steadily boosting grain demand. Meanwhile, the growing demand for water in the cities is often satisfied by diverting irrigation water from agriculture. With every country in the region facing irrigation water shortages, grain imports are climbing. Last year, the water required to produce the grain and other farm commodities imported into the region was roughly equal to the annual flow of the Nile River.[39]

In effect, water scarcity is crossing national borders through the international grain trade. While the world has responded easily to the rising demand for imported grain in North Africa and the Middle East, expanding grain imports into China and India as their aquifers are depleted simultaneously could destabilize the world grain market. China, which has a huge balance-of-trade surplus with the United States, can easily afford to import massive quantities of grain. Stated simply, falling water tables in China could lead to rising food prices for the world.[40]

A similar situation exists for oceanic fisheries. Given the capacity of fishing fleets to roam the oceans today accompanied by factory processing ships, scarcity can quickly move from one fishery to another. This is why, regardless of local demand for seafood, fisheries everywhere are being fished at or beyond capacity.

Given increasing economic integration, trends tend to emerge simultaneously in many countries, whether it be the acceleration of population growth that began a half-century ago in developing countries or the depletion of aquifers, a more recent phenomenon. In the first case, the international availability of basic medical technology, such as vaccines, and of advancing agricultural technology led to the nearly simultaneous fall in mortality rates after mid-century in scores of developing countries. This in turn led to rapid population growth when fertility rates did not follow suit. Similarly, it is the near universal availability of diesel and electrically powered pumps in both industrial and developing countries that enables farmers everywhere to pump water from aquifers faster than they can recharge.

Many countries that have failed to rein in their population growth are already facing potentially unmanageable problems. When developing countries embarked on the development journey a half-century ago, they followed one of two demographic paths. In the first, illustrated by South Korea, Taiwan, and Thailand, early efforts to shift to smaller families set in motion a positive reinforcing cycle of higher savings, rising living standards, and falling fertility rates. (See Table 1–1.) These countries are now approaching population stability.[41]

In the second category, which prevails in sub-Saharan Africa and the Indian subcontinent, fertility has remained relatively high, setting the stage for a potentially dangerous downward spiral in which rapid population growth reinforces poverty. After several decades of incessant population growth, some countries are simply being overwhelmed. Governments struggling with the simultaneous challenge of educating growing numbers of children coming of school age, of creating jobs for swelling ranks of young job seekers, and of dealing with the environmental effects of population growth are stretched to the limit. When a major new threat arises—such as a new infectious disease or the collapse of an ecosystem—governments often cannot cope.[42]

In countries with continuing rapid pop-

Table 1–1. Current and Projected Populations in Selected Developing Countries, 1999 and 2050

Country	1999	2050	Growth from 1999 to 2050	
	(million)	(million)	(percent)	
Developing Countries Where Population Growth is Approaching Stability				
South Korea	46	51	5	+ 11
Taiwan	22	25	3	+ 14
Thailand	61	74	13	+ 21
Developing Countries Where Rapid Population Growth Continues				
Ethiopia	61	169	108	+177
Nigeria	109	244	135	+124
Pakistan	152	345	193	+127

SOURCE: United Nations, *World Population Prospects: The 1998 Revision* (New York: December 1998).

ulation growth, three of the trends cited earlier—the spread of HIV, falling water tables, and shrinking cropland per person—either are already replacing progress with decline or have the potential to do so. Problems routinely managed in industrial societies are becoming full-scale humanitarian crises. For example, all industrial countries have managed to hold the HIV infection rate in their adult populations below 1 percent, but in several sub-Saharan African countries, as noted earlier, more than 20 percent of adults are HIV-positive. The resulting high mortality trends are more reminiscent of the dark ages than the bright new millennium so many had hoped for. Tragically, they suggest that some countries may already have crossed a deterioration/decline threshold.[43]

In sub-Saharan Africa, home to some 800 million people, the HIV epidemic is devastating society. Death rates are rising, infant mortality is rising, and life expectancy—perhaps the most basic measure of economic development—is falling. Before the on-slaught of AIDS, life expectancy in Zimbabwe was 65 years. In 1998, it was 44 years. By 2010, it is projected to fall to 39 years. The figures for Kenya are 66 years before AIDS and 44 years expected by 2010. For Namibia, the decline is 65 years to 39 years in 2010; for South Africa, 65 years to 48 years; and for Botswana, 61 years to 38 years.[44]

In these countries where the HIV epidemic is spiraling out of control, several other negative trends are set in motion, trends that collectively are reversing the process of development itself. AIDS patients are overwhelming health care systems in many sub-Saharan societies. In some hospitals in South Africa, 70 percent of the beds are occupied by AIDS patients. In Zimbabwe, half of the health care budget now goes to care for AIDS patients. The risk is that this unanticipated pressure on health care systems will deprive many people of even the most basic health services, no doubt raising death rates further.[45]

The loss of so many working-age adults in rural Africa is adversely affecting food production. The number of able-bodied workers is reduced not only by deaths, but also by the number who are sick and unable to work, as well as by those who are obligated to care for the sick. Godfrey Ssewankambo of Uganda observes that in rural areas, "from the time one adult family member is bedridden, AIDS compromises the nutrition and food security of the whole family."[46]

Governments are having difficulty grasping the dimensions of the epidemic. Despite the extraordinarily damaging effects of HIV in Zimbabwe, where 25 percent of the adult population is HIV-positive, the government is spending $70 million a month on the war in the Democratic Republic of the Congo, but only $1 million on prevention of AIDS.[47]

Companies operating in countries with high infection rates find their health insurance costs doubling, tripling, or even quadrupling. As a result, many companies that were operating comfortably in the black suddenly find themselves in the red. The combination of rising costs of health care insurance, the heavy loss of workers to AIDS, and the associated need to recruit and train new workers is making investment in these areas increasingly unattractive, which means that capital inflows are likely to decline and may even dry up altogether. What began as an unprecedented social tragedy is translating into an economic disaster.[48]

To make matters worse, in Africa it is often the better educated, more socially mobile populations who have the highest infection rate. Africa is losing the agronomists, the engineers, and the teachers it needs to sustain its economic development. In South Africa, for instance, at the University of Durban-Westville, where many of the country's future leaders are trained, 25 percent of the student body is HIV-positive.[49]

The long-term social consequences of the HIV epidemic remain to be seen, but for the first time in the modern era, progress is turning into a sustained decline for an entire region. With the virus continuing to spread, there is no reversal of the situation in prospect.

One of the results of the failure of governments to protect their people—whether it is from a deadly infectious disease or from aquifer depletion—may be the loss of political legitimacy. When this happens, the ability of governments to actually govern is impaired, making it even more difficult to respond to new threats to their people.

Two Keys to Regaining Control of Our Destiny

The overriding challenges facing our global civilization as the new century begins are to stabilize climate and stabilize population. Success on these two fronts would make other challenges, such as reversing the deforestation of Earth, stabilizing water tables, and protecting plant and animal diversity, much more manageable. If we cannot stabilize climate and we cannot stabilize population, there is not an ecosystem on Earth that we can save. Everything will change. If developing countries cannot stabilize their populations soon, many of them face the prospect of wholesale ecosystem collapse.

The exciting thing about the climate and population challenges is that we already have the technologies needed to succeed at both. Restructuring the energy economy to stabilize climate requires investment in climate-benign energy sources. It is the greatest investment opportunity in history. Stabilizing population, though it requires additional investment in reproductive health services and in the education of young women in developing countries, is more a matter of behavioral change—of couples having fewer children and investing more in the health and education of each.

Stabilizing climate means shifting from a fossil-fuel or carbon-based energy economy to alternative sources of energy. Nuclear power, once seen as an alternative to fossil fuels, has failed on several fronts. Within a few years, the closing of aging power plants is expected to eclipse the new plants still coming online, setting the stage for the phaseout of nuclear power. Electricity from the power source that was once described as "too cheap to meter" has now become too costly to use. The issue is no longer whether it is economical to build nuclear power plants but—given the high operating costs—whether it even makes economic sense in many situations to continue using those already built.

The only feasible alternative is a solar/hydrogen-based economy, one that taps the various sources of energy from the sun, such as hydropower, wind power, wood, or direct sunlight. (See also Chapter 8.) The transition to a solar/hydrogen economy has already begun, as can be seen in energy use trends from 1990 to 1998. (See Table 1-2.) Coal burning, for example, did not increase at all during this period. Meanwhile, wind power and photovoltaic cells—two climate-benign energy sources—were expanding at 22 percent and 16 percent a year, respectively. But the transition is not moving fast enough to avoid potentially disruptive climate change.

Although the use of all sources of energy that derive from the sun directly and indirectly will probably expand, wind and solar cells are likely to be the cornerstones of the new energy economy. Already Denmark gets 8 percent of its electricity from wind. For Schleswig-Holstein, the northernmost

state in Germany, the figure is 11 percent. Navarra, a northern industrial state in Spain, gets 20 percent of its electricity from wind. In the United States, wind generating capacity is moving beyond its early stronghold in California as new wind farms come online in Minnesota, Iowa, Oregon, Wyoming, and Texas, dramatically broadening the industry's geographic base.[50]

Within the developing world, India, with 900 megawatts of generating capacity, is the unquestioned leader. With the help of the Dutch, China began operation in 1998 of its first commercial wind farm, a 24-megawatt project in Inner Mongolia, a region of vast wind wealth.[51]

The world wind energy potential can only be described as enormous. Today the world gets over one fifth of all its electricity from hydropower, but this is dwarfed by the wind power potential. For example, China is richly endowed with wind energy and could double its national electricity generation from wind alone. An inventory of wind resources in the United States by the Department of Energy indicates that

Table 1-2. Trends in Global Energy Use, by Source, 1990-98[1]

Energy Source	Annual Rate of Growth
	(percent)
Wind power	22
Solar photovoltaics	16
Geothermal power	4
Hydroelectric power[1]	2
Oil	2
Natural gas	2
Nuclear power	1
Coal	0

[1] 1990–97 only.
SOURCE: Worldwatch estimates based on UN, BP, DOE, EC, Eurogas, PlanEcon, IMF, LBL, IAEA, BTM Consult, and *PV News*.

three states—North Dakota, South Dakota, and Texas—have enough harnessable wind energy to satisfy national electricity needs.[52]

With the costs of wind electric generation dropping from $2,600 per kilowatt in 1981 to $800 in 1998, wind power is fast becoming one of the world's cheapest sources of electricity, in some locations undercutting coal, traditionally the cheapest source. Once cheap electricity is available from solar sources, it can be used to electrolyze water, producing hydrogen—an ideal means of both storing and transporting solar energy.[53]

In 1998, sales of solar cells, the silicon-based semiconductors that convert sunlight into electricity, jumped 21 percent, reaching 152 megawatts. This growth reflected the sharp competition emerging among major industrial countries in the solar cell market as the world looks for clean energy sources that will not destabilize climate. The development of a solar cell roofing material in Japan has set the stage for even more rapid future growth in solar cell use. With this technology, the roof becomes the power plant for the building.[54]

In Japan, nearly 7,000 rooftop solar systems were installed in 1998. The German government announced in late 1998 the goal of 100,000 solar roofs in that country. In response, Royal Dutch Shell and Pilkington Solar International jointly are building the world's largest solar cell manufacturing facility in Germany. Italy joined in with a goal of 10,000 solar rooftops.[55]

While wind and solar cell use are soaring, the worldwide growth of oil use has slowed to less than 2 percent a year and may peak and turn downward as early as 2005. The burning of natural gas, the cleanest of the three fossil fuels, is growing by 2 percent per year. It is increasingly seen as a transition fuel, part of the bridge from the fossil-fuel-based energy economy to the solar/hydrogen energy economy.[56]

The goal is to convert small positive growth rates for fossil fuels into negative rates, as they are phased down, and to boost dramatically the growth in wind power and solar cells. Because wind energy is starting from such a small base, and because the urgency of stabilizing climate is mounting, it should perhaps be growing at triple-digit annual rates, not just in the double digits. If coral reefs are dying and if the Antarctic ice cap is beginning to break up because Earth's temperature is rising, maybe wind-generating capacity should be doubling each year, much as the number of host computers linked to the Internet did each year from 1980 to 1995.[57]

One way of dramatically boosting the growth in wind power would be to reduce income taxes and offset them with a carbon tax on fossil fuels, one that would more nearly reflect the full costs associated with air pollution, acid rain, and climate disruption. Such a move would raise investment not only in wind power, but also in solar cells and energy efficiency. It could push wind power growth far above the current rates, greatly accelerating the shift to a solar/hydrogen energy economy.

Sharply accelerating the wind power growth rate depends on restructuring tax systems to reduce taxes on income and wages while increasing those on environmentally destructive activities, such as carbon emissions from fossil fuel burning. Some countries have already begun to do this, including Denmark, Finland, the Netherlands, Spain, Sweden, and the United Kingdom. And in late 1998, the new coalition government in Germany announced the first step in a massive restructuring of the tax system, one that would simultaneously reduce taxes on wages and raise them on energy use. In April 1999, the first of four annual tax shifts was implemented. This ecological tax shift of some $14 billion—the largest yet contemplated by any government—was taken unilaterally, not bogged down in the politics of the global climate treaty or contingent on steps taken elsewhere. The framers of the new tax structure justified it primarily on economic grounds, mainly the creation of additional jobs. It would also help reduce carbon emissions.[58]

This bold German initiative is setting the stage for tax restructuring in other countries. If the world is going to make the economic changes needed in the time available, tax restructuring must be at the center of the effort. No other set of policies can bring about the needed changes quickly enough. In an article in *Fortune* magazine, which argued for a 10-percent reduction in U.S. income taxes and a 50¢-per-gallon hike in the tax on gasoline, Professor N. Gregory Mankiw of Harvard noted: "Cutting income taxes while increasing gasoline taxes would lead to more rapid economic growth, less traffic congestion, safer roads, and reduced risk of global warming—all without jeopardizing long-term fiscal solvency. This may be the closest thing to a free lunch that economics has to offer."[59]

While stabilizing climate is largely a matter of investing in new energy sources, stabilizing population is more a matter of changing reproductive behavior. The annual addition to world population increased steadily from 38 million in 1950 to the historical peak of 87 million in 1989. After that, it dropped to 78 million in 1998. While the annual additions in many developing countries have been increasing, they have been declining elsewhere. Some 32 countries—virtually all of industrialized Europe, from the United Kingdom to Russia, plus Japan and Canada—have succeeded in stabilizing their population size. Births and deaths are essentially in balance, as they must be in a sustainable society. This group of countries contains some 15 percent of the world population.[60]

To make the economic changes needed in the time available, tax restructuring must be at the center of the effort.

Another, much larger group of countries has reached replacement level of fertility of 2.1 children per couple, but this does not immediately translate into population stability because of the disproportionately large number of young people moving into their reproductive years. This group, containing over 40 percent of the world's people, includes two of the most populous countries: China and the United States. In each of these, population is growing at just under 1 percent a year.[61]

One of the keys to the needed changes in reproductive behavior is information that will help people understand the consequences of not shifting quickly to smaller families. Few people intentionally want their children or grandchildren to be deprived of adequate water supplies or of education because they themselves have too many children. Thus information is vital. Governments can provide this information through national carrying capacity assessments—studies to determine how many people the cropland, water, grassland, and forest resources of a country can sustain. This also involves a tradeoff between the size of population and the level of consumption. The carrying capacity calculations provide the information needed for that choice.

The key to stabilizing world population is for national governments to formulate strategies for stabilizing population humanely rather than waiting for nature to intervene with its inhumane methods, as it is in Africa. Once these strategies are developed, it is in the interest of the international community to support the stabilization effort.

In two countries that curbed the spread of HIV after it reached epidemic proportions, the heads of state led the campaign.

At the U.N. Conference on Population and Development in Cairo in 1994, it was estimated that the annual cost of providing quality reproductive health services to all those in need in developing countries would be $17 billion in the year 2000. By 2015, this would climb to $22 billion. Industrial countries agreed to provide one third of the funds, with developing countries providing the remainder. While developing countries have largely honored their commitments, industrial countries—including the United States—have regrettably reneged on theirs. And, almost unbelievably, in late 1998 the U.S. Congress withdrew all funding for the U.N. Population Fund, the principal source of international family planning assistance.[62]

Fortuitously, the same family planning services that provide reproductive health counseling and that supply the condoms to help slow population growth also help to check the spread of HIV. Investment in efforts to slow population growth can thus also help to check the spread of the virus.

In stabilizing climate and stabilizing population, there is no substitute for leadership. Examples of this abound in both initiatives. Denmark, for instance, has simply banned the construction of coal-fired power plants. Meanwhile, it has adopted a series of economic incentives for investment in wind power that has fostered the development of the world's largest wind turbine manufacturing industry. As a result, in 1998 wind turbines of Danish design accounted for half of all turbines installed worldwide. Though scarcely a major industrial power, Denmark has a commanding position in this fast-expanding new industry.[63]

In every developing country where population growth has slowed dramatically, family planning programs have enjoyed strong government support. The same is true for containing the HIV epidemic. In the two countries that have successfully curbed the spread of HIV after it reached epidemic proportions—Uganda and Thailand—the heads of state led the containment campaign. In Uganda, President Museveni personally led the effort and continuously referred not only to the dangers of the virus but also to the behavioral changes needed to check its spread. In Thailand, Prime Minister Anand Punyarachun both provided the personal leadership to direct the campaign and was instrumental in raising the appropriations for containing the virus from $2.6 million in 1990 to $80 million in 1996.[64]

Leadership and time are the scarce resources. The world desperately needs more of both. Saving the planet, including the stabilization of climate and the stabilization of population, is a massive undertaking by any historical yardstick. This is not a spectator sport. It is something everyone can participate in. Few activities offer more satisfaction.

We can participate not only as individuals, but also in an institutional sense. All of society's institutions—from organized religion to corporations—have a role to play.

Although many individuals and corporations want to do something about the environment, few recognize the need for systemic change. Corporations take pride in listing in their annual reports the steps they have taken to help protect the environment. They will cite gains in office paper recycling or reductions in energy use. These are obviously moves in the right direction. And they are to be applauded. But they do not deal with the central issue, which is the need to restructure the global economy quickly.

This is not likely to happen unless corporations use some of their political leverage with governments to actively support tax restructuring.

There is no middle path. The challenge is either to build an economy that is sustainable or to stay with our unsustainable economy until it declines. It is not a goal that can be compromised. One way or another, the choice will be made by our generation, but it will affect life on Earth for all generations to come.

Chapter 2

Anticipating
Environmental "Surprise"

Chris Bright

If there is any comfort to be found in the prospect of environmental decline, it lies in the idea that the process is gradual and predictable. All sorts of soothing cliches follow from this notion: Even if we have not turned the trends around, our children will rise to the challenge. There's time. We're constantly learning; you can see plenty of progress already.

But this way of thinking is sleepwalking. To understand why, you have to look at decline close up. Here, for instance, is how it has happened in one small country, with big implications. Honduras, in the early 1970s, was caught up in a drive to build agricultural exports. Landowners in the south increased their production of cattle, sugarcane, and cotton. This more intensive farming reduced the soil's water absorbency, so more and more rain ran off the fields and less remained to evaporate back into the air. The drier air reduced cloud cover and rainfall. The region grew warmer—a lot warmer. The local weather station recorded an increase in the median annual tempera-

ture of 7.5 degrees Celsius between 1972 and 1990, by which time it had exceeded 30 degrees.[1]

The hotter, drier landscape was poor habitat for the kind of mosquitoes that carry malaria, so the mosquitoes largely died off and malaria infection declined. But of course the land was also becoming less productive, so people began to leave. Many found work on big plantations that were being carved out of the rainforests to the north. The plantations were growing export crops too, primarily bananas, melons, and pineapples. But it is difficult to mass-produce big, succulent fruits in a rainforest—even a badly fragmented rainforest—because there are so many insects and fungi around to eat them. So the plantations came to rely heavily on pesticides. From 1989 to 1991, Honduran pesticide imports increased more than fivefold, to about 8,000 tons.[2]

This steaming, ragged forest was perfect habitat for malaria mosquitoes. Around the plantations, the insecticide drizzle suppressed them for a time, but they eventually acquired resistance to a whole spectrum of chemicals, and that basically released them from human control. When their populations bounced back, they encountered a landscape stocked with their favorite prey: people. And since these people were from an area where malaria infection had become rare, their immunity to the disease was low. Malaria rapidly reasserted itself: from 1987 to 1993, the number of cases in Honduras jumped from 20,000 to 90,000.[3]

The situation was brought to light in 1993 by a group of researchers concerned about the public health implications of environmental decline. But their primary interest was not in what had already happened—it was in what might happen next. Some very nasty surprises might be tangled up somewhere in this web of pressures. They argued, for example, that deforestation and changing patterns of disease had made the country "especially vulnerable to climatic change and climate instability."[4]

They were right. In October 1998, Hurricane Mitch slammed into the Gulf coast of Central America and stalled there for four days. Nightmarish mudslides obliterated entire villages; half the population of Honduras was displaced and the country lost 95 percent of its agricultural production. Mitch was the fourth strongest hurricane to enter the Caribbean this century, but much of the damage was caused by deforestation: had forests been gripping the soil on those hills, fewer villages would have been buried in mudslides. And in the chaos and filth of Mitch's wake there followed tens of thousands of additional cases of malaria, cholera, and dengue fever.[5]

It is hard to shake the feeling that "normal change"—even change for the worse—should not happen this way. In the first place, too many trends in this scenario are spiking. Instead of gradual change, the picture is full of discontinuities—very rapid shifts that are much harder to anticipate. There is a rapid warming in the south, then an abrupt expansion in deforestation in the north, as plantations are developed. Then malaria infections jump. Then those mudslides, in addition to killing thousands of people, cause a huge increase in the rate of topsoil loss.

There also seem to be too many overlapping pressures—too many synergisms. The mudslides were not the work of Mitch alone; they were caused by Mitch plus the social conditions that encouraged the farming of upland forests. The malaria emerged not just from the mosquitoes, but from the movement of a low-immunity population into a mosquito-infested area, and from heavy pesticide use.

Such discontinuities and synergisms frequently catch us by surprise. (See Tables 2–1 and 2–2.) They tend to subvert our sense of the world because we so often assume that a trend can be understood in isolation. It is tempting, for example, to believe that a smooth line on a graph can be used to see into the future: all you have to do is extend the line. But the future of a trend—any trend—depends on the behavior of the entire system in which it is embedded. When we isolate a phenomenon in order to study it, we may actually be preventing ourselves from knowing the most important things about it.

This fragmented form of inquiry is becoming increasingly dangerous—and not just because we might miss problems in small, poor countries like Honduras. After all, there is nothing special about the pressures in the Honduran predicament. Deforestation, climate change, chemical contamination—these and many other forms of environmental corrosion are at work on a

Table 2–1. Three Types of Environmental Surprise

Type of Surprise	Example
A discontinuity is an abrupt shift in a trend or a previously stable state. The abruptness is not necessarily apparent on a human scale; what counts is the time frame of the processes involved.	Overfishing has pushed some fish species into a population crash rather than a gradual decline. As recently as the 1970s, for example, the white abalone occurred along the coast of northern Mexico and southern California at densities of up to 10,000 per hectare. Today its total population is probably no more than a few dozen and its extinction is imminent.
A synergism is a change in which several phenomena combine to produce an effect that is greater than would have been expected from adding up their effects taken separately.	The monstrous 1998 flood of China's Yangtze River did $30 billion in damages, displaced 223 million people, and killed another 3,700. The damage was a synergism caused not just by heavy rains, but by extremely dense settlement of the floodplain and by deforestation—the Yangtze basin has lost 85 percent of its forest cover.
An unnoticed trend, even if it produces no discontinuities or synergisms, may still do a surprising amount of damage before it is discovered.	In the United States, where natural areas are monitored with much greater attention than in most parts of the world, aggressive non-native weeds may still have to be in the country for 30 years or have spread to more than 4,000 hectares before they are even discovered. In the United States and elsewhere, such weeds displace native plants, upset fire and water cycles, and do billions of dollars in agricultural damage every year.

SOURCE: Callum M. Roberts and Julie P. Hawkins, "Extinction Risk in the Sea," *TREE* (Trends in Ecology and Evolution), June 1999; Janet N. Abramovitz and Seth Dunn, "Record Year for Weather-Related Disasters," Vital Signs Brief 98-5 (Washington, DC: Worldwatch Institute, 27 November 1998); U.S. Congress, Office of Technology Assessment, *Harmful Nonindigenous Species in the United States* (Washington, DC: September 1993).

global scale. Each has engendered its own minor research industry. But even as the publications pile up, we may actually be missing the biggest problem of all: what might the inevitable convergence of these forces do?

"When one problem combines with another problem, the outcome may be not a double problem, but a super-problem." That is the assessment of Norman Myers, an Oxford-based ecologist who is one of the most active pioneers in the field of environmental surprise. We have hardly begun to identify those potential super-problems, but in the planet's increasingly stressed natural systems, the possibility of rapid, unexpected change is pervasive and growing.[6]

You can find this potential nearly anywhere, but this chapter will consider three of the most important theaters of surprise—tropical rainforests, coral reefs, and the atmosphere.

Tropical Rainforests: The Inferno Beneath the Canopy

Eight thousand years ago, before people began to clear land on a broad scale, more than 6 billion hectares, or around 40 percent of the planet's land surface, were covered with forest. Today, Earth's tattered cloak of natural forests (as opposed to tree

Table 2–2. Some Major Causes of Discontinuities and Synergisms

Cause	Example
A synergism can produce a discontinuity.	In eastern Canada, a mild long-term drought has reduced streamflow into some of the region's lakes. The weakened streams are washing in less organic debris, so the lakes have grown clearer. The increase in water clarity has combined synergistically with the deterioration of the ozone layer to produce a major discontinuity in ultraviolet (UV) light penetration. Previously, UV light penetrated only the upper 20–30 centimeters of water. In some lakes, it now penetrates to 1.5 meters. Acid rain could clarify the water even further, allowing UV penetration to reach 3 meters. (UV light can injure aquatic organisms just as it injures humans.)
A discontinuity can produce a synergism.	The use of pesticides against the sweet potato whitefly eventually caused a huge discontinuity: when a pesticide-resistant strain of the fly emerged in the 1980s, its populations exploded. In South America, the resistant strain has combined with various plant viruses that the fly can transmit to produce a synergism: fly-virus combinations have forced the abandonment of more than 1 million hectares of cropland.
A positive feedback loop can produce a discontinuity. (A positive feedback loop is a cycle of change that amplifies itself.)	Global warming appears to be shrinking the Arctic ice pack, which is 5 percent smaller than it was in 1996 and considerably thinner. Ice absorbs less than half the sunlight that hits it, but open ocean absorbs some 90 percent. So as the ice dwindles, the oceans will warm more rapidly, which will in turn accelerate the loss of the ice. If the ice pack continues to thin, a major discontinuity is inevitable: an Arctic ocean that is largely free of ice in the summer. That, of course, would cause the oceans to warm even faster.
Cascading effects can lead to multiple discontinuities and synergisms. (Cascading effects occur when a change in one component of a system produces change in another component, which in turn changes yet another component, and so on.)	In the waters off the Alaskan coast, a decline in perch and herring seems to have depressed populations of sea lions and seals, which prey on the fish. Killer whales, which normally prey on the sea lions and seals, turned instead to otters, forcing a collapse in their populations. The otter collapse caused a population explosion of their favorite prey, sea urchins, which in turn demolished the kelp forests on which they feed. The disappearance of the kelp jeopardizes a large number of species—fishes, marine invertebrates, marine mammals, and birds.

SOURCE: David W. Schindler et al., "Consequences of Climate Warming and Lake Acidification for UV-B Penetration in North American Boreal Lakes," *Nature*, 22 February 1996; whitefly from Lori Ann Thrupp, *Bittersweet Harvests for Global Supermarkets* (Washington, DC: World Resources Institute, 1995); Richard Monastersky, "Sea Change in the Arctic," *Science News*, 13 February 1999; J.A. Estes et al., "Killer Whale Predation on Sea Otters Linking Oceanic and Nearshore Ecosystems," *Science*, 16 October 1998.

plantations) amounts to 3.6 billion hectares at most. Every year, at least another 14 million hectares are lost—and maybe considerably more than that. This is an enormous evolutionary tragedy. Among the many thousands of species that are believed to go extinct every year, the overwhelming majority are forest creatures, primarily tropical insects, who have been denied their habitat. That, anyway, is the best estimate, but the forests are vanishing far more rapidly than they can be studied. We really don't even know what we are losing.[7]

Currently, well over 90 percent of forest loss is occurring in the tropics—on a scale so vast that it might appear to have exceeded its capacity to surprise us. In 1997 and 1998, fires set to clear land in Amazonia claimed more than 5.2 million hectares of Brazilian forest, brush, and savanna—an area nearly 1.5 times the size of Taiwan. In Indonesia, some 2 million hectares of forest were torched during 1997 and 1998. All this is certainly news, but if you are interested in conservation, it is the kind of dreadful news you have come to expect.[8]

The damage may not look dramatic, but another tract of forest may be doomed by a positive feedback loop of fire and drying.

And yet our expectations may not be an adequate guide to the sequel, assuming the destruction continues at its current pace. A substantial portion of the damage is "hidden"—it does not show up in the conventional analysis. But once you take the full extent of the damage into account, you can begin to make out some of the surprises it is likely to trigger.

Consider, for example, the destruction of Amazonia. Over half the world's remaining tropical rainforest lies within the Amazon basin, where more forest is being lost than anywhere else on Earth. Deforestation statistics for the area are intended primarily to track the conversion of forest into farms and ranches. Typically, the process begins with the construction of a road, which opens up a new tract of forest to settlement.

In June 1997, for instance, some 6 million hectares of forest were officially released for settlement along a major new highway, BR-174, which runs from Manaus, in central Amazonia, over 1,000 kilometers north to Venezuela. Ranchers and subsistence farmers clearcut patches of forest along the road and burn the slash during the July-November dry season. (The farmers generally have few other options: Brazil has large numbers of poor, land-hungry people, and the plots they cut from the forest usually lose their fertility rapidly, so there is a constant demand for fresh soil.)[9]

But the damage to the forest generally extends much farther than the areas that are "deforested" in this conventional sense, because of the way fire works in Amazonia. In the past, major fires have not been a frequent enough occurrence to promote any kind of adaptive "fire proofing" in the region's dominant tree species. Some temperate-zone and northern trees, by contrast, are "fire-adapted" in one way or another—they may have especially thick bark, for example, or the ability to resprout after burning. The lack of such adaptations in Amazonian trees means that even a small fire can begin to unravel the forest.[10]

During the burning season, the flames often escape the cuts and sneak into neighboring forest. Even in intact forest, there will be patches of forest floor that are dry enough at that time of year to allow a small "surface fire" to feed on the dead leaves. Surface fires do not climb trees and become crown fires. They just crackle along the forest floor, here and there, as little patches of flame, going out at night, when the temperature drops, and rekindling the next day. They will not kill the really big trees, and they do not cover every bit of ground in a burned patch. But they are fatal to most of the smaller trees they touch. Overall, an initial surface fire may kill perhaps 10 percent of the living forest biomass.[11]

The damage may not look all that dramatic, but another tract of forest may already be doomed by an incipient positive feedback loop of fire and drying. After a surface fire, the amount of shade is reduced from about 90 percent to around 60 percent, and the dead and injured trees rain debris down on the floor. So a year or two later, the next fire in that spot finds more tinder, and a warmer, drier floor. Some 40 percent of forest biomass may die in the second fire. At this point, the forest's integrity is seriously damaged; grasses and vines invade and contribute to the accumulation of combustible material. The next dry season may eliminate the forest entirely. Once the original forest is gone, the scrubby second growth or pasture that replaces it will almost certainly burn too frequently to allow the forest to restore itself.[12]

New roads admit not just settlers and ranchers but loggers as well. Commercial logging involves a form of damage that is in some ways similar to the surface fires. Unlike its temperate-zone counterparts, the Amazonian timber industry does not generally clearcut. Most of its operations are what foresters call "high grading"—the best specimens of the most desirable species are cut and hauled out. The result is not outright deforestation, but the forest loses its largest trees and suffers extensive collateral damage from the felling and hauling. An intricate network of logging roads undermines the canopy—a human termite's nest through the wood. An Amazonian timber operation typically kills 10–40 percent of the forest biomass, thereby reducing canopy coverage by up to 50 percent. The forest floor grows warmer and drier; the forest becomes increasingly flammable.[13]

As the main mechanisms of deforestation, the logging and burning are hardly surprising per se, but a great deal of the resulting damage is still obscure. Deforestation estimates are derived primarily from satellite photos, and while the photos show a great deal of detail, the mapping process is generally designed to give a picture of gross canopy destruction. Damage that leaves most of the canopy intact, or that has been masked by second growth, is usually not factored in to the estimates.[14]

In a recent survey, researchers crosschecked satellite maps with field observations and concluded that conventional deforestation estimates for Brazilian Amazonia were missing some 1–1.5 million hectares of severe forest damage done by logging every year. Surface fire damage is harder to quantify, but the same researchers did a fire survey and found that the amount of standing forest that had suffered a surface fire in 1994 and 1995 was 1.5 times the area fully deforested in those years. Overall, they suggested, the area of Amazonian forest attacked by surface fire every year may be roughly equivalent to the area deforested outright. And in some parts of the basin, the extent of this cryptic damage is so great that the conventional measurements may no longer be all that useful. In one region, around Paragominas in eastern Amazonia, the researchers found that although 62 percent of the land was classified as forested, only about one tenth of this consisted of undisturbed forest.[15]

Apart from these direct losses, the logging and burning are likely to trigger various forms of second-order damage through fragmentation. Cutting the forest up into smaller and smaller pieces renders the surviving tracts increasingly vulnerable to "edge effects." Near the edge of a major clearing, competing vegetation often invades forest, choking out saplings. Higher winds dry out the soil and sometimes topple trees. In Amazonia, these effects may extend a kilometer from the edge itself.[16]

As the Amazonian forest dwindles, a more surprising second-order effect may emerge as the hydrological cycle changes. Because trees exhale so much water vapor, a forest to some degree creates its own climate. Much of this water vapor condenses out below the canopy and drips back into the soil. Some of it rises higher before falling back in as rain—researchers estimate that most of the Amazon's rainfall comes from water vapor exhaled by the forest. Widespread deforestation will therefore tend to make the region substantially drier, and that will accelerate the feedback loop created by the fires.[17]

Some degree of deforestation-induced drought already appears to have affected other parts of the humid tropics—parts of Central America, for instance, Côte d'Ivoire, and peninsular Malaysia, where the drying has been severe enough to force the abandonment of some 20,000 hectares of rice paddy. It may not be possible to define the point at which such a drought takes hold in Amazonia, but about 13 percent of the Brazilian Amazon has now been deforested outright. If you add to that the tracts of forest that have been seriously degraded—by logging, surface fires, and fragmentation—that fraction could rise to more than a third.[18]

The fire feedback loop is also likely to gain momentum from forces outside the region. Over the past two decades, Amazonia has seen several unusually intense dry seasons, during which the burning was far worse than normal. These periods corresponded with recent El Niño weather events (in 1982–83, 1992–93, and 1997–98). El Niño events appear to be growing longer, more intense, and more frequent. Many climatologists regard this

trend as a likely effect of climate change—the change in the behavior of Earth's climate system caused by increasing atmospheric concentrations of carbon dioxide (CO_2) and other greenhouse gases. Climate change, in other words, may be accelerating the Amazonian fire cycle. By burning large amounts of coal and oil, the United States, China, and other major carbon-emitting countries may in effect be burning the Amazon.[19]

Other kinds of surprises are lurking in tropical forests as well. As developing countries industrialize, some forest maladies better known in the industrial world are likely to appear in these countries too. Acid rain, for example, is already reported to be affecting the forests of southern China. In parts of South Asia, Indonesia, South America, and West Africa, this form of pollution is bound to increase as industrialization proceeds and cities enlarge. The soil in these areas tends to be fairly acidic already, which would make it incapable of buffering large doses of additional acid. At least in some of these places, acid-induced decline may therefore be much more abrupt than in the temperate zone.[20]

Other development pressures may have no exact temperate-zone analog. Increased hunting pressure, for instance, has emptied out a good number of tropical forests. Many forest-dwelling peoples have armed themselves with shotguns and rifles, which are far more lethal than traditional weapons. And logging is bringing additional hunters into forest interiors. Hunting often supplements the logging: it feeds the loggers, and the surplus is sold as bush meat in towns and cities farther from the frontier. Such hunting is typically very indiscriminate; almost any creature of any size is potential game. Hundreds—even thousands—of animals may fall prey to just a single logging camp. In the Republic of Congo, for example, the annual take in a single camp of 648 people was found to be 8,251 animals, amounting to 124 tons of meat. In tropical forests the world over, mammals, birds, and reptiles play critical roles in pollinating trees, dispersing their seeds, and eating other creatures that prey on the seeds. If hunting continues at its present rate, some tropical tree species are liable to disappear along with the animals themselves.[21]

Recent research suggests that the Central African bushmeat trade may have sparked the AIDS epidemic. It is likely that the original host of the HIV virus was a population of chimpanzees in Cameroon and Gabon. Chimps in that area are commonly hunted for their meat; now, apparently, one of their diseases is hunting us. The appearance of such "rainforest diseases" as AIDS and Ebola suggests one other surprise that could result from tropical deforestation. Would you be willing to bet that humans have already encountered all the potentially lethal pathogens that haunt these forests? And yet that is a bet that our species is collectively making every time another patch of forest is felled.[22]

Coral: Death in the Warming Seas

Coral reefs are perhaps the greatest collective enterprise in nature. Reefs are born from the slow accretion of the skeletons of millions of coral—small, sedentary, cup-shaped animals that form a living veneer over the limestone remnants of their forebears. Reefs form in shallow tropical and warm temperate waters, and host huge numbers of plants and animals. The reef biome is a tiny part of the ocean—it includes only about 0.3 percent of total ocean area. But it is the richest type of ecosystem in the oceans and the second richest on Earth, after tropical rainforests. One out of every four ocean species thus far identified is a reef-dweller, including at least 65 percent of marine fish species.[23]

Coral is extremely vulnerable to heat stress, and the unusually high sea surface temperatures (SSTs) of the past two decades may have damaged this biome just as badly as the unusual fires have damaged tropical forests. Elevated SSTs appear to be a feature of global warming. Atmospheric concentrations of heat-trapping CO_2 are now higher than they have been in 420,000 years; the average annual global temperature had risen from 13.8 degrees Celsius in 1950 to 14.6 degrees by 1998. That is warmer than Earth has been in at least 1,200 years, and most climate scientists expect a further warming, in the range of 1 to 3.5 degrees by 2100, depending on the rate at which CO_2 and other greenhouse gases continue to build up in the atmosphere.[24]

All this is very bad news for coral, which has a fairly limited temperature comfort zone. Heat tolerance varies somewhat from one species to the next, but if the warm-season SSTs climb past 28 degrees Celsius or so, the coral polyp may expel the algae that live within its tissues. This action is known as "bleaching" because it turns the coral white. Coral usually recovers from a brief bout of bleaching, but if the syndrome persists it is generally fatal, because the coral depends on the algae to help feed it through photosynthesis. Published records of bleaching date back to 1870, but show nothing comparable to what began in the early 1980s, when unusually warm water caused extensive bleaching throughout the Pacific. Coral bleached over millions of hectares. By the end of the decade, mass bleaching was occurring in every coral reef region in the world. (See Figure 2–1.) The full spectrum of coral species was affected in these events—a phenomenon that had never been observed before.[25]

In the second half of the 1990s, SSTs set new records in much of the coral's range—in some areas pushing past 30 degrees—and the bleaching became even more intense. Over a vast tract of the Indian Ocean, from the African coast to southern India, 70 percent of the coral appears to have died. Some scientists think that a shift from episodic events to chronic levels of bleaching is now under way. And the problem is not limited to corals; other creatures that live in symbiosis with algae are also bleaching—sea anemones, sponges, even certain types of mollusks.[26]

A huge discontinuity has apparently been triggered by what might seem to be a fairly modest amount of warming. Like most surprises, this one has an air of improbability about it. Coral reefs are among the world's oldest communities, so surely, you might think, they have weathered temperature shifts in the remote past. What could account for their sensitivity now? But the main problem is not so much the size of the temperature shift—it is the fact that the shift is so rapid (in terms of life history) and so pervasive. It is occurring nearly everywhere in the coral's range at once. Reefs will not have an opportunity to "reassemble" themselves in more favorable sites.

Coral reefs are coming under other pressures as well. Because they are shallow-water communities, they generally occur in coastal zones, where they are likely to be exposed to agricultural runoff and sewage. The nitrogen and phosphorus in this runoff fertilizes algae. Along the world's coastlines, algal blooms are growing increasingly common; the

number of low-oxygen "dead zones" created by these blooms is thought to have tripled worldwide over the past 30 years. Out on the reefs, floating algae can starve corals for light; macro-algae—"seaweeds"—can colonize the reefs themselves and displace the coral directly. Pollution-enriched algae is the likely reason Jamaica's reefs never recovered from Hurricane Allen in 1980; 90 percent of the reefs off the island's northwest coast are now just algae-covered lumps of limestone.[27]

The decay of neighboring ecosystems is exacerbating the nutrient pollution and burdening reefs with yet another pressure: siltation. The stretch of shallow, protected water between a reef and the coast often nurtures beds of seagrass. When they are healthy, these beds filter out sediment and effluent that would injure the reefs. But along many tropical coastlines, the seagrass beds themselves are silting up under tons of sediment from development, logging, mining, and the construction of shrimp farms. They are suffocating under algal blooms in nitrogen-polluted waters; they are being poisoned by herbicide runoff.[28]

If you follow the seagrass-choking sedi-

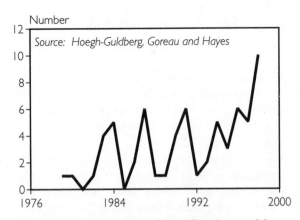

Number

Source: Hoegh-Guldberg, Goreau and Hayes

Figure 2–1. Number of Reef Provinces with Moderate to Severe Bleaching, 1979–98

ment back the way it came, you are increasingly likely to find a shoreline denuded of mangroves—stilt-rooted trees that stabilize coastlines and trap sediment as well. The mangroves' importance as a sediment filter is perhaps greatest in the center of reef diversity, the Australasian region. About 450 coral species are known to grow there; the Caribbean, by comparison, contains just 67 species. Australasia is correspondingly rich in fish too: a quarter of the world's fish species inhabit these waters. But throughout the region, logging and shrimp farming are obliterating mangrove forests and releasing a tremendous burden of silt. Southeast Asia has lost half its mangrove stands over the past half-century. A third of the mangrove cover is gone from Indonesian coasts, three quarters from the Philippines.[29]

In addition to the sedimentation and nutrient pollution, the demise of the seagrass and the mangroves is impoverishing the reef communities in another way as well, through loss of fish habitat. Tropical seagrass provides crucial cover for young fish; it is the major nursery for many species that spend their adult lives out on the reefs. Perhaps 70 percent of all commercially important fish spend at least part of their lives in the seagrass. The mangrove ecosystem is incredibly productive as well—mangrove canopies are home to all sorts of terrestrial animals; their stilt roots harbor a rich array of marine life. [30]

This complex of overlapping stresses is the setting—and perhaps in part the cause—for a series of spectacular epidemics. The first of these that scientists noticed is something now called black-band disease, which emerged in the Caribbean in 1973, off the coast of Belize. Black band is caused by a three-layer complex of blue-green "algae" (actually, cyanobacteria), each layer consisting of a different species; the bottom layer secretes highly toxic sulfides, which kill the coral. The complex creeps very slowly over a head of coral in a narrow band, leaving behind only a bare white skeleton.[31]

Black band has since been joined by a whole menagerie of other diseases. Simply finding common names for them is beginning to tax the lexicon of marine biology: there's white band, yellow band, red band, patchy necrosis, white pox, white plague type I and II, rapid wasting syndrome, dark spot. The modes of action are as various as the names. White pox, for example, is caused by an unknown pathogen that almost dissolves the living coral tissue. Infected polyps disintegrate into mucous-like strands that trail off into the water, and bare, dead splotches appear on the reefs, giving them a kind of underwater version of the mange. Rapid wasting syndrome most likely starts with infection by a fungus that for some reason is extremely attractive to coral-chewing spotlight parrotfish. The fish attack the coral much more aggressively than they would otherwise.[32]

On the reefs off Florida, the number of diseases has jumped from 5 or 6 to 13 during the past decade. In 1996, 9 of the 44 coral species occurring on these reefs were diseased; a year later, the number of infected species had climbed to 28. Coral epidemics are also turning up here and there throughout the Pacific and Indian Oceans, in the Persian Gulf, and in the Red Sea. As with the bleaching, the scale of these epidemics has no clear precedent.[33]

For most of these diseases, a pathogen has yet to be identified; some of them may be syndromes that have nothing to do with pathogens at all. But it is unlikely that the diseases are "new" in the sense of being caused by pathogens that have recently evolved. It is much more likely that the coral's vulnerability to them is new. And in

31

some cases, at least, that could be a kind of synergistic effect. Black band, for example, seems to be particularly virulent in nutrient-polluted water.[34]

Or consider *Aspergillus sydowii*, a pathogen that is killing sea-fan coral around the Caribbean. *A. sydowii* belongs to a very common genus of terrestrial fungi. The last time you threw something out of your refrigerator because it was moldy, you could well have been looking at an *Aspergillus* species. In a very bizarre form of invasion, *A. sydowii* breached the land-sea barrier and found a second home in the ocean. But it evidently took the plunge decades ago and has only been killing sea-fans for some 15 years or so. Why? Part of the answer may be the higher SSTs: *A. sydowii* seems to do better in warmer water. But the reasons for the sudden appearance of other diseases are not yet known.[35]

Whatever those reasons may be, the interplay between coral die-off and climate change is liable to generate additional surprises as both processes gather momentum. Thomas Goreau, president of the Global Coral Reef Alliance, reports that of the 207 coral reef sites the Alliance tracks worldwide, nearly 75 percent experienced SSTs that were high enough to inflict some bleaching in 1998. More than half the affected sites showed bleaching intense enough to kill most of the corals. But the rising SSTs are not the only effects of climate change that the corals will have to contend with. As atmospheric CO_2 increases, more and more of the gas will move from the air into the oceans, where it will work a chemical change that reduces the amount of calcium available for reef-building. A doubling of atmospheric CO_2, expected by 2065, will lead to a 30-percent drop in the amount of calcium that tropical oceans can hold. Even on reefs that escape

the bleaching, coral growth may eventually be stunted for lack of this basic nutrient.[36]

Extensive coral death will begin to jeopardize the very structures of reefs, since the coral will not be able to repair damage from storms and from coral-chewing predators. And because reefs buffer many coasts from wave action, any loss of reef structure is likely to increase coastal erosion, especially from storm surges. This effect will be exacerbated by yet another consequence of climate change: rising sea levels. (Sea levels will rise because warming water expands and because runoff from melting glaciers will increase.) Higher sea levels alone may cause some coral atolls to disappear beneath the waves. And if the reefs begin to crumble, the rising seas may inflict a substantial increase in storm damage on many tropical and warm temperate coasts. Mangrove communities are likely to disintegrate even faster; more silt will settle on the seagrass. Yet more fish habitat will disappear.[37]

About one sixth of the world's coasts are shielded by reefs, and some of these coasts, like those in South and Southeast Asia, support some of the densest human populations in the world. Reef fish account for perhaps 10 percent of the global fish catch; one estimate puts their contribution to the catch of developing countries at 25 percent. The prospect of reef decline gives new meaning to the term "natural disaster," but it is also a social disaster in the making.[38]

The Atmosphere: An Invisible Confluence of Poisons

The delicate membrane of gases that makes up our atmosphere is as thin, comparatively speaking, as the skin of an onion. The atmosphere's outer border is very diffuse—a faint scattering of gas molecules extends into space for hundreds of kilometers—but 90 percent of those molecules lie within 16 kilometers of sea level. Every ecosystem on Earth is linked to the chemistry of the membrane that separates us from outer space, and that chemistry is changing in many ways. Levels of some "trace gases," such as sulfur dioxide, nitrogen oxides, and carbon dioxide, are increasing. Many novel compounds have entered the brew as well—for instance, chlorofluorocarbons, the ozone-destroying chemicals used as refrigerants. From this immense potential for change, look at just three basic phenomena: acid rain, nitrogen pollution, and increasing levels of CO_2.[39]

Fossil fuel combustion is the source of acid rain (which falls not just as rain but also as dry particles). Acid rain is composed in large measure of sulfuric acid, which derives from the sulfur dioxide released by coal-burning power plants and metal smelters. (Sulfur is a common contaminant of coal and metal ores.) Smoke stack "scrubbers" and a growing preference for low-sulfur coal and natural gas have helped reduce sulfur dioxide emissions in North America and Western Europe, although high sulfur emissions are still common elsewhere. The other primary constituent of acid rain is nitric acid, which is generated from the nitrogen oxides in fossil fuel emissions. Unfortunately, nitric acid is likely to be more difficult to control than sulfuric acid, since a substantial share of the nitrogen oxides comes from gasoline burned in the world's expanding fleets of cars. (See Figure 2–2.)[40]

Acid rain can travel downwind for hundreds of kilometers—then fall on forests and farmland, where the idea of air pollution may seem quite incongruous. The acid can dam-

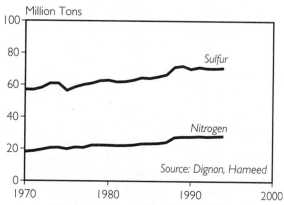

Figure 2–2. World Sulfur and Nitrogen Emissions from Fossil Fuels, 1970–94

age plant tissues directly, but its worst effects come as a series of discontinuities that are much harder to see. As the acid drips into the soil, decade after decade, it tends to leach out the stock of calcium and magnesium, both essential plant nutrients. Depending upon how nutrient-rich a soil is to begin with, this process may or may not be an immediate concern, but if it persists, the nutrient decline will eventually cross a threshold of scarcity: it will begin to cripple plant growth.[41]

A second discontinuity will occur once soil calcium has grown scarce. Without calcium to neutralize it, the incoming acid will just build up in the soil—soil acidification will increase abruptly, even if the amount of incoming acid remains constant. The growing acidity will work another change, by releasing aluminum from its mineral matrix. Aluminum is a common soil constituent; when bonded to other minerals, it is biologically inert, but free aluminum in acid conditions is toxic to both plants and animals. In plants, acidity plus aluminum damages the fine roots. That could affect water uptake, and thereby increase susceptibility to drought. Root damage will also cripple a plant's ability to absorb whatever nutrients remain in the soil.[42]

Acid lowers the calcium level, which allows the acid to build up, which releases aluminum, which interferes with calcium uptake. There is a cascade of chemical effects here, reinforcing the nutrient starvation. Other kinds of second-order effects may emerge as well. In the United States, for example, recent research over a swath of the Midwest from southern Illinois to southern Ohio has uncovered a correlation between increasing acidity in forest soils and a decline of soil organisms—earthworms, beetles, and so on. As biological activity in these soils has dropped off, decay of woody debris appears to have slowed radically, so the calcium "locked up" in the dead wood is not being released back into the soil. The nutrient cycle has apparently been constricted to some degree.[43]

Acid-induced decline may unfold for decades as a hidden process that escapes casual notice.

Because they are chronic, these changes in soil chemistry also create many opportunities for synergisms. Aluminum, for example, is not the only metal that acid tends to "mobilize." Toxic heavy metals such as cadmium, lead, and mercury may also be present in the soil, or they may arrive in trace amounts, on the same winds that brought in the acid. (Like sulfur, heavy metals are common contaminants of coal and ores, although usually in much smaller amounts.) Increasing acidity will tend to make these metals more soluble and toxic as well.[44]

Even though it involves discontinuities, acid-induced decline may still unfold for decades as a hidden process that largely escapes casual notice. Does a tract of forest have fewer large trees than it once did, or fewer species that need more alkaline soil?

Even if it does, it may still be perfectly green. It may still show vigorous growth, but the growth may be concentrated in younger plants, and in acid-tolerant species. It may be on its way to becoming a kind of "acid thicket."

In the eastern United States, over a large portion of Appalachia, the death rate of oaks appears to have doubled and that of hickories to have nearly tripled from 1960 to 1990; a recent review found a strong correlation between these declines and exposure to acid rain and ozone pollution. In the mountains of New Hampshire, at the Hubbard Brook Experimental Forest, acid-induced leaching of minerals has been identified as the main reason vegetation there has shown virtually no net growth since the mid-1980s. Here and there throughout the U.S. Northeast, acidity may be a factor in the failure of the sugar maple, one of the region's dominant tree species, to generate new seedlings. Acid rain is a threat elsewhere in the country as well—in the southern Appalachians, for example, and in the mountains of Colorado.[45]

The worst acid rain, however, is in Asia, particularly in China, which gets 73 percent of its energy from burning coal. The vast quantities of sulfur dioxide released in the process are reportedly now affecting some 270 million hectares of land—more than a quarter of the country's land mass. The acid is reaching Japan and the Koreas as well; Japan, for instance, currently receives more than a third of its sulfur deposition from China. In a recent study, scientists built a computer model of China's energy development and concluded that if the country does not curb its appetite for coal, then over the next two decades acid rain could overwhelm many of the region's soils.[46]

The study is not wholly pessimistic: under its best case scenario, state-of-the-art pollution controls are installed at all of China's coal-burning power plants and factories, causing sulfur dioxide emissions to fall to 31 percent of their 1990 value by 2020. But under a more realistic scenario—with state-of-the-art pollution control installed only in new power plants and some fuel-switching to cleaner energy—sulfur dioxide emissions actually increase by 40 percent over the same time frame. China's coal use is therefore an invitation to widespread discontinuity—not just on an ecosystem level, but also, because of its potential for poisoning cropland, on a social level as well.[47]

Acid rain overlaps with another, broader form of global change: the alteration of the planet's "nitrogen cycle." Nitrogen is an essential plant nutrient and the main constituent of the atmosphere: 78 percent of the air is nitrogen gas. But plants cannot metabolize this pure, elemental nitrogen directly. The nitrogen must be "fixed" into compounds with hydrogen or oxygen before it can become part of the biological cycle. In nature this process is accomplished by certain types of algae and bacteria, as well as by lightning strikes, which fuse atmospheric oxygen and nitrogen into nitrogen oxides.[48]

Humans have radically amplified this process. Farmers boost the nitrogen level of their land through fertilizers and the planting of nitrogen-fixing crops (actually, symbiotic microbes do the fixing). The burning of forests and the draining of wetlands releases additional quantities of fixed nitrogen that had been stored in vegetation and organic debris. And fossil fuel combustion produces still more fixed nitrogen, in the form of nitrogen oxides. Natural processes probably incorporate around 140 million tons of nitrogen into the terrestrial nitrogen cycle every year. (The ocean cycle is largely a mystery.) Thus far, human activity has at least doubled that amount.[49]

Fixed nitrogen is often a "limiting nutrient" in terrestrial ecosystems: it is in high demand and relatively short supply, so its availability determines the amount of plant growth. If you add more, you get more growth—at least in some species. That is why fertilizer is mostly nitrogen. But if you keep adding more, you run into trouble. The excess nitrogen becomes a kind of poison that may interact synergistically with acid rain and other pressures. (Of course, since nitrogen compounds often contribute to the acidity, the processes are not entirely distinct.)[50]

In forests, for example, excess nitrogen tends to inhibit fine root growth, just as the acid-aluminum combination does. Acidity plus nitrogen pollution could therefore deal a double blow to trees' ability to withstand drought and to take in calcium and magnesium. Above ground, excess nitrogen may stimulate extra growth, but it is likely to be produced faster than the tree can absorb those mineral nutrients. The new growth will therefore tend to be weak—essentially, malnourished. This effect also can be exacerbated by acid rain, since acid leaches out those minerals in the first place.[51]

The weakness of the new growth is not just physical—there can be chemical weaknesses too. Since nitrogen may also be a limiting nutrient for insects and other small organisms that feed on trees, nitrogen-rich foliage is likely to be very attractive to pests. And the low mineral concentrations within the tree's tissues can interfere with its ability to produce the chemicals that make up its "immune system"—compounds that, for example, inhibit infection or make foliage less palatable to insects. Other physiological effects are probably at work as well; excess nitrogen, for instance, appears to lower a tree's ability to cope with cold weather.

Combinations of various effects like these may eventually produce substantial discontinuities. In one monitoring experiment in the U.S. Northeast, researchers found that by increasing the nitrogen in pine and spruce-fir stands, they induced declines in growth and increases in tree mortality over a period of just six years.[52]

Nor is it just forests that are at risk from excess nitrogen. In prairies and heaths, too much nitrogen can favor the terrestrial equivalent of algae—whatever fast-growing, weedy species happen to be present—at the expense of slower-growing species that do not have the adaptations necessary to use the extra fertilizer. Several field experiments have shown that this process can have a dramatic homogenizing effect. In one such study, a highly diverse prairie in Minnesota dissolved into a luxuriant patch of fast-growing, aggressive grass. The main beneficiaries of this process are often likely to be non-native "exotic" species, since the ability to grow fast (and therefore to capitalize on any extra nitrogen) will tend to make a plant a good invader in the first place. In this way, nitrogen pollution could converge synergistically with bioinvasion, the spread of exotic species.[53]

Now factor in climate change. Although the processes of climate change are too complex to permit accurate prediction of local effects, the higher latitudes are generally expected to warm much more than the tropics. In the north, the warming is likely to proceed faster than forests can respond by "migrating" further north (where the soil and water conditions permit such movement). Unless carbon emissions are reduced, the result is likely to be substantial forest decline. The immediate causes of decline will likely vary from place to place, but will often involve drying and changes in the freeze-thaw regime during winter and early spring. (Such changes can cause trees to start growing too early in the season.) Both types of change invite an overlap with acid rain and nitrogen pollution, which can make trees less drought- and frost-tolerant. The biggest potential for such an overlap may be in Siberia, where air pollution has degraded vast areas and destroyed some 1 million hectares of forest outright.[54]

In northern forests, unusually warm years often provoke massive defoliation from insects. Recent warming in Alaska, for example, apparently underlies a spruce budworm attack that had chewed up some 20 million hectares of forest by the end of 1998. In parts of northern Europe, southeastern Canada, and the U.S. Northeast, any such climate-driven insect response could combine with the tendency of nitrogen pollution to promote insect damage.[55]

The warming may trigger less direct stresses as well. In the north, climate change is likely to weaken the stratospheric ozone layer, thereby allowing more harmful ultraviolet radiation to reach Earth's surface. (Greenhouse gases are keeping more heat in the lower atmosphere, so the stratosphere is cooling, and air currents are likely to exacerbate that effect in the north. Stratospheric cooling affects the ozone layer because ozone-destroying substances are more effective at lower temperatures.) The extra ultraviolet radiation will damage the foliage of many trees—another overlap with air pollution stress. Chronic damage to foliage slows growth and tends to increase susceptibility to other pressures, such as drought and pests.[56]

Drought stress, pollution, insect attack, ultraviolet light—the critical issue here is not whether any particular synergism will occur, it is the increase in the aggregate risk of a major surprise. As the pressures build, so does the chance of triggering some unanticipated "super-problem."

An Agenda for the Unexpected

Human pressures on Earth's natural systems have reached a point at which they are more and more likely to engender problems that we are less and less likely to anticipate. Dealing with this predicament is obviously going to require more than simply reacting to problems as they appear. We need to forge a new ethic for managing our relationship with nature—one that emphasizes minimal interference in the lives of wild beings and in the broad natural processes that sustain all living things. Such an ethic might begin with three basic principles.

First, nature is a system of unfathomable complexity. Our predominant response to that complexity has been specialization, in both the sciences and public policy. Learning a lot about a little is a form of progress, but it comes at a cost. Expertise is seductive: it is easy for specialists to get into the habit of thinking that they understand all the consequences of a plan. But in a complex, highly stressed system, the biggest consequences may not emerge where the experts are in the habit of looking. This inherent unpredictability condemns us to some degree of error, so it is important to err on the side of minimal disruption whenever possible.[57]

Second, nature gives away nothing for free. You cannot get an appreciable quantity of anything out of nature without sacrificing something in the process. Even sustainable resource management is a trade-off—it's simply one we regard as acceptable. In our dealings with nature, as with any other sort of transaction, we need to know the full cost of the goods before deciding whether they are worth the price, or whether there is a better way to pay for them.[58]

Third, nature has no reset button. Environmental corrosion is not just killing off individual species—it is setting off system-level changes that are, for all practical purposes, irreversible. Even if, for example, all the world's coral reef species were miraculously to survive the impending bout of rapid climate change, that does not mean that our descendents will be able to reconstruct reef communities. The near impossibility of restoring complex systems to some previous state is another strong argument for minimal disruption.[59]

These are basic features of the natural world: we will never understand it completely, it will not do our bidding for free, and we cannot put it back the way it was. A policy ethic sensitive to these facts of life might emphasize the following themes.

Nature has no reset button.

Monoculture technologies are brittle, so plan for diversity. Huge, uniform sectors generally exhibit a kind of superficial efficiency because they generate economies of scale. You can see this in fossil-fuel-based power grids, megadam projects, even in woodpulp plantations. But because they are beholden to vast quantities of invested capital—both financial and political—industrial monocultures are extremely difficult to reform when their hidden costs begin to catch up with them. More diverse technologies are liable to be more adaptable because their investors are not all "betting" on exactly the same future. And whether the goal is the production of irrigation water, paper, or electricity, a more adaptable system is likely to be more durable over the long term. (See Chapters 3, 6, and 8.)[60]

Direct opposition to a natural force is generally counterproductive, so plan to work with nature. Heavy-handed approaches sometimes exacerbate a problem, as when intensive pesticide use causes a population explosion of resistant pests. But successes in this genre are often worse than the outright failures. Dams and levees, for instance, may end up controlling a flood-prone river—and largely killing the riverine ecosystem in the process. Sound management often tends to be more "oblique" than direct. Restoration of water-absorbing floodplain ecosystems can make for more effective flood control than dams. Cropping systems that mimic natural floral diversity make it harder for any particular pest to dominate a field.[61]

You can never have just one effect, so plan to have several. Thinking through the likely "ripple effects" of a plan will help locate not just the risks, but additional opportunities. Encouraging organic agriculture in the U.S. Midwest, for example, could help reduce nutrient leakage into the Mississippi. That, in turn, could ease the stress on Caribbean reefs. Reef conservation may therefore "overlap" with agricultural reform; it might even be possible to extend this overlap to include reform of the North American diet. (See Chapter 4.) Environmental policy is full of such latent positive synergisms. In many countries, for instance, there may be a powerful overlap between the need for meaningful employment and the need to replace the "throwaway" economy with one that emphasizes durable goods. (See Chapter 9.)[62]

Solutions are almost never permanent, so plan to keep on planning. In the 1950s, organochlorine pesticides were hailed as a permanent "fix" for insect pest problems; given the pervasive ecological damage that these chemicals are now known to cause,

the idea of a permanent chemical solution to anything may seem rather naive today. (See Chapter 5.) But because our relationship with nature is in a constant state of flux, even realistic fixes will need regular revision. The Montreal Protocol is not a permanent patch for the ozone layer, in part because climate change will probably exacerbate ozone loss. The Green Revolution is not a permanent answer to world hunger, in part because conventional agriculture is overtaxing aquifers. (See Chapter 3.) The growing strain on Earth's natural systems will probably force an increase in the tempo of policy revision—so it makes sense to take full advantage of the powerful new information and communications technologies. (See Chapter 7.) Because of their ability to bring together enormous quantities of data from different areas and disciplines, such technologies could help counter the blinkering effects of specialization.

None of us may find the answer alone, but together we probably can. In social as well as natural systems, there is a potent class of properties that exists only on the system level—properties that cannot be directly attributed to any particular component. In a political system, for example, institutional pluralism can create a public space that no single institution could have created alone. One of the most important policy activities may therefore be to encourage innovation outside policy institutions. (See Chapter 10.) Policy may need to become increasingly a matter of creating not so much solutions per se as the conditions from which solutions can arise. In the face of the unexpected, our best hopes may lie in our collective imagination.

Chapter 3

Redesigning Irrigated Agriculture

Sandra Postel

Six thousand years ago, settlers in Mesopotamia embarked upon a new way of growing food. They had migrated south from the Mesopotamian highlands to the drier plains between the Tigris and Euphrates Rivers, the area that is now southern Iraq. In their new homeland, crops would sprout and grow but then wither from dryness before harvest time. These migrants, who came to be known as Sumerians, devised a simple remedy for this dilemma that had profound and lasting effects. They dug a ditch and diverted some water to their fields from the Euphrates River—and, in so doing, gave rise to the practice of irrigation.

Irrigation transformed both the land and human society like no other activity had before. It created large food surpluses that had to be stored and distributed, which in turn led to new forms of centralized management. Those surpluses also freed many people to pursue nonfarm activities. Historians credit the Sumerians with the development of writing and the wheel as well as the creation of sailboats, water-lifting devices, and yokes for harnessing animals to plows.

Over time, as the range of activities widened, these early societies became more stratified. Populations increased in size and density, leading to the first true cities. The Sumerians were followed by a number of other irrigation-based civilizations in Mesopotamia, including the famed Babylonians, who gave the world magnificent palaces and Hammurabi's historic code of law. Outside this region, early irrigation societies also arose in the Nile River valley of Egypt, the Indus River valley of present-day Pakistan, the Yellow River basin of north-central China, and, much later, in several river valleys in North and South America. Each of these produced unique cultural and scientific advances. On the whole, irrigation established a new foundation from which civilizations blossomed and

Sandra Postel is Director of the Global Water Policy Project in Amherst, Massachusetts, and a Senior Fellow with Worldwatch. This chapter is based on her book, *Pillar of Sand* (New York: W.W. Norton & Company, 1999).

profoundly shaped societal development.

At the same time, history tells us that in the long run most irrigation-based societies fail. The inherent environmental instability of irrigated agriculture can weaken seemingly advanced cultures, rendering them less able to cope with political and social disturbances. Salt can poison the soil, sapping its fertility and causing crop production to decline. Dams, canals, and levees can fall into disrepair, be wiped out by floods, or be destroyed by enemies. Competition over irrigated land and water can contribute to militarism and regional warfare. All told, history suggests that the irrigation base, upon which everything else rests, requires constant vigilance; if it is neglected, a cascade of destabilizing effects can unfold.[1]

As we begin a new century and millennium, these lessons are much more than a historical curiosity. Irrigation underpins our modern society just as it did so many failed societies of the past. Because irrigated farms typically get higher yields and can grow two or three crops per year, the spread of irrigation has been a key driver in this century's rise in food production. Many nations, including China, Egypt, India, Indonesia, and Pakistan, rely on irrigated land for more than half of their domestic food output. Today some 40 percent of the world's food comes from the 17 percent of cropland that is irrigated. India, China, the United States, and Pakistan together account for over half of the world's irrigated land; the top 10 countries collectively account for two thirds of the global total. (See Table 3–1.)[2]

Worldwide, after a remarkable period of growth, the pace of irrigation's spread slowed substantially toward the end of the twentieth century. Between 1982 and 1994, global irrigated area grew at an average rate of 1.3 percent a year, down from an annual rate of 2 percent between 1970 and 1982. Irrigation expansion began to reach diminishing returns. In most areas, the best and easiest sites were already developed; bringing irrigation water to new sites was more difficult and costlier.[3]

Over the next couple decades, the global irrigation base is unlikely to grow faster than 0.6 percent a year, and even this may turn out to be optimistic because of worsening soil salinization and shortages of irri-

Table 3–1. Irrigated Area in the Top 20 Countries and the World, 1995

Country	Irrigated Area	Share of Cropland That is Irrigated
	(million hectares)	(percent)
India	50.1	29
China	49.8	52
United States	21.4	11
Pakistan	17.2	80
Iran	7.3	39
Mexico	6.1	22
Russia	5.4	4
Thailand	5.0	24
Indonesia	4.6	15
Turkey	4.2	15
Uzbekistan	4.0	89
Spain	3.5	17
Iraq	3.5	61
Egypt	3.3	100
Bangladesh	3.2	37
Brazil	3.2	5
Romania	3.1	31
Afghanistan	2.8	35
Italy	2.7	25
Japan	2.7	62
Other	52.4	—
World	255.5	17

SOURCE: U.N. Food and Agriculture Organization, *1996 Production Yearbook* (Rome: 1997).

gation water. By 2020, per capita irrigated area will likely be 17–28 percent below the peak, which occurred in 1978.[4]

As the twenty-first century opens, the need for new approaches to irrigated agriculture is clear. Sixty percent of our irrigation base is less than 50 years old, yet threats to the continued productivity of much of it have surfaced with unsettling speed. One out of five hectares of irrigated land is damaged by salt, the silent scourge that plagued ancient Mesopotamian societies. More and more rivers are running dry for parts of the year, leaving agriculture vulnerable to the reallocation of water to burgeoning cities and industries. And damage to fisheries and aquatic habitats from excessive water use is creating pressure to leave more water in rivers, streams, and lakes.[5]

At the same time, the cultural, economic, and political forces that shaped the modern irrigation era and underpinned the impressive rise of irrigation around the world are being realigned. From western urbanites and recreation enthusiasts in the United States to environmentalists and human rights advocates in India, new voices have begun to question irrigation's ends and means. The upshot is the need for a redesign of irrigated agriculture to make it more efficient, equitable, and environmentally sound.

Mounting Water Deficits

Of all the vulnerabilities characterizing irrigated agriculture today, none looms larger than the depletion of underground aquifers. Across large areas, farmers are pumping groundwater faster than nature is replenishing it, causing a steady drop in water tables. Just as a bank account dwindles if withdrawals routinely exceed deposits, so will an underground water reserve decline if pumping routinely exceeds recharge. During the last several decades, as the number of groundwater wells skyrocketed, aquifer depletion has spread from isolated pockets to large areas of irrigated cropland. The problem is now widespread in central and northern China, northwest and southern India, parts of Pakistan, much of the western United States, North Africa, the Middle East, and the Arabian Peninsula.

In India, the situation is so severe that in September 1996 the Supreme Court directed one of the nation's premier research centers to examine the problem of declining groundwater levels. The National Environmental Engineering Research Institute (NEERI) found that "overexploitation of ground water resources is widespread across the country" and that water tables in critical agricultural areas are sinking "at an alarming rate."[6]

Nine Indian states are now running major water deficits, which in the aggregate total just over 100 billion cubic meters a year, and those deficits are growing. The situation is particularly serious in Punjab and Haryana, India's principal breadbaskets. Village surveys found that water tables are dropping 0.6–0.7 meters per year in parts of Haryana and half a meter per year across large areas of Punjab. In the western state of Gujarat, 87 out of 96 observation wells showed declining groundwater levels during the 1980s, and aquifers in the Mehsana district are now reportedly depleted. Overpumping in this western coastal state has also caused salt water to invade freshwater aquifers, contaminating drinking water supplies. David Seckler, Director General of the International Water Management Institute in Sri Lanka, estimates that a quarter of India's grain harvest could be in jeopardy from groundwater depletion.[7]

Farmers usually face problems long before their wells run dry. At some point, the cost of pumping gets too high or their well yields drop too low to continue irrigating at the same level. They then have several options. They can take irrigated land out of production, eliminate one or more harvests, switch to less water-intensive crops, or adopt more-efficient irrigation practices. Apart from shifting out of thirsty nonfood crops like sugarcane or cotton, improving efficiency is the only option that can sustain food production while lowering water use. Yet investments in efficiency are minuscule relative to the challenge at hand.[8]

Besides constraining future food production, groundwater overpumping is widening the income gap between rich and poor in some areas. As water tables drop, farmers must drill deeper wells and buy larger pumps to lift the water to the surface. The poor cannot afford these technologies. In parts of Punjab and Haryana, for example, wealthier farmers have installed deeper, more-expensive tubewells at a cost of about 125,000 rupees ($2,890). As the shallower wells dry up, some of the small-scale, poorer farmers end up renting their land to the larger well owners and becoming laborers on these larger farms.[9]

Other countries are facing similar problems. In Pakistan's province of Punjab, the country's leading agricultural region, groundwater is being pumped at a rate that exceeds recharge by an estimated 27 percent. In Bangladesh, groundwater use is about half the level of natural replenishment on an annual basis. During the dry season, however, when irrigation is most needed, heavy pumping causes many wells to go dry. On about a third of Bangladesh's irrigated area, water tables routinely drop below the suction level of shallow tubewells during the dry months. Monsoon rains

recharge these aquifers and water tables rise again later in the year, but farmers run out of water when they need it most. Again, the greatest hardships befall poor farmers, who cannot afford to deepen their wells or buy bigger pumps.[10]

In China, the world's largest grain producer, groundwater conditions are unsettling as well. Northern China is running a chronic water deficit, with groundwater overpumping amounting to some 30 billion cubic meters (bcm) a year. Across much of the north China plain, which produces roughly 40 percent of China's grain, the water table has been dropping 1–1.5 meters a year, even as water demands continue to increase.[11]

How the Chinese respond to this predicament—whether they take land out of irrigated agriculture, switch to less thirsty crops, or irrigate more efficiently—will make a big difference to China's grain outlook. The projected 2025 water deficit for the Hai and Yellow River basins is roughly equal to the volume of water needed to grow 55 million tons of grain—14 percent of the nation's current annual grain consumption and more than a quarter of current global grain exports.[12]

In the United States, farmers are overpumping aquifers in several important crop-producing regions. California is overdrafting groundwater at a rate of 1.6 bcm a year, equal to 15 percent of the state's annual groundwater use. Two thirds of this depletion occurs in the Central Valley, which supplies about half of the nation's fruits and vegetables.[13]

By far the most serious case of depletion, however, is in the region watered by a geologic formation called the Ogallala—one of the planet's greatest aquifers. The Ogallala spans portions of eight states, and prior to exploitation held a volume of water equivalent to more than 200 years of Colorado River flow. Today, the Ogallala alone waters one fifth of U.S. irrigated land. Particularly in its southern reaches, this aquifer gets very little replenishment from rainfall, so almost any pumping diminishes it. Currently it is being depleted at a rate of about 12 bcm a year. Total depletion to date amounts to some 325 bcm, a volume equal to the annual flow of 18 Colorado Rivers. More than two thirds of this depletion has occurred in the Texas High Plains.[14]

Driven by falling water tables, increased pumping costs, and historically low crop prices, many farmers who depend on the Ogallala have already abandoned irrigated agriculture. At its peak in 1978, the total area irrigated by the Ogallala in Colorado, Kansas, Nebraska, New Mexico, Oklahoma, and Texas reached 5.2 million hectares. Less than a decade later, this had fallen to 4.2 million hectares. A long-range study of the region, done in the mid-1980s, suggested that more than 40 percent of the peak irrigated area would come out of irrigation by 2020; if this happens, another 1.2 million hectares will either revert to dryland farming or be abandoned over the next two decades.[15]

Much of North Africa and the Arabian Peninsula, where it rarely rains, is underlain by "fossil aquifers"—remnants of ancient climates that were much wetter than current conditions. These aquifers get little or no replenishment from current rainfall, so pumping from them permanently depletes the supply, much as extracting oil depletes an oil reserve.

Saudi Arabia sits atop several aquifers estimated to contain (prior to large-scale exploitation) some 2,000 cubic kilometers of water—just over half as much as the Ogallala. A massive two-decade experiment with desert agriculture drew heavily on these sources for the production of wheat. In its peak years of grain production during the early 1990s, the nation ran a water deficit of some 17 bcm a year, using up more than 3,000 tons of water for each ton of grain produced in the hot, windy desert. At that rate, underground water reserves would have run out by 2040. In the mid-1990s, when budgetary problems forced King Fahd's government to slash agricultural subsidies, grain production fell by 60 percent within two years. The groundwater depletion rate has dropped back as well, but the Saudis are likely still running a water deficit of some 6 bcm a year.[16]

Northern China is running a chronic water deficit, with groundwater overpumping of some 30 billion cubic meters a year.

Africa's northern tier of countries—from Egypt to Morocco—also relies heavily on fossil aquifers, with estimated depletion running at 10 bcm a year. Nearly 40 percent of this occurs in Libya, which is now pursuing a massive water scheme known as the Great Man-Made River Project. The $25-billion project aims at pumping water from desert aquifers in the south and transferring it 1,500 kilometers north through a vast system of concrete pipe. As of early 1998, it was delivering 146 million cubic meters a year to Tripoli and Benghazi. If all stages are completed, the scheme will eventually transfer up to 2.2 billion cubic meters a year, with 80 percent of it slated for agriculture. As in Saudi Arabia, however, the greening of the desert will be short-lived: some engineers say the wells would likely dry up in 40–60 years.[17]

The upshot of this survey of groundwater use is that many food-producing regions are sustained by the hydrologic equivalent of deficit financing. Irrigators are drawing down water reserves to support today's production, racking up large water deficits that at some point will have to be balanced. Collectively, estimated annual water depletion in India, China, the United States, North Africa, and the Arabian Peninsula adds up to about 160 billion cubic meters a year—equal to the annual flow of two Nile Rivers. (See Table 3–2.) Factoring in Australia, Pakistan, and other areas for which comparable data are not available would likely raise this figure by at least 10 percent.

The vast majority of this overpumped groundwater is used to irrigate grain, the staples of the human diet. Since it takes about 1,000 tons of water to produce one ton of grain (and a cubic meter of water weighs one ton), some 180 million tons of grain—roughly 10 percent of the global harvest—is being produced by depleting water supplies. This finding raises an unsettling question: If so much of irrigated agriculture is operating under water deficits now, where are farmers going to find the

additional water that will be needed to feed the more than 2 billion people projected to join humanity's ranks by 2030?

The New Water Wars

There is an old saying in the western United States that "water flows uphill toward money." In the popular movie *Chinatown*, Hollywood capitalized on the drama of Los Angeles sucking farms dry in the Owens Valley. American writer and humorist Mark Twain captured the West's tension over water with his famous quip that "whiskey's for drinking, water's for fighting about." As water becomes more scarce, competition for it is increasing—not just in the western United States, but in many parts of the world.[18]

Rapidly growing cities and industries are increasingly looking to irrigated agriculture as the last big pool of available water. Mark Rosegrant and Claudia Ringler of the International Food Policy Research Institute in Washington, D.C., project that annual water demands by households and industries in developing countries will climb by 590 billion cubic meters between 1995 and 2020—a volume equivalent to the annual flow of seven Nile Rivers. They project that the share of water going to these activities will more than double—from 13 percent of total water use to 27 percent.[19]

Some portion of these increased residential and industrial demands will be met by shifting irrigation water out of agriculture. What is not known is how much water irrigators will transfer, whether they will transfer that water voluntarily, and to what extent crop production will decline as a result of the transfers. If, for the sake of illustration, half of the projected rise in urban and industrial demand by 2020 is met by shifting irri-

Table 3–2. Water Deficits in Key Countries and Regions, Mid-1990s

Country/Region	Estimated Annual Water Deficit
	(billion cubic meters per year)
India	104.0
China	30.0
United States	13.6
North Africa	10.0
Saudi Arabia	6.0
Other	unknown
Minimum Global Total	163.6

SOURCE: Various references cited in the text and author's estimates.

gation water to these users, and little improvement is made in irrigation efficiency, grain production could drop by some 300 million tons—one sixth of the current global harvest. As Rosegrant and Ringler conclude, the way the farm-to-city reallocation of water is managed "could determine the world's ability to feed itself."[20]

In China, the number of cities has climbed from 130 in 1949 to more than 600 today. About half of them are already short of water, and there is increasing pressure to pull supplies away from agriculture to narrow the urban water deficits. A mid-1990s planning study by China's State Statistical Bureau and Ministry of Water Resources concluded that 40 percent of the projected demand gap in 2000 could be met by shifting water out of agriculture. Moreover, these gaps will widen over the next couple of decades. The United Nations projects that more than half of China's people will live in cities by 2025, up from a third at present.[21]

In China, as elsewhere, both politics and economics drive water's reallocation. A cubic meter of water used in China's industries generates more jobs and about 70 times more economic value than the same quantity used in agriculture. As supplies tighten, water will shift to where it is more highly valued. Already in parts of north China, reservoirs that had supplied irrigation water to farms are now used almost exclusively to supply households and factories. Farmers in Daxing County, about 50 kilometers south of Beijing, for example, no longer receive irrigation water that used to be shipped in by canal from the city's reservoirs.[22]

Farmers in India face mounting competition over water as well. India is expected to add some 340 million people to its cities between 1995 and 2025, more than the current populations of the United States and Canada combined. Reallocations are reportedly occurring to increase supplies for the cities of Madras, Coimbatore, and Tirupur and for a number of smaller towns. Many farmers within 35 kilometers of Tirupur have abandoned farming and instead sell their groundwater to urban and industrial users.[23]

Rice farmers in parts of the Indonesian island of Java are losing water supplies to textile factories, even though Indonesian law gives agriculture higher priority for water. A study of one irrigated region in West Java found that factories often take more water than their permits allow and also take it directly out of irrigation canals, leaving too little for the farms. Some factories buy or rent rice fields from farmers just to get access to the irrigation water, leaving the fields fallow. Lacking legally enforced rights to the water they have been accustomed to using, the farmers have little recourse. Researchers Ganjar Kurnia, Teten Avianto, and Bryon Bruns note that "many farmers, suffering from lost production and insecurity of water supplies, feel they have no choice but to sell their land."[24]

Overall, cities in industrial countries will likely pull less water out of agriculture because their water demands are rising more slowly. To the extent residential and industrial users improve their own water-use efficiency—for example, by installing more-efficient fixtures, fixing leaks, recycling wastewater, and replacing thirsty lawns with native grasses and shrubs—the need to shift water out of agriculture will be lessened. But in rapidly urbanizing, water-short areas, such as the western United States, the city-farm competition is heating up. Cities are buying water, water rights, or land that comes with water rights in parts of Arizona, California, Colorado, and elsewhere.

Although these trades are voluntary, they

are not without controversy. The biggest ones so far have involved the Imperial Irrigation District (IID) in southern California. Thanks to a century-old deal with the federal government, IID gets about one fifth of the Colorado River's annual flow essentially for free. The irrigation district is within striking distance of urban areas that are home to 16 million people and still growing. Not surprisingly, these cities look to IID for water.

A decade ago, IID agreed to a trade with the Metropolitan Water District (MWD) of Los Angeles, the water wholesaler for southern California cities. Under this agreement, no cropland needs to come out of production. MWD is investing in efficiency improvements within IID in exchange for the water those investments save. The trade will shift up to 106,000 acre-feet (130.8 million cubic meters) a year from farm to urban uses for 35 years. MWD benefits because the cost of the conserved water will be less than 10¢ per cubic meter, much lower than its next best supply option. IID benefits from the cash payments and an upgraded irrigation network, while keeping its land in production.[25]

In 1998, IID struck a deal with San Diego to transfer up to 246.8 million cubic meters of water a year at an initial price of 20–27¢ per cubic meter. Again, if most of the water that is transferred results from increased efficiency and shifts to less thirsty crops, farmers will not necessarily need to take land out of production.[26]

Water transfers often affect people not involved directly in the sale, which makes a full accounting of costs and benefits hard to achieve. But the costs to so-called third parties, who rarely have a place at the negotiating table, can be substantial—especially where water trades do result in land coming out of production. These costs can also be cumulative, affecting employment, the tax base, the stability of rural communities, and the environment. Because poor farm laborers may be the ones to lose jobs, even economically efficient water trades may worsen inequities.

Without a doubt, cities in water-short areas will continue to siphon water away from agriculture. At the same time, there is growing pressure in parts of Australia, Central Asia's Aral Sea basin, the United States, and elsewhere to transfer some water from human uses—especially irrigation—back to the natural environment. Citizens, scientists, and political leaders concerned about the loss of fisheries, wetlands, lakes, and other ecological assets are shifting the balance of power governing water use.

The states in Australia's Murray-Darling river basin, for example, have agreed to allocate 25 percent of the river's natural flow to maintaining the system's ecological health. A basinwide freeze has been placed on withdrawals for irrigation. The basin commission has also recommended capping allocations to major cities and towns at projected year-2000 levels of water use and has suggested that cities meet any additional demands by purchasing water from irrigators. A number of proposals and plans are afoot in the western United States as well to return water to the natural environment. A federal law passed in 1992, for instance, dedicates about 10 percent of the water from the Central Valley Project in California to maintaining fish and wildlife habitat.[27]

Competition for water is also increasing between countries, as populations continue to grow rapidly in some of the most water-short regions. In five of the world's hot spots of water dispute—the Aral Sea region, the Ganges, the Jordan, the Nile, and the Tigris-Euphrates—the population of the nations within each basin is projected to climb between 32 percent and 71 percent by 2025. (See Table 3–3.) In the absence of water-sharing agreements, this competition could lead to regional instability or even conflict.

When a country's renewable water supplies drop below about 1,700 cubic meters per capita (what some analysts call the water-stress level), it becomes difficult for that country to mobilize enough water to satisfy all the food, household, and industrial needs of its population. Countries in this situation typically begin to import grain, reserving their water for household and industrial uses. As noted, each ton of grain represents about 1,000 tons of water, so

countries in effect balance their water books by purchasing grain rather than growing it themselves.[28]

At present, 34 countries in Africa, Asia, and the Middle East are classified as water-stressed, and all but two of them—South Africa and Syria—are net importers of grain. Collectively, these water-stressed countries import nearly 50 million tons of grain a year—about a quarter of the total traded internationally. By 2025, the number of people living in water-stressed countries is projected to climb from 470 million to 3 billion—more than a sixfold increase. The vast majority of water-stressed populations will be in Africa and South Asia, where

Table 3–3. Populations in Selected Hot Spots of Water Dispute, 1999, with Projections to 2025

River Basin/Countries	Total 1999 Population	Projected 2025 Population	Change
	(million)		(percent)
Aral Sea[1] Kazakhstan, Kyrgyzstan, Tajikistan, Turkmenistan, Uzbekistan	56	74	+ 32
Ganges Bangladesh, India, Nepal	1,137	1,631	+ 43
Jordan Gaza, Israel, Jordan, Lebanon, Syria, West Bank	34	58	+ 71
Nile Burundi, Democratic Republic of Congo, Egypt, Eritrea, Ethiopia, Kenya, Rwanda, Sudan, Tanzania, Uganda	307	512	+ 67
Tigris-Euphrates Iraq, Syria, Turkey	104	156	+ 50

[1]Excluding Afghanistan and Iran, which hydrologically are part of the basin.
SOURCE: Global Water Policy Project; population projections from Population Reference Bureau, *1999 Population Data Sheet*, wallchart (Washington, DC: 1999).

the deepest pockets of poverty and hunger are today. It therefore seems dangerous to presume that there will be enough exportable grain to meet increased import demands at a price the poorer countries can afford. And with world food aid at its lowest level since the mid-1950s, relying on the generosity of grain-surplus nations to fill food gaps is a risky strategy.[29]

The Productivity Frontier

For the last half-century, agriculture's principal challenge has been raising land productivity—getting more crops out of each hectare of land. As we move into the twenty-first century, the new frontier is boosting water productivity—getting more benefit from every liter of water devoted to crop production.

More than half of the water removed from rivers and aquifers for irrigated agriculture never benefits a crop. Because water performs many functions as it travels through the landscape toward the sea, however, it is important to think systemically about where water goes once it comes under human management. Is it actually lost through evaporation to the atmosphere? If so, reducing evaporative losses will increase the available supply—a true savings of water. Or does the water seep through a canal and recharge an underlying aquifer, which another user then taps into with a well? If so, reducing those canal losses may allow more water to remain in the river or aquifer from which it was taken, but it may also reduce the groundwater supply someone else depends on. By tracking water's flow through the landscape and thinking about its varied roles and functions, creative new ways may be found of getting more benefit out of each liter we extract from nature.[30]

There is a long and growing list of measures that can increase agricultural water productivity. (See Table 3–4.) The key is to custom-design strategies to fit the farming culture, climate, hydrology, crop choices, water use patterns, environmental conditions, and other characteristics of each particular area. Successful strategies almost always involve a synergistic mix of measures. Farmers will not invest in efficient technologies, for example, if they have no incentive to do so. And these technologies will only improve water productivity if accompanied by good management practices.

Drip irrigation ranks near the top of measures with substantial untapped potential. Consisting of a network of perforated plastic tubing installed on or below the soil surface, drip systems deliver water directly to the roots of individual plants. In contrast to a flooded field, which allows a large share of water to evaporate without benefiting a crop, drip irrigation results in negligible evaporation losses. When combined with soil moisture monitoring or other ways of assessing crops' water needs accurately, drip irrigation can achieve efficiencies as high as 95 percent, compared with 50–70 percent for more conventional flood or furrow systems.[31]

Besides saving water, drip irrigation usually increases crop yield and quality—a result of maintaining a nearly ideal moisture environment for the plants. In countries as diverse as India, Israel, Jordan, Spain, and the United States, studies have consistently shown drip irrigation to cut water use by 30–70 percent and to increase crop yields by 20–90 percent—often leading to a doubling of water productivity.[32]

Over the last two decades, the area of land irrigated by drip and other micro-irrigation methods has risen 50-fold, to an estimated 2.8 million hectares. Nonetheless, this total represents just over 1 percent of all irrigated

Table 3–4. Menu of Options for Improving Irrigation Water Productivity

Category	Option or Measure
Technical	• Land leveling to apply water more uniformly • Surge irrigation to improve water distribution • Efficient sprinklers to apply water more uniformly • Low-energy precision application sprinklers to cut evaporation and wind drift losses • Furrow diking to promote soil infiltration and reduce runoff • Drip irrigation to cut evaporation and other water losses and to increase crop yields
Managerial	• Better irrigation scheduling • Improving canal operations for timely deliveries • Applying water when most crucial to a crop's yield • Water-conserving tillage and field preparation methods • Better maintenance of canals and equipment • Recycling drainage and tail water
Institutional	• Establishing water user organizations for better involvement of farmers and collection of fees • Reducing irrigation subsidies and/or introducing conservation-oriented pricing • Establishing legal framework for efficient and equitable water markets • Fostering rural infrastructure for private-sector dissemination of efficient technologies • Better training and extension efforts
Agronomic	• Selecting crop varieties with high yields per liter of transpired water • Intercropping to maximize use of soil moisture • Better matching crops to climate conditions and the quality of water available • Sequencing crops to maximize output under conditions of soil and water salinity • Selecting drought-tolerant crops where water is scarce or unreliable • Breeding water-efficient crop varieties

SOURCE: Amy L. Vickers, *Handbook of Water Use and Conservation*, prepublication manuscript (Amherst, MA: Amy Vickers & Associates, 1999); J.S. Wallace and C.H. Batchelor, "Managing Water Resources for Crop Production," *Philosophical Transactions of the Royal Society of London: Biological Sciences*, vol. 352 (1997), pp. 937–47.

land worldwide. A few recent developments suggest, however, that drip's share could expand markedly in the years ahead.[33]

First, researchers and farmers have been getting good results with drip irrigation of cotton and sugarcane—two thirsty and widely planted crops. In addition, some Kansas farmers who pump water from the dwindling Ogallala aquifer are drip-irrigating corn—the third most widely planted crop worldwide. They have typically reduced their water use by half compared with flood irrigation and by 20–30 percent compared with conventional center-pivot sprinklers. Second, there now exists a spectrum of different drip systems keyed to various income levels and farm sizes that makes the technology affordable and practical for many more farmers. R.K. Sivanappan, a researcher in Coimbatore who is often called the father of drip irrigation in India, believes that drip's potential there may exceed 10 million hectares—an area equal to a fifth of the nation's current irrigated land.[34]

Better information about how much irrigation water crops really need can also raise water productivity. For instance, California's Department of Water Resources operates the California Irrigation Management Information Service (CIMIS), a network of more than 100 automated and computerized weather stations in key agricultural areas of the state. Each station hourly collects local climate data—including solar radiation, wind speed and direction, relative humidity, rainfall, and air and soil temperature—and then transmits this information to a central computer in Sacramento. For each station site, the computer then calculates the reference evapotranspiration rate, which is the amount of water that grass would transpire at that particular location. Using this benchmark, farmers can calculate the evapotranspiration rates of their particular crops and determine quite accurately when and how much to irrigate.[35]

A 1995 survey conducted by the University of California at Berkeley found that, on average, farmers using CIMIS to schedule irrigation experienced an 8-percent yield increase and a 13-percent reduction in water use. Economic benefits ranged from $99 per hectare of alfalfa to $927 per hectare of lettuce.[36]

Similar agricultural weather networks are operating in Arizona, in Washington state, and in the Pacific Northwest region overall. U.S. farmers who own computers can also time their irrigations more accurately by using one of several popular computer programs, including the U.S. Department of Agriculture's *NRCS Scheduler*. At the global level, the International Water Management Institute has developed a computerized tool for irrigation and crop planning called the World Water and Climate Atlas. Available on CD-ROM and through the Internet, the atlas integrates agricultural weather data from 56,000 weather stations around the world for the period 1961 to 1990. It can be used to better match crops and crop varieties to local climate conditions as well as to determine whether rainfall is adequate to grow certain crops in certain locations.[37]

Despite the myriad technologies and measures available to raise agricultural water productivity, there are few examples of countries or even large farming districts that have implemented effective programs. Israel began pushing for greater water efficiency soon after it was founded a half-century ago. Nationwide, water use per hectare dropped from an annual average of 8,200 cubic meters in 1951 to about 5,200 cubic meters in 1985—a 37-percent drop. Though impressive, these gains were mainly achieved with vegetables, fruits, and other high-value crops—not with the widely planted crops that consume the bulk of the world's irrigation water.[38]

In dry and drought-prone northwest Texas, where irrigation depends on the Ogallala aquifer, farmers and researchers have made substantial progress with cereal grains and cotton. Spearheaded by the High Plains Underground Water Conservation District in Lubbock, which oversees water management in 15 counties of northwest Texas, the effort has involved agricultural researchers, extension agents, district officials, and farmers in a major upgrade of the region's irrigation systems.

Several technologies have been key to the program's success. Time-controlled surge valves that distribute water more uniformly down the parallel furrows of gravity systems along with new low-pressure sprinkler designs have produced water savings of 25–37 percent compared with conventional furrow irrigation. (See Table 3–5.) The most efficient sprinkler, known as low-energy precision application (LEPA), is equipped with drop tubes extending vertically from the sprinkler arm that deliver water much closer to the plants, thereby reducing evaporation losses in the hot, windy plains. Under LEPA irrigation, corn yields have increased about 10 percent and cotton yields by 15 percent, while water use has dropped by 15–35 percent.[39]

The Texas High Plains program has also included extension work to help farmers adopt good water practices—including, for example, the construction of small earthen ridges in the furrows of fields so as to capture both rain and irrigation water and to promote infiltration into the soil. The district also uses weather and crop water use information to develop irrigation schedules for farmers. Overall, the Texas water district's program, bolstered by state-funded low-interest loans for efficiency improvements, has achieved impressive results. Over the last two decades, growers have raised the water productivity of cotton, which accounts for about half the cropland area, by 75 percent.[40]

The list of measures and strategies that can boost water productivity—whether on a farm, within an irrigation district, or more broadly within a region or river basin—is too long to describe here in its entirety. But together with dietary shifts away from heavy meat intake and other actions to reduce consumption (see Chapter 4), they offer the hope of producing enough food for all in a world of increasing water scarci-

Table 3–5. Efficiencies of Selected Irrigation Methods, Texas High Plains

Irrigation Method	Typical Efficiency	Water Application Needed to Add 100 Millimeters to Root Zone	Water Savings Over Conventional Furrow[1]
	(percent)	(millimeters)	(percent)
Conventional Furrow	60	167	—
Furrow with Surge Valve	80	125	25
Low-Pressure Sprinkler	80	125	25
LEPA Sprinkler	90–95	105	37
Drip	95	105	37

[1]Data do not specify what portion of savings result from reduced evaporation versus runoff and seepage.
SOURCE: Based on High Plains Underground Water Conservation District (Lubbock, Texas), *The Cross Section*, various issues.

ty. Collectively, they can move us toward a modern irrigation-based society that is more efficient, productive, ecologically sound, and potentially lasting.

Expanding Irrigation to Poor Farmers

The benefits of irrigation, like so many of society's advances, have not been shared equally. Tens of millions of the world's poorest farmers do not have access to irrigation because the technologies available are too big for their small plots and too expensive. Typically the cheapest way to irrigate with groundwater, for example, is to install a diesel pump on a tubewell, which costs upwards of $350—out of reach for farmers who earn barely this much in a year.

Partly because so many farmers lack access to irrigation, poverty and hunger remain entrenched in large pockets of the countryside. Like trickle-down economics, trickle-down food security does not work well for the very poor. India, for example, is self-sufficient in grain, yet tens of millions of Indians suffer from hunger and malnutrition. The production and trade of more "surplus" food will not solve their problems of hunger because many of them are too poor to buy that food, even at today's historically low prices. The surest and most direct way of reducing hunger among the rural poor is to raise their productive capacities directly. Access to irrigation, in turn, is one of the best ways to boost small-farm productivity. With a secure water supply, farmers can get higher yields and harvest an additional crop or two each year.[41]

A spectrum of irrigation technologies— custom-designed for small plots and affordable by the poorest farmers—has begun to open the door to irrigation's benefits for millions of farm families, from the Gangetic plains of South Asia to the terraced hills of Nepal to the drylands of sub-Saharan Africa. (See Table 3–6.) Collectively, these little-known technologies could lift productivity on tens of millions of hectares of farmland and raise the incomes of many of the world's poorest people.

Over the past 16 years, for example, large areas of Bangladesh have been transformed by a human-powered irrigation device called a treadle pump. To an affluent westerner, it resembles a Stairmaster exercise machine, and is operated much the same way. Grasping a horizontal arm on the bamboo frame, the user pedals up and down on two long poles, or treadles. On the upward stroke, groundwater is sucked up into a pair of cylinders while water from the previous stroke is expelled directly into a field channel. The volume of water that can be pumped in an hour depends on the distance to the water table, the diameter of the pump cylinders, and the energy expended by the operator. In general, the treadle pump is practical for lifting water up to 6 meters below the surface.[42]

Since its introduction in Bangladesh in the early 1980s, the treadle pump has quietly revolutionized small-farmer agriculture there. Each pump irrigates roughly a fifth of a hectare (half an acre) and costs about $35, including installation of the tubewell. Net returns have averaged more than $100 per pump, so farmers recoup their investment in less than a year, and often in one season.[43]

So far, Bangladeshi farmers have purchased 1.2 million treadle pumps, which are raising the productivity of a quarter-million hectares of farmland and injecting an additional $325 million a year into the poorest parts of the Bangladeshi economy. Thanks to creative and effective marketing efforts

Table 3–6. Low-Cost Irrigation Methods for Small Farmers

Technology or Method	General Conditions Where Appropriate	Example Locations
Cultivating wetlands, delta lands, valley bottoms; flood-recession cropping; rising-flood cropping	Seasonally waterlogged floodplains or wetlands	Niger and Senegal river valleys; *fadama* of northern Nigeria; *dambos* of Zambia and Zimbabwe; other parts of sub-Saharan Africa
Treadle pump, rower pump; pedal pump; rope pump; swing basket; Archimedian screw; shadouf or beam and bucket; hand pump	Very small (less than 0.5 hectare) farm plots underlain by shallow groundwater or near perennial shallow streams or canals in dry areas or areas with a distinct dry season	Eastern India; Bangladesh; parts of southeast Asia; valley bottoms, *dambos*, *fadama*, and other wetlands of sub-Saharan Africa
Persian wheel; bullock and other animal-powered pumps; low-cost mechanical pumps	Similar to those above, but where average size of farm plots is roughly 0.5–2.0 hectares	Those above, in addition to parts of North Africa and Near East
Various forms of low-cost micro-irrigation, including bucket kits; drip systems; pitcher irrigation; as well as microsprinklers	Areas with perennial but scarce water supply; hilly, sloping, or terraced farmlands; tail-ends of canal systems; can apply to farms of various sizes, depending on the micro-irrigation technique	Much of northwest, central, and southern India; Nepal; Central Asia, China; Near East; dry parts of sub-Saharan Africa; dry parts of Latin America
Tanks; check dams; percolation ponds; terracing; bunding; mulching; other water-harvesting techniques	Semiarid and/or drought-prone areas with no perennial water source	Much of semiarid South Asia, including parts of India, Pakistan, and Sri Lanka; much of sub-Saharan Africa; parts of China

SOURCE: Global Water Policy Project, Amherst, MA.

orchestrated by International Development Enterprises (IDE), a Colorado-based non-profit organization with offices in many developing countries, the treadle pump has spread entirely through the private sector, with no government subsidies. The technology is now supported by a network of more than 70 manufacturers, 830 dealers, and 2,500 installers—creating jobs and raising incomes in urban areas as well. IDE President Paul Polak estimates that the total

market for treadle pumps in the developing world may number 10 million—including 6 million in India, 3 million in Bangladesh, and another 1 million spread among a number of Asian and African countries.[44]

Lack of access to affordable technologies for small farm plots has also been a major constraint on irrigation's spread in sub-Saharan Africa. In Malawi, Tanzania, Zambia, and many other sub-Saharan African countries, small-scale farmers account for

80–85 percent of the farming population. Most are subsistence farmers just scraping by, and they face many barriers to irrigation. Equipment often costs 2–10 times more than in Asia. Many farmers lack secure tenure to their land, making it risky to invest in improvements. Many smallholders—especially the women, who do much of the farm work in Africa—lack access to credit to purchase supplies and equipment. And poor transportation and marketing facilities often make it difficult to profit from agricultural investments.[45]

Few resources and little support have been devoted to raising the productive potential of rain-fed lands using indigenous, small-scale methods.

Nonetheless, small-scale irrigation schemes have the potential to make a huge contribution in the region. Many of them build on Africa's long and diverse tradition of indigenous irrigation, including such practices as cultivating wetlands, flood-recession farming, and a variety of techniques to capture rainfall and runoff and to channel more moisture into the soil. These water management practices are typically not counted as "irrigation" in official statistics, even though they provide more water to crops than natural rainfall would alone. In addition to 6.3 million hectares of irrigated land in sub-Saharan Africa, the U.N. Food and Agriculture Organization reports that there are more than 1 million hectares of cultivated wetlands and valley bottoms and another 1 million hectares of flood-recession cropping. Including these lands increases the total area under some form of water management in sub-Saharan Africa by a third.[46]

Compared with conventional irrigation methods, few resources and little support have been devoted to raising the productive potential of rain-fed lands using these indigenous and small-scale methods. Greater efforts to do this, however, seem warranted.

Malawi, Zambia, and Zimbabwe are among the sub-Saharan African countries with substantial areas of seasonal wetlands called *dambos*. Crops grown on *dambos* typically get 85–90 percent of their moisture through capillary action from the underlying shallow groundwater. Farmers laboriously use buckets or handpumps to get water from shallow wells to meet the remainder of the plants' water needs. With careful management, small human-powered devices—such as swing baskets, rope pumps (made locally from a bicycle wheel and rope), and very shallow treadle pumps—can be ideal for easing the watering burden, raising the productivity of these lands, and spreading the cultivated area, while at the same time sustaining ecological functions. In Zimbabwe, about 20,000 hectares of *dambos* are cultivated out of a total *dambo* area of 1.3 million hectares.[47]

Substantial areas of valley bottomlands are suited for irrigation with low-cost pumps as well. In Kenya, for example, a local nongovernmental organization called Approtech reconfigured the Asian treadle pump into a lighter, portable device called a pedal pump for irrigating small valley-bottom plots. Nicknamed "the moneymaker," the pedal pump sells for about $70 and is actively marketed in Kenyan towns and villages.[48]

Modern enhancements to traditional practices of flood-recession agriculture also show promise in parts of Africa. As the phrase implies, flood-recession farming involves planting crops after a river's seasonal flood recedes. The moisture stored in the floodplain soils supports the plants throughout the growing season. Farmers practicing this method typically trade higher yields for greater drought resistance and early maturation of crops, because if the soil moisture runs out before harvest time, they can lose their entire crop.

In good years, between 300,000 and 500,000 hectares are used for flood-recession cultivation in the Senegal, Niger, and Lake Chad basins. Judged by grain yields alone, this cultivation method appears considerably less productive than modern intensive irrigated agriculture. Typical cereal yields range between 400 and 800 kilograms per hectare, compared with 2,000–4,000 kilograms per hectare on irrigated lands with good water control. But crop yields alone fail to capture other highly productive elements of a flood-recession system—especially livestock herding, fishing, and vegetable production—that together contribute to higher water productivity and greater food and income security.[49]

For example, a research team studying the effects of the Manantali Dam on Mali's Bafing River (a major tributary of the Senegal River) found that the dam could store enough water for both the planned hydroelectric production and the release of a controlled flood. Moreover, according to researcher Michael Horowitz, "the net returns per unit land from the total array of traditional production—flood-recession farming, herding, and fishing—actually exceed those from more-conventional irrigation, without taking into account the latter's huge start-up and recurrent operating and maintenance costs."[50]

In areas with a perennial but scarce water supply, a variety of low-cost micro-irrigation systems can help raise production, as noted earlier. But in semiarid and drought-prone areas with no year-round water

source, farmers need methods to capture rainfall and runoff in the wet season for use during the dry season. There are millions of such farmers in semiarid parts of South Asia, sub-Saharan Africa, Latin America, and China. Because of unreliable rains, many of them suffer crop failures in one out of three years.

A set of practices known as water harvesting offers the best hope of raising crop production and food security in many of these regions. These measures aim at capturing and channeling a greater share of rainfall into the soil, and conserving moisture in the root zone where crops can use it. Researchers in India, for example, are reviving interest in tanks—small reservoirs of varying sizes that store rainfall or runoff from the wet season for use during the dry season. Used for centuries, some 2 million tanks dot India's landscape, but many now lie in disrepair.[51]

In many watersheds, constructing a small dam across a gully can trap large amounts of runoff, which can either be channeled directly to a field, stored for later use in a tank or small reservoir, or allowed to percolate through the soil to recharge the underlying groundwater. Often called check dams, temporary structures can be made of loose rocks, earth, woven wire, or other locally available materials. They typically last two to five years and cost $200–400, depending on the materials used, the size of the gully, and the dam's height. More-permanent check dams made of stone, brick, or cement typically cost $1,000–3,000.[52]

These descriptions only hint at the rich diversity of small-scale irrigation and water harvesting methods that have substantial untapped potential. The enhanced food security and rising incomes of large numbers of small-farm families can be a powerful engine of economic growth in the world's poorest regions, and stem the exodus of rural dwellers to burgeoning cities. But that engine must be jump-started. Especially as large-scale water projects come under closer scrutiny for their environmental and social costs, a coordinated effort to provide access to small-scale irrigation methods that poor farmers will invest in and use is a critical addition to the portfolio of irrigation advances.

The Policy Challenge

From the wealthiest farms in California to the poorest ones in Bangladesh, irrigation involves many players, each of whom behaves according to a set of rules and incentives. These players include farmers, irrigation districts, water user organizations, state or provincial water agencies, national ministries, development banks, aid agencies, private voluntary organizations, engineering firms, politicians, and taxpayers.

The rules, by and large, have been stacked against efficiency, equity, and environmental sustainability. Large government subsidies—an estimated $33 billion a year worldwide—keep water prices artificially low, discouraging farmers from investing in efficiency improvements. Inflexible laws and regulations have discouraged the marketing of water, leading to inefficient water allocation and use. The absence of rules to regulate groundwater use has led to the overpumping and depletion of aquifers, and has worsened inequities between the rich, who can afford to deepen their wells, and the poor, who cannot. And the failure to place a value on freshwater ecosystem services—including maintenance of water quality, flood control, and the provision of fish and wildlife habitat—has left far less water in natural systems than is socially optimal.[53]

Correcting these policy failings is no easy task. In most cases it requires bucking entrenched and powerful interests. But if society is to redesign irrigated agriculture to make it both productive and sustainable in an era of water scarcity, there is little choice but to take up the challenge.

More than 25 countries are now in the process of devolving responsibility for irrigation systems from central government to local farmers' groups or other private organizations. Most are driven as much by the need to cut government expenditures as by the desire to improve irrigation performance. The hope is that, in reducing subsidies, such management transfers will accomplish both aims.[54]

The largest shift of this kind has occurred in Mexico, where the management of more than 85 percent of the 3.3 million hectares of publicly irrigated land has been turned over to farmers' associations. With the large reduction in government subsidies, water fees have risen to cover costs. The irrigation districts are now about 80 percent financially self-sufficient, up from 37 percent prior to the management transfer. The cost of irrigation water to farmers, although higher, is still generally within 3–8 percent of total production costs, a typical range for irrigated agriculture. Other important aspects of Mexico's efforts—in particular, whether the turnover is promoting greater equity and more efficient use of scarce water—have yet to be assessed.[55]

In many areas, raising water prices can be a political high-wire act. But a spectrum of options exists between full-cost pricing, which could put farmers out of business, and a marginal cost of nearly zero to the farmer, which is a clear invitation to waste water. One option, for example, is a tiered pricing structure. Farmers are charged the rate they are accustomed to paying for up to, say, 80 percent of their average water use, a much higher price for the next 10 percent, and the full marginal cost for the last 10 percent. This gives farmers a strong incentive to reduce their water use so as to avoid the higher unit rates.

The Broadview Water District in California instituted a tiered pricing structure when faced with the need to reduce drainage into the highly polluted San Joaquin River. For each crop, the district determined the average volume of water used in 1986–88, and then applied a rate of $16 per acre-foot (1.3¢ per cubic meter), which was the rate farmers were accustomed to paying, to 90 percent of this amount. Any deliveries above that level were charged at a rate of $40 per acre-foot—2.5 times higher. Even though they were still paying much less than the real cost of their irrigation water, the farmers had incentive to conserve. On average, cotton growers used 25 percent less water over the period 1990–93 compared with 1986–89. Similarly, water use on tomatoes fell by 9 percent, on cantaloupes by 10 percent, on wheat by 29 percent, and on alfalfa seed by 31 percent. Crop yields either held steady or increased.[56]

Lifting barriers to water marketing can also help promote more efficient use and allocation of water, although checks are needed to guard against worsening inequities. The ability to sell some of their water gives farmers an incentive to improve their efficiency so as to profit from the resulting savings. Formal water markets only work where farmers have legally enforced rights to their water (either private or communal), and where those rights can be traded. Australia, Chile, Mexico, and many western states in the United States now have laws and policies that allow for water markets.

As with pricing, a variety of marketing options exists. Farmers can sell their water on a seasonal basis in response to short-term conditions, or they can enter into long-term contracts (say for 5, 10, or 30 years). They can also sell their water rights to another user, in which case the legal water entitlement permanently changes hands.

Another option—and one that can be particularly helpful during droughts—is the trading of water through a water bank. A farmer (or other water user) "deposits" water in the bank in hopes that another user will rent that water at a profit-yielding price. The benefit of a water bank, which is typically run by a regional or state water authority, is that it gives farmers extra flexibility in the face of uncertainty. They can choose to deposit water one year, but then do something different the next. In 1991, during the fourth consecutive year of drought, a water bank operated by California's Department of Water Resources purchased 820,000 acre-feet (about a billion cubic meters) of water, half of it coming from farmers who had decided to stop irrigating temporarily.[57]

Groundwater overpumping ranks among the most serious threats to the world's food supply, yet no government has made a concerted effort to curtail the practice. To the contrary, India encourages overpumping in many areas by charging a flat fee for electricity. This policy makes the marginal cost of water nearly zero for many farmers, discouraging efficiency improvements. Likewise, Texas irrigators get a break on their federal income taxes for depleting the Ogallala aquifer: they receive a depletion allowance, much as oil companies do for depleting oil reserves.

In addition to reducing such subsidies, an important step for governments is to commission credible assessments of the replenishment rates for every major groundwater basin or aquifer. Only after these are determined can a plan be devised to balance pumping with recharge. Where landowners have de facto rights to pump as much water as they like from beneath their land, including in India and some parts of the United States, legislatures and the courts may need to intervene in order to elevate the public interest over private rights. Some scholars have recommended use of the public trust doctrine to deal with India's severe groundwater problem. This doctrine asserts that governments hold certain rights in trust for the public and can take action to protect those rights from private interests.

If gains in efficiency are used merely to support ever more consumption, we will get no closer to a sustainable world.

Once a legal basis for regulating groundwater use is established, the challenge is to develop an equitable plan for making that use sustainable. The council overseeing Mexico's Lerma-Chapala River basin, an area that contains 12 percent of Mexico's irrigated land, is taking a constructive approach. It is setting specific regulations for each aquifer in the region. Technical committees are responsible for devising plans to reduce overpumping. Because these committees are composed of a broad mix of players, including groundwater users, they lend legitimacy to both the process and the outcome.[58]

Technologies and policies of the sort described in this chapter can go a long way toward making irrigated agriculture more

efficient and moving the global economy toward more sustainable water use overall. However, they are not by themselves sufficient. If gains in efficiency are used merely to support ever more consumption, we will get no closer to a sustainable world. Meeting the challenges posed by Earth's finite supply of fresh water also requires a scaling back of our individual and collective pressures on natural systems.

One way to reduce the pressure is to eat lower on the food chain. This can be just as effective at saving water as the use of better irrigation practices. The typical American diet, with its large share of animal products, requires twice as much water to produce as do nutritious but less meat-intensive diets common in some Asian and European countries. If Americans reduced their meat intake to these lower (and often healthier) levels, the same volume of water could feed two people instead of one, leaving more water in rivers, streams, and aquifers.[59]

Further reductions in population growth would ease pressures on fresh water as well. Stepped-up support for family planning services, women's reproductive health programs, and educational and economic opportunities for women would have the double benefit of bettering women's lives and lowering birth rates. Coupled with reductions in high-end consumption, slower population growth would make the goal of adequately feeding all people while keeping natural ecosystems intact far more achievable.

Like the rulers of ancient Mesopotamia, today's political leaders seem unaware of the dangers of a faltering irrigation base. Yet history strongly warns against complacency when it comes to the sustainability of our agricultural foundation. It is a lesson our irrigation society has yet to heed and act upon.

Chapter 4

Nourishing the Underfed and Overfed

Gary Gardner and Brian Halweil

In a large desert tent in war-torn Ajeip, Sudan, doctors and nurses from the aid agency Médecins sans Frontières are hard at work. They hover over starving Sudanese children, administering high-calorie biscuits and sugar water in an effort to revive them. These children have traveled from miles around, most arriving by foot. All of them are emaciated, with skin stretched tightly across their bones, and with stomachs that, ironically, are clearly distended—a condition of calorie and protein deficiency known as "kwashiorkor" that disturbs the water balance of the abdomen.

Half a world away, patients arrive at the cardiac ward of Mt. Sinai hospital in Manhattan, many complaining of chest pains. Doctors and nurses prepare some for open-heart surgery, schedule others for angioplasties to open clogged arteries, and administer anti-clotting medications to the remainder. Like the Sudanese children, many of these patients are noteworthy for their swollen abdomens—in this case, of course, the product of too much food. In

contrast to the Sudanese children, many of whom will not pull through, most of the New Yorkers will survive, only to return later with similar symptoms.

The disparity in wealth and in health care between the Ajeip and Manhattan centers is clear. Perhaps less obvious is the condition common to both, one shared by a large and diverse portion of the human family: malnutrition. The World Health Organization (WHO) estimates that roughly half the population in all nations—wealthy and poor—suffers from poor nutrition of one kind or another. This includes 1.2 billion hungry people, those whose diets are lacking in calories. It also includes their polar opposite on the nutritional spectrum—those who eat too much—at least 1.2 billion people and probably the fastest-growing group of malnourished. A large third group of several billion people overlaps the first two; dubbed the "hidden hungry," these people appear to be adequately fed, but are nonetheless debilitated by a lack of essential vitamins and minerals. (See Table 4–1.)[1]

Table 4–1. Types and Effects of Malnutrition, and Number Affected Globally, 2000

Type of Malnutrition	Effect on Diet	Minimum Number Affected Globally
		(billion)
Hunger[1]	Deficiency of calories and protein	at least 1.2
Micronutrient Deficiency	Deficiency of vitamins and minerals	2.0
Overconsumption[1]	Excess of calories often accompanied by deficiency of vitamins and minerals	at least 1.2

[1]Indicated by the number of underweight or overweight people.
SOURCE: See endnote 1.

Malnutrition is both parent and child of a society's development process. It stifles children's mental and physical development and increases susceptibility to disease and death, which in turn burdens health care systems and hampers economic advancement. On the other hand, malnutrition itself is shaped by societal trends. What appears on a dinner plate and what is missing from it are closely related to a smorgasbord of societal issues, including income levels, women's status, migration to cities, civil strife, land distribution, and exposure to advertising, to name a few. The central place of nutrition in development makes it an issue that policymakers ignore at their country's peril.

A Malnourished World

Today our food supply is nothing less than cornucopian: more food and more varieties of food are available to more people than ever before. Yet more people and a greater share of the world today are malnourished—hungry, deficient in vitamins or minerals, or overfed—than ever in human history. To be sure, the three faces of malnutrition are tak-

ing different directions globally. But overall, trends in nutrition and in the way we get and consume food are headed down unhealthy paths. (See Table 4–2.)

Malnutrition is an imbalance—a deficiency or an excess—in a person's intake of nutrients and other dietary elements needed for healthy living. Carbohydrates, protein, and fat—the macronutrients—provide the basic building blocks for cellular growth. They are also the body's only source of energy, or calories: each gram of carbohydrate or protein provides about four calories; each gram of fat, about nine. Micronutrients are vitamins and minerals, such as iron, calcium, and vitamins A through E; these provide no energy and are consumed in small quantities. But they are no less important nutritionally. Vitamins and minerals help macronutrients to build and maintain the body. Cholesterol, fiber, and other components of food also affect nutrition and health, although these are not defined as nutrients.[2]

Foods vary widely in their nutrient composition. Whole grains are mostly carbohydrates, but they also provide fiber, protein, and small amounts of fat and micronutrients. Meat's profile is just the opposite: it

Table 4–2. Share of Children Who Are Underweight and Adults Who Are Overweight, Selected Countries, Mid-1990s

Country	Share Underweight	Country	Share Overweight
	(percent)		(percent)
Bangladesh	56	United States	55
India	53	Russian Federation	54
Ethiopia	48	United Kingdom	51
Viet Nam	40	Germany	50
Nigeria	39	Colombia	43
Indonesia	34	Brazil	31

SOURCE: U.N. Food and Agriculture Organization, *The State of Food Insecurity in the World* (Rome: 1999); World Health Organization, *Obesity: Preventing and Managing the Global Epidemic*, Report of a WHO Consultation on Obesity (Geneva: 1997); K.M. Flegal et al., "Overweight and Obesity in the United States, Prevalence and Trends, 1960–1994," *International Journal of Obesity*, August 1998.

provides large quantities of fat, cholesterol, protein, and some micronutrients, but no carbohydrates or fiber. Meanwhile, fruits and vegetables tend to be low in fat, but rich in fiber and micronutrients. And other foods, such as refined sugar, provide "empty calories": energy without any additional nutritional value. Striking a healthy nutrient balance requires eating from each of these categories, but in different quantities. A joint research effort by the Harvard School of Public Health and Oldways Preservation & Exchange Trust found that traditional diets associated with good health and lengthy adult life expectancies are generally plant-based—rich in whole grains, vegetables, fruits, and nuts—supplemented by sparing amounts of animal products.[3]

Hunger, the worst of the three forms of malnutrition because it often cuts short lives that have barely begun, has been reduced modestly in recent decades. The U.N. Food and Agriculture Organization (FAO) reports that 790 million people—roughly one out of five people in the developing world—are chronically hungry, down from 918 million in 1970. While this esti-mate captures the trend, it likely underestimates the number of hungry, because it is based on calculations of calories per person in developing countries—a method that does not account for the unequal distribution of food found virtually everywhere. For example, FAO estimates that 21 percent of India's population is chronically undernourished. But recent, on-the-ground surveys paint a more accurate and desperate picture: 49 percent of adults and 53 percent of children in India are underweight, a proxy measurement for hunger.[4]

Based on the share of children who are underweight, hunger in the developing world has fallen by 10 percent over the past two decades: 26 million fewer children are estimated to be underweight in 2000 than in 1980. (See Table 4–3.) The greatest absolute progress in reducing hunger was made in Asia, while the greatest relative reduction came in Latin America, where the population of children underweight was cut in half. Africa, on the other hand, saw a worsening in the number of children who are underweight. There, the number in this condition nearly doubled—making the continent a focus of particular concern.

Despite the progress in some regions, hunger remains stubbornly rooted across much of the developing world. The greatest concentration of chronically hungry people is in South Asia and sub-Saharan Africa. Some 44 percent of South Asia's children are underweight, with the shares in India, Bangladesh, and Afghanistan are well above this average. In sub-Saharan Africa, 36 percent of children are underweight, but the figure climbs toward 50 percent in nations such as Somalia, Ethiopia, and Niger. Even in Latin America, with a relatively small share of hungry people, particular countries and subregions, such as Haiti and Central America, still have high levels of hunger.[5]

Hunger is found in pockets of the industrial world as well. The U.S. Department of Agriculture (USDA) estimates that in 1998 people in some 10 percent of U.S. households were "food-insecure"—hungry, on the edge of hunger, or worried about being hungry. These households are home to near-

Table 4–3. Underweight Children in Developing Countries, 1980 and 2000

Region	Underweight Children		Share of Children Who Are Underweight	
	1980	2000	1980	2000
	(million)		(percent)	
Africa	22	38	26	29
Asia	146	108	44	29
Latin America & Caribbean	7	3	14	6
All developing countries	176	150	37	27

SOURCE: United Nations, Administrative Committee on Coordination, Sub-Committee on Nutrition, "Fourth Report on The World Nutrition Situation" (draft) (New York: July 1999).

ly one in five American children. In Europe, Australia, and Japan, where social safety nets are more widespread, hunger is rarer.[6]

While the ranks of the hungry are shrinking modestly, overeating is more prevalent today than ever, according to WHO, which has called the swift spread of being overweight and its extreme form, obesity, "one of the greatest neglected public health problems of our time." Worldwide, the number of overweight people now rivals the number who are underweight. (Overweight and obesity are defined using body-mass index [BMI], a scale calibrated to reflect the health effects of weight gain. A healthy BMI ranges from 19 to 24; a BMI of 25 or above indicates "overweight" and brings increased risk of illnesses such as cardiovascular disease, diabetes, and cancer. A BMI above 30 signals "obesity" and even greater health risks.)[7]

The United States has been at the leading edge of this overeating wave. Today it is more common than not for American adults to be overweight: 55 percent have a BMI over 25. The share of American adults who are obese has climbed from 15 to 23 percent just since 1980. And one out of five American children is overweight or obese, a 50-percent increase in the last two decades. Meanwhile, the incidence of being overweight or obese in Europe has shown similar growth in the last decade, although from somewhat lower levels. The prevalence of obesity in England, for example, has doubled in the last 10 years to 16 percent.[8]

Significantly, the condition of being overweight or obese is no longer found only in wealthy nations. A 1999 survey for the United Nations found obesity at measurable levels in all developing regions, and growing rapidly, even in countries where hunger persists. In China, for example, the share of adults who are overweight jumped by more than half—from 9 percent to 15 percent—between 1989 and 1992. In several Latin American nations, such as Brazil and Colombia, the prevalence of overweight people—at 31 and 43 percent, respectively—approaches the share in some of Europe. Today, the overweight population in many developing nations exceeds the number who are underweight.[9]

The third malnourished population—those with inadequate intakes of vitamins and minerals—partially overlaps with both the hungry and the overfed. Micronutrient deficiencies typically result from a lack of dietary variety—three bowls of rice and little else each day, for example, or heavy dependence on hamburgers and french fries. Low intake of three micronutrients—iodine, vitamin A, and iron—is of particular concern globally because it is so widespread.

The human body needs only one teaspoon of iodine over a lifetime, consumed in tiny amounts. Yet iodine deficiency disorder (IDD), the world's leading preventable cause of mental retardation, affects 740 million people—some 13 percent of the world population—in at least 130 developing countries. (The prevalence of IDD stood at double this level in the mid-1990s, before a worldwide salt-iodization initiative.) Another 2 billion people are reportedly at risk. Vitamin A deficiency (VAD), which leads to blindness and even death in children, is less well documented, but data from problem countries in Latin America and Africa indicate that the prevalence ranges from 9 to 50 percent of the population.[10]

The most commonly lacking micronutrient, however, is iron—a mineral found in whole grains, green leafy vegetables, and meat. A 1999 U.N. report estimates that 5 billion people, more than 80 percent of the human family, suffer from varying degrees of iron deficiency. Among the 2 billion most severely affected, largely women and children in poor countries, lack of iron leads to anemia and cognitive disability.[11]

Changes in hunger, overeating, and micronutrient intake have been linked to changes in eating patterns, as traditional diets and food sources are replaced by less nutritious alternatives. On every continent, for instance, the age-old practice of breast-feeding is being replaced by more "modern" alternatives, even though breast milk is the best source of nutrition for infants and protects babies from some illnesses. In only three countries in Latin America is exclusive breast-feeding—the UNICEF recommendation for infants under the age of six months—practiced more than half the time. And rates are even lower in the industrial world: only 14 percent of mothers in the United States practice exclusive breast-feeding.[12]

Consumption of fat and sugar has surged far beyond earlier levels as people eat more livestock products and as oil and sugar are added to foods of all kinds. In Europe and North America, fat and sugar account for more than half of caloric intake and have squeezed complex carbohydrates, such as grains and vegetables, to just a third of total calories—a near complete reversal of the diet of our hunter-gatherer forebears. Moreover, whole grain products have been largely replaced with refined grains, which are stripped of fiber and of many vitamins and minerals; just 2 percent of the wheat flour consumed in the United States is unrefined. And fast-food items, which often surpass government guidelines for daily intake of fat, sugar, cholesterol, and sodium in a single meal, may be displacing nutritious dark green and yellow vegetables: one fifth of the "vegetables" Americans eat are french fries and potato chips. These trends are rapidly spreading beyond industrial nations as the fast-food culture goes global.[13]

The Roots of Hunger

Most people would correctly identify hunger as the most critical form of malnourishment. But few can describe its causes. The myth persists today that hunger results from scarce food supplies, and that poor harvests, usually because of poor rainfall or barren soils, are the root cause of hunger. The reality is that hunger is the product of human decisions—especially decisions about how a society is organized. Whether people have a decent livelihood, what status is accorded to women, and whether governments are accountable to their people—these have far more impact on who eats and who does not than a country's agricultural endowment does.

Nobel Laureate Amartya Sen argues persuasively that poverty—rather than food shortages—is frequently the underlying cause of hunger. Indeed, nearly 80 percent of all malnourished children in the developing world in the early 1990s lived in countries that boasted food surpluses. The more important common feature among these countries is pervasive poverty, which limits people's access to food in the market or to land and other inputs needed to produce food. Poverty also means poor access to health care, education, and a clean living environment, which increases the likelihood of hunger. Diseases like diarrhea, for instance, prevent a child from absorbing available nutrients, while a poor education often means poor job prospects, with adverse consequences for income, and, in turn, for nutrition.[14]

Poverty often strikes hardest among women, the nutritional gatekeepers in many families. FAO estimates that more than half of the world's food is produced by women, and in rural areas of Africa, Latin America, and Asia, the figure soars to 80 percent. Yet women have little or no access to landownership, credit, agricultural training, education, and social privileges in general. In five African nations, for example—Kenya, Malawi, Sierra Leone, Zambia, and Zimbabwe—where at least 40 percent of the people are chronically hungry, women receive less than 10 percent of the credit awarded to smallholder farmers and just 1 percent of agricultural credit overall, which limits their ability to increase income and boost nutrition.[15]

Moreover, women in developing countries reinvest nearly all of their earned income to meet household food and other needs, whereas men often set aside up to a quarter of their income for alcohol, sex, and other nonhousehold expenses. And women impoverished to the point of hunger bear hungry children, and are less able to care for their children and to breast-feed, conditions that perpetuate hunger across generations. In sum, societies that abandon women to poverty are weakening one of their key defenses against malnutrition.[16]

Poverty and hunger also result from a range of misguided government policies, from inequitable distribution of land and other resources to poor management of foreign debt. Where governments have allowed land to be concentrated in a few hands, rural poverty and hunger tend to be most severe. And many governments saddled by heavy debts have reduced public spending, cut food aid, and eliminated subsidies for staple crops, often under pressure from international donor organizations such as the International Monetary Fund—all of which is likely to increase the ranks of the poor.[17]

Many indebted countries turn to the export of coffee, bananas, flowers, and other cash crops, often at the expense of their own domestic food supply, in an effort to earn foreign exchange to pay down debt. The main beneficiaries of this export orientation are typically large corporations, foreign investors, and large landowners, not the rural poor. Indeed, the export boom in Latin America and Southeast Asia has tended to squeeze out smallholder farmers and to divert resources away from household food production.[18]

The growing emphasis on free trade in agricultural products brings additional nutritional vulnerabilities. Current trade arrangements, such as the Agreement on Agriculture under the World Trade Organization, permit the industrial farmers of Europe and North America to sell subsidized grain, oils, and other commodity surpluses cheaply in developing nations, undercutting local farmers and forcing many off the land—their source of nutrition security.[19]

Meanwhile, dependence on foreign markets for staple foods leaves importing countries vulnerable to price fluctuations and currency devaluations that can increase the price of food substantially. Mexican consumers, for example, initially benefited from cheap imports of corn from the United States under the North American Free Trade Agreement. But when the peso was devalued in 1995, the price of corn doubled in one year. Some hungry Mexicans—including farmers undercut by the flood of American corn—took to looting rail freight cars to obtain food.[20]

While hunger can spark social unrest, conflict can often produce or worsen hunger. From Angola to Kosovo, millions are hungry because warring factions use food as a weapon, choking distribution channels or forcing people from their farmland. War can exacerbate hunger by destroying crops, eliminating jobs, and driving up food prices. And the destruction of agricultural infrastructure, schools, hospitals, and factories means that war's damage to nutrition persists long after the conflict has ended. In Afghanistan, landmines planted in conflicts since 1979 are estimated to prevent between half and two thirds of the nation's arable land from being plowed. A rash of civil wars and resulting refugee flows in central Africa in the 1990s are at the center of the worsening hunger there.[21]

The Creation of Overeating

Even as hunger persists in much of the developing world, a large share of humanity has shifted from a traditional diet dominated by grains and vegetables to one with foods heavy in fats and sugars. This "nutrition transition" did not emerge spontaneously from farms and kitchens worldwide. Instead, societal changes—from rising incomes to urban migration and the emancipation of women—shaped eating habits in two important ways. First, they unleashed the natural human desire for fat and sugar. And second, they created a relatively open operating space for food companies to meet this demand. In the absence of regulation of the food industry, and with government providing only minimal dietary guidance for consumers, overeating and the consumption of nutrition-poor foods were nearly inevitable.

In part, the shift to chronic overeating was a physiological response. The desire for fatty and sugary food seems to be innate—a vestige of our hunter-gatherer roots, when energy-dense foods were scarce but crucial for weathering lean times. And the sugary and fatty foods we prefer are also easily overconsumed: the human body has weak satiation mechanisms for fat and sugar, in contrast to foods high in fiber and complex carbohydrates, such as whole grain bread and potatoes, which leave us feeling full. But these physiological traits were largely corralled until this century, when a host of social forces transformed the food environment and allowed our desires to run largely unconstrained.[22]

The mechanization of agriculture and the breeding of high-yielding crops, for example, boosted the availability of all foods, including meat, vegetable oils, dairy products, sugar, and other commodities previously reserved for the affluent or for holidays. For those with money, food simply became far more available. And more people had more money, as industrialization boosted incomes. The fourfold increase in per capita income in China since 1978, for instance, correlated with a drop in consumption of rice and starchy roots and a rapid increase in the consumption of high-fat foods, such as pork. Moreover, as the price of vegetable oils fell in China—and in much of the rest of the world—fried foods became more popular, making a relatively high-fat diet available even to people who could not yet afford meat. Indeed, increased global vegetable oil consumption since the 1960s alone has added roughly 30 grams of fat to the average daily diet. (See Figure 4–1.) In short, the twentieth century saw more high-fat foods become available to more people at lower prices than ever before.[23]

Not only were these foods available and cheap, but new lifestyles spawned by urbanization made them still more attractive. As the urban share of global population rose from 10 percent in 1900 to nearly 50 percent today, time for food preparation was squeezed as women began to work outside the home—and because men do not typically share domestic responsibilities. Breastfeeding also began to decline. Urban consumers began to demand convenience foods, which are often high in fat, cholesterol, and sugar and low in fiber, vitamins, and minerals. And these foods are generally cheaper and more widely available in cities than in rural areas, at least in developing countries. Not surprisingly, a study of 133 countries has demonstrated that in poorer nations, migration to the city—without any changes in income—will generally double per capita intake of sweeteners and increase fat intake by at least 25 percent.[24]

Even as more calories become available to more people in cities, fewer are needed for daily living. As jobs of all kinds become mechanized, work in the manufacturing and service sectors demands less physical exertion than farming. In the developing world, the share of the population engaged in agriculture has declined from roughly 80 percent in 1950 to 55 percent. And electri-

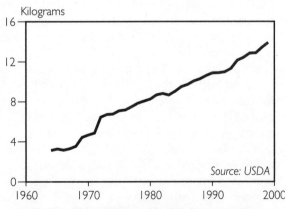

Kilograms

Source: USDA

Figure 4–1. World Vegetable Oil Consumption Per Person, 1964–99

fied, labor-saving amenities at home, motorized transportation, and sedentary leisure activities further reduce energy expenditure.[25]

As urbanization has broken the close connection between farmer and consumer, and as changing lifestyles encourage more processed, packaged, and fast foods, the food industry has assumed a powerful place in the food chain—catering to and influencing consumer demands. To maximize revenues, food sellers invest heavily in the creation of a "food environment" that makes unhealthy food and its promotion ubiquitous in modern life.

The most powerful tool used by companies to shape this environment is advertising. In the United States, food companies spend more on advertising than any other industry—an estimated $30 billion annually. Food is also the top ad category in Austria, Belgium, and France. Food ad expenditures in the developing world are lower, but are growing fast as incomes increase rapidly. In Southeast Asia, for example, food ad expenditures tripled between 1984 and 1990, from $2 billion to $6 billion.[26]

Unfortunately, the most heavily advertised foods tend to be of dubious nutritional value. A 1996 Consumers International study found that candy, sweetened breakfast cereals, and fast-food restaurants accounted for over half of all food ads in Australia, Norway, the United States, and 10 European Union nations. In the United States, fast-food restaurants alone account for one third of food advertising expenditures. And Coca-Cola and McDonald's are among the top 10 ad spenders worldwide. Moreover, food advertisers disproportionately target children—the least savvy consumers—with troubling implications over the long run. While a sweet tooth is a natural human

characteristic, research shows that repeated exposure to sweet and fatty foods in childhood results in a life-long craving for these unhealthy items.[27]

The food industry has assumed a powerful place in the food chain—catering to and influencing consumer demands.

In theory, companies could promote nutritional foods, such as apples and carrots, instead of potato chips and cookies. But two factors compel food companies to keep their eyes on sweet and fatty foods. They know that consumers will always have a special interest in these products, given our predisposition for them. And processed foods are most likely to have "added value"—the alterations or packaging that allow a company to earn higher profits. Doughnuts, for example, will fetch a greater profit than if the flour, oil, eggs, and sugar in them were sold separately. Part of the profit is reinvested in advertising the product, which keeps the cycle of promotion and sales in constant motion.

Not all processed foods are nutritionally poor, of course. But when the added "value" is sugar, salt, fats, or oils, the result is usually a tasty and profitable but nutritionally suspect product that is irresistible for consumers and companies alike. In fact, sugar and salt are the two most widely added ingredients to food, and fats and oils are also added in large doses. Consumers are often unaware of the quantities of these additives in what they are eating, especially where nutritional labeling is missing or unclear. In 1909, for example, when the range of processed foods was much smaller, two thirds of discretionary sugar consumed in the United States was added in the household. Today, three quarters is added during the processing of food, out of the consumer's sight.[28]

Value added generates the revenues needed to fund other marketing strategies in addition to advertising, including product placement, distribution, and efforts to expand the reach of junk food. Since 1990, for instance, soda companies have offered millions of dollars to cash-strapped school districts in the United States for exclusive rights to sell their products in the schools. Indeed, the marketing prowess of fast-food firms makes them a growing presence in U.S. schools. More than 5,000 schools—13 percent of the U.S. total—have contracts with fast-food establishments to provide either food service, vending machines, or both.[29]

Among the newest marketing strategies is the "supersizing" of french fries, popcorn, pizzas, soda, and other fast-food items—at little extra cost to the consumer. The standard 6.5-ounce soda container of the 1950s, for example, has been supplanted by 20-ounce bottles, and one U.S. convenience store pushes a 64-ounce, 600-calorie "Double Gulp" soda bucket. The extra food content costs manufacturers very little, since the ingredients account for only a small share of the sale price; the consumer is paying mostly for the brand name, packaging, and marketing. To the consumer, such a "good value" is highly enticing. But supersizing may be skewing peoples' conception of what a "normal" serving is: surveyed Americans consistently label as "medium" portions that are double or triple the size of USDA's "medium" guidelines. And consumer behavior studies show that people consistently eat more from larger containers.[30]

Whatever the marketing strategy, in most nations its presence dwarfs efforts at nutri-

tion education. In the United States, the billions of dollars of advertising by fast-food restaurants and the myriad advertisements for snack foods, soda, candy, and sugary breakfast cereals—Kellogg's spends $40 million to promote Frosted Flakes alone—makes USDA's $333-million budget for nutrition education look like a pittance. This imbalance in information and power between industry, consumers, and government results in what Kelly Brownell, a Yale University psychologist, has labeled a "toxic food environment": unprecedented access to high-calorie foods that are low in cost, promoted heavily, and good-tasting.[31]

As these strategies run their course in increasingly saturated industrial markets, and as incomes rise in developing nations, many food companies are looking overseas for greater profits. Mexico recently surpassed the United States as the top per capita consumer of Coca-Cola, for example. And that company's 1998 annual report marvels at Africa's rapid population growth and low per capita consumption of carbonated beverages, noting that "Africa represents a land of opportunity for us." American fast-food restaurants are growing rapidly overseas: four of the five new restaurants that McDonald's opens every day are found outside the United States. From Cali to Calcutta, urbanites in the developing world are tempted by the same foods that are jeopardizing health in the industrial world.[32]

Despite these troubling nutritional trends, the emergence of a countertrend—perhaps the first wave of a new nutrition transition—offers some hope for healthier eating. High-income and well-educated people in Brazil, Chile, South Korea, the United States, and elsewhere are beginning to spurn junk foods, improve exercise habits, and embrace nutritional basics—a diet low in fat and added sugar and high in whole grains, fruits, and vegetables. In these countries, the traditional relationship between obesity and socioeconomic levels is being inverted: the wealthiest often have the best eating habits, while the poorest, with little access to healthy food and nutrition education, are increasingly likely to be overweight. The question for coming decades is whether better eating will spread from elites to entire societies, or whether obesity will join hunger as a poor person's affliction worldwide.[33]

Diet and Health

Hunger, deficiencies of micronutrients, and overeating are all closely linked to health, suppressing or promoting diseases from the common cold to cancer. Sometimes dietary influence on disease is direct, as with the capacity of iodine deficiency to cause men-

tal retardation. Other times diet interacts with lifestyle patterns to set off a chain reaction: increased fat and sugar consumption, for example, combined with more sedentary activities typically leads to obesity, which raises the risk of developing heart disease, diabetes, and hypertension. Whatever the path, researchers are increasingly confident that a large share of many diseases can be blamed on poor nutrition. (See Table 4–4.)[34]

In some ways, the undernourished and overnourished face similar health risks: both are more susceptible to disease and both have reduced life expectancies. But differences exist as well. Hunger and the lack of vitamins and minerals do their greatest damage early in life, while overnourishment degrades the body gradually, with heart disease, cancer, and other chronic ailments appearing typically in middle and old age. Moreover, the effects of undernourish-

Table 4–4. Health Problems That Could Be Avoided Through Dietary Change

Disease or Condition	Share of Cases Preventable Through a Change in Diet[1] (percent)	Public Initiatives That Could Lower the Risk of Disease or Condition
Cancer	30–40	Promote the consumption of fruits and vegetables; promote low-fat diets and higher levels of physical activity
Coronary Heart Disease	17–25	Promote low-fat diets and higher levels of physical activity
Childhood Blindness	25–50	Establish vitamin A supplementation programs; ensure access to vegetables and other food rich in micronutrients
Mental Retardation	33–43	Iodize salt
Adult-Onset Diabetes	64–74	Promote low-fat diets and higher levels of physical activity

[1]The share of preventable cases of cancer, coronary heart disease, and adult-onset diabetes is substantially greater in high-risk groups such as obese people.

SOURCE: See endnote 34.

ment, such as stunted growth and blindness, are often irreversible, while illnesses from overnourishment can be corrected through changes in diet and lifestyle.

An assessment of disease worldwide undertaken early in the 1990s by the World Bank and Harvard University captured the broad impact of malnutrition on health. The study measured deaths due to various risk factors, as well as "years of healthy life lost" because of death or disability. Hunger and micronutrient deficiencies are now regarded as the top two causes for loss of a year of healthy life: updated analysis by Dr. John Mason, who worked on the original study, finds hunger to be responsible globally for 22 percent of these losses, and micronutrient deficiencies for 18 percent. The study did not measure the contribution of being overweight. But Dr. Alan Lopez, head of the Harvard team that contributed to the study, is confident that overeating is responsible for at least as large a share of global illness as hunger. Thus even considering some overlap between micronutrient deficiency and hunger or overeating, the three forms of malnutrition would likely account for more than half of the global burden of disease.[35]

Some of hunger's worst effects are sown even before a child is born. A 1999 U.N. report on nutrition estimates that each year some 20 million infants—16 percent of all live births—are underweight in the womb, the result of malnutrition in the mother. These children are scarred for life, suffering from impaired immune systems, neurological damage, and retarded physical growth. Underweight *in utero*, for example, translates in adulthood to five centimeters' loss of height and five kilograms' loss of weight.[36]

The consequences of malnutrition in the womb may be broader, however. Epidemiologist David Barker of the University of Southampton hypothesizes—and many nutritionists agree—that *in utero* malnutrition may predispose a person to chronic disease in later life. Barker's theory is that malnutrition in a fetus triggers metabolic and physiologic responses that help it to survive hunger. The same adaptive responses later in life, when food is more plentiful, can lead to high blood pressure, cardiovascular disease, and glucose intolerance, a risk factor for diabetes. The terrible irony of Barker's hypothesis is that a child born hungry might escape the "diseases of poverty" only to be at increased risk of dying from a "disease of affluence."[37]

Many underweight infants, however, do not survive to see adulthood. The risk of death for an infant weighing two thirds of normal is 10 times greater than for a healthy baby. Barker observes soberly that 60 percent of all newborns in India would be in intensive care had they been born in California. And WHO estimates that of the five leading causes of child death in the developing world, 54 percent of cases have malnutrition as an underlying condition.[38]

Hungry infants—and older children and adults as well—are also more susceptible to infectious disease because they lack the energy needed to fight such illness. Infectious disease, in turn, often suppresses appetite and prevents a hungry body from absorbing the few nutrients it does get. Diarrhea, for example, sweeps away nutrients before they are used and can leave a body hungrier than before. Thus hunger and sickness chase each other in a vicious circle known as the malnutrition-infection cycle.[39]

Even if a person gets sufficient calories, micronutrient deficiencies can take a heavy toll on development and health. Iodine deficiency, as noted earlier, can stunt physical and mental growth, leading to mental retardation in severe cases. And vitamin A deficiency is the world's leading cause of preventable blindness. It is also a killer—researchers have learned that eliminating VAD can reduce infant mortality by an average of 25 percent. Meanwhile, iron deficiency is a particular concern for women: some 44 percent of all women in developing countries, and 56 percent of pregnant women there, suffer from anemia, compared with 33 percent of men. Most micronutrient deficiencies in expectant mothers will damage the unborn child. Severe anemia, for example, is a major cause of mental impairment *in utero*.[40]

Some micronutrient deficiencies are widespread in industrial nations as well, especially among poorer people. As calorie-rich junk foods squeeze healthy items from the diet, obesity often masks nutrient starvation. In the United States, iron deficiency affects nearly 20 percent of all premenopausal women—and 42 percent of all poor, pregnant African-American women. And the exorbitance of industrial-nation diets also adversely affects micronutrient levels: extremely high protein intake—characteristic of diets dominated by animal products—tends to leach calcium from bones, heightening the risk of osteoporosis.[41]

As a country grows wealthier, and as people begin to overconsume calories and to eat nutrition-poor foods, a major shift occurs in health patterns. Hospitals see fewer cases of infectious disease and an increase in chronic illness, including cardiovascular disease, diabetes, and cancer. The nutrition transition, in other words, spawns an epidemiological transition—essentially trading diseases of dietary deficiency for diseases of dietary excess.[42]

Two long-term surveys have helped establish the link between dietary habits and the prevalence of chronic disease. The China Health Survey has tracked 6,500 Chinese undergoing the nutrition transition since 1983, and found strong correlations between high intake of fat and protein, particularly from animal sources, and the incidence of heart disease, stroke, and colorectal and breast cancer. Meanwhile, the Framingham Heart Study followed more than 5,000 residents of Framingham, Massachusetts, since the 1950s and found that chronic illness is not a normal or inevitable consequence of aging but is closely tied to modifiable dietary and exercise habits. The long-term nature of these studies and their broad coverage make them particularly strong chroniclers of the effect of poor eating on health.[43]

Sixty percent of all newborns in India would be in intensive care had they been born in California.

Diets high in calories and fat encourage obesity, which raises the risk of heart disease, stroke, diabetes, and various cancers. These four sets of disease are responsible for more than half of all deaths in the industrial world. High-calorie, high-fat diets also promote high blood pressure and clogged arteries, which are additional risk factors for a variety of degenerative diseases. Dr. Graham Colditz at the Harvard School of Public Health has estimated that among obese Americans, slimming to a healthy weight and maintaining it could prevent 96 percent of diabetes cases, 74 percent of hypertension, 72 percent of coronary heart disease, 32 percent of colon cancers, and 23 percent of breast cancers. Indeed, researchers at the World Cancer Research Fund and the American Institute for Cancer Research (WCRF/AICR) report that changes in diet alone could prevent 30–40 percent of all cancers worldwide—at least as many cases as could be prevented by a cessation of smoking, a more familiar cause of cancer.[44]

People who are overweight also suffer disproportionately from a range of nonlethal but debilitating conditions, including osteo-arthritis, hormonal disorders, asthma, sleep apnea, back pain, and infertility. And like hunger, obesity raises the susceptibility to infection by impairing immune function.[45]

The most dramatic health effect of the surge in global obesity is the parallel rise in diabetes. The global population with adult-onset diabetes—the type that is associated with being overweight—jumped nearly five-fold between 1985 and 1998, from 30 million to an estimated 143 million. And the affected population is broadening rapidly as well. Adult-onset diabetes was once rare in people under 40. But up to 20 percent of new pediatric patients at Columbia University's Naomi Berrie Diabetes Center had this condition in 1998, compared with less than 4 percent in the early 1990s—a trend confirmed at clinics around the nation. As obesity spreads to younger populations, it is likely that other "adult" diseases—from heart disease to stroke and cancer—will also strike young people more frequently.[46]

Because of an aging population, unhealthy diets, obesity, and a sedentary lifestyle, developing countries face a growing wave of chronic disease. The International Diabetes Federation estimates that the number of diabetics worldwide will double to 300 million by 2025, with three quarters of the growth coming in developing nations. And the share of deaths from cancer is projected to double in developing countries to 14 percent by 2015, while holding steady in industrial countries at 18 percent. Of even greater concern is recent epidemiological evidence that non-Caucasian populations, particularly Pacific Islanders, American-Indians, Asian-Indians, and Mesoamericans and other Hispanics, appear to be susceptible to some chronic diseases, especially adult-onset diabetes, at lower levels of overweight and at younger ages than Caucasians.[47]

Societal Costs of Poor Diet

The toll of malnutrition on society is broader but perhaps more subtle than its impact on individual health. Children's performance at school, worker productivity, the size of the national health budget—all of these are affected by malnutrition. Most studies of this effect are partial, analyzing hunger but not obesity, for example, or the impact on children but not adults. But these hint at a huge societal cost and yield a clear conclusion: malnutrition hampers a country's development—in poor nations, by making them less productive and less prosperous; in wealthy nations, by paring back hard-won developmental gains, such as less disability and longer life spans.

Nutritional deficiencies often limit children's capacity to learn. The World Bank reports that undernutrition and micronutrient deficiencies together result in a 5- to 10-percent loss in learning capacity. And a set of nine studies on nutrition's role in educational performance suggests that early malnutrition can affect school aptitudes, concentration, and attentiveness and can lead to delays in starting school. In Spain and Indonesia, children in areas with high levels of iodine deficiency reportedly average three fewer years in school than their peers in comparable but nondeficient communities. In the Philippines, a 12-year study of more than 2,000 children linked stunting during infancy with a marked increase in the dropout rate, later enrollment in school, and poorer school performance.[48]

Meanwhile, hunger among adults reduces their strength and physical stamina, which lowers productivity at work. For five South Asian nations, economist Susan Horton of the University of Toronto calculates that productivity losses due to hunger range from 2 to 6 percent for the moderately undernourished and from 2 to 9 percent for the severely malnourished. Fatigue induced by iron deficiency is especially devastating: for light blue-collar work, productivity losses are measured at 5 percent, and for heavy manual labor, at 17 percent.[49]

Horton also expressed these productivity reductions in terms of lost wages. She calculates that productivity losses from hunger and micronutrient deficiencies cost developing countries 1–2 percent of their gross domestic product each year in lost wages. A modest-sounding sum, these losses nevertheless amount to more than $5 billion annually for these five countries alone—equal to the total government health budgets of these nations. If Horton's estimates hold for all developing countries, between $64 billion and $128 billion is drained from developing-country economies just from productivity losses. Horton's work is pathbreaking yet conservative: it does not measure the costs associated with child deaths, premature births, stunting, disruption to the immune system, and less quantifiable consequences of undernutrition.[50]

Malnutrition takes a great toll on productivity in industrial countries as well. In a 1994 study of 80,000 Americans, obese participants were found to account for a disproportionate share of health-related absences from work. For example, they were half again as likely as other study participants to take sick days in bed. And a 1995 study from Sweden estimated that obesity accounted for 7 percent of lost productivity due to sick leave and disability,

and that obese people were twice as likely as the general population to take long-term sick leave. These results are also conservative, because they do not include analysis of overweight (as distinguished from obese) populations in these countries.[51]

Overeating is costly to a nation's health care system too. Unlike the diseases of undernutrition, which often kill the young or inflict permanent and untreatable damage such as stunting or blindness, treatment of chronic illness often involves frequent use of the health care system. Overweight and obese people in the Netherlands visit their physicians 20 percent and 40 percent more, respectively, than people of healthy weight. And obese people there were 2.5 times more likely to take drugs prescribed for cardiovascular and circulation disorders.[52]

The economic cost of these burdens to health care appears to be enormous. Comparing the prevalence of hypertension, heart disease, cancer, diabetes, gallstones, obesity, and food-borne illness among vegetarians and meat eaters in the United States, the Physicians Committee for Responsible Medicine estimated total annual medical costs in 1995 related to meat consumption of between $29 billion and $61 billion. The costs would likely have been higher if stroke and other arterial diseases had been studied as well.[53]

Meanwhile, Dr. Graham Colditz of the Harvard School of Public Health has calculated the direct costs (hospital stays, medicine, treatment, and visits to the doctor) and indirect costs (reduced productivity, missed work-days, disability pensions, and other nonmedical costs) of obesity in the United States to be $118 billion annually, or nearly 12 percent of the nation's health care expenditures. (This is more than double the $47 billion in direct and indirect costs attributable to cigarette smoking in the United States.) Add to this enormous sum the $33 billion spent on diet drugs and weight loss programs, together with the unmeasurable psychological costs from the social exclusion associated with being overweight, and the full cost of overeating begins to emerge.[54]

Poor nutrition—like excessive military spending—has the potential to constrain or unravel a nation's development gains.

Whichever figure is used, such huge sums are a drag on a nation's development. Money spent to counteract obesity or to reverse the damage done by meat consumption—health effects that are largely avoidable—cannot be used to address other social problems. In this sense, poor nutrition—like pollution, crime, or excessive military spending—has the potential to constrain or even unravel a nation's development gains.

The high cost of malnutrition poses particularly acute challenges for developing countries, which must tackle large caseloads of both infectious and chronic diseases on shoestring budgets. Many developing countries spend less than $25 per person annually on health care, compared with $4,000 in industrial nations. India and China, for example, spend $17 and $31, respectively, on health care.[55]

Yet treating chronic diseases will cost a bundle: WCRF/AICR researchers warn that without effective policies for cancer prevention, the cost of treating cancer in developing countries will rise 25-fold between 1985 and 2015. The cost of cancer alone may exceed the entire national budget for health care in many developing

countries by then. Because chronic disease usually afflicts the wealthy—at least initially—in developing countries, and because the wealthy often wield disproportionate political clout, the pressure to give priority to treatment of chronic diseases will probably be strong. Those suffering from infectious disease—typically the poor—are likely to be left untreated.[56]

The broad costs of malnutrition extend still further, since malnutrition tends to reinforce the social inequalities that spawn poor eating in the first place. Where hunger leads to high rates of infant mortality, for instance, population growth is often high, as couples have large families to offset likely losses of children. Such overcompensation further burdens poor families. At the other end of the spectrum, obesity can also keep people in poverty. Obese women in the United States generally have less education and less income than other women, and obese young adults are less likely to reach the same social class level as their peers.[57]

Nutrition First

Half of the world is malnourished, roughly half of the world's burden of disease is likely linked to poor diet, and some of these diseases are spreading at an epidemic clip. The world is in the midst of a nutrition crisis, and its toll on human development is large, but also largely unrecognized. Reversing the crisis requires that nutrition, long treated as an afterthought by many national leaders, become a clear policy priority. And it requires that policymakers see hunger and overeating as the product of human decisions—often societal but sometimes individual—that government policies can influence.

The most urgent task facing policymakers is to attend to those who are currently hungry or deficient in micronutrients. Distribution of food aid, supplementary feeding programs, and fortification of foods can all be set up in the short run, often with great success. UNICEF and WHO, for example, set out early this decade to eliminate iodine deficiency disorder, largely by persuading 47 countries to establish programs to iodize salt, thereby doubling the number of countries with this capacity. As a result, the global population at risk of IDD was slashed in half between 1994 and 1997, from 29 percent to 13 percent.[58]

A commitment like this could be made concerning iron, vitamin A, and other micronutrients, many of which can easily be added to foods. Such fortification is generally quite affordable, at around 10¢ per beneficiary per year. (See Table 4–5.) Feeding programs require more fiscal effort, costing some $70–100 per beneficiary annually, but the alternative—in terms of loss of life, loss of productivity, and increased health care costs—is likely to make such investments look like a bargain.

Perhaps a greater bargain, and another initiative that can be quickly implemented, is the promotion of breast-feeding. Researchers calculated in the 1980s that during a baby's first six months, effective campaigns could reduce illness from diarrhea by 8–20 percent and death from diarrhea by 24–27 percent. Some governments have gone a step further in encouraging breast-feeding: for example, Papua New Guinea allows the sale of feeding bottles only with a prescription, and the Baby-Friendly Hospital Initiative, a joint UNICEF and WHO program, facilitates breast-feeding by prohibiting the promotion of breast-milk substitutes in hospitals and permitting mothers and infants to room together.[59]

Table 4–5. Costs of a Range of Nutrition Interventions

Intervention	Cost per Beneficiary per Year (dollars)
Micronutrient fortification	0.05–0.15
Micronutrient supplementation	0.20–1.70
Mass media education	0.20–2.00
Other education programs, such as breast-feeding promotion	5
Community-based programs such as home gardening and growth monitoring	5–10
Feeding programs (per 1,000 calories a day)	70–100

SOURCE: Susan Horton, "The Economics of Nutritional Interventions" (draft, April 1999), in R.D. Semba, ed., *Nutrition and Health in Developing Countries* (Totowa, NJ: Humana Press, forthcoming).

Interventions such as these, however, are not sufficient to end hunger. Eradicating hunger requires elimination of its root cause, which is generally poverty. Job creation, affordable food pricing, micro-credit, women's education, land redistribution—any number of diverse measures that reduce poverty and improve social services are likely to improve nutrition. A 1999 study of malnutrition in 63 countries by the International Food Policy Research Institute found social factors—health environment, women's education, and women's status—to account for nearly three quarters of the reduction in malnutrition in these countries, with food availability per person an important fourth factor. Indeed, where governments are committed to improving social safety nets and ensuring gender equity, hunger has largely been contained.[60]

In South Korea and Taiwan, for example, strong commitment to public education and other social infrastructure meant wide access to economic opportunities, while land redistribution and heavy investment in agriculture—the leading economic sector in many developing countries—created a level playing field for rural enterprise. The result has been rapid and shared economic growth that has dampened income inequalities and made poverty, as well as hunger, virtually nonexistent. Especially in rural areas, where most of the world's hungry now live, improving agricultural credit, marketing, and distribution and targeting agricultural investment to small- and middle-size farms are often the fastest routes to eliminating poverty.[61]

Where governments are committed to improving social safety nets and ensuring gender equity, hunger has largely been contained.

Even where economic growth is not robust, attention to poverty reduction can reduce malnutrition substantially. In the Indian state of Kerala, for instance, where per capita incomes and the rate of economic growth are below the Indian average, a strong social safety net is credited with keeping the share of hungry children at 32 percent—well below the national average of 53 percent. Subsidized food shops, health care, and education guarantee acess to all, including women and those in rural areas— two groups that often have the lowest incomes.[62]

The international community shares responsibility for eliminating hunger by assisting national governments and by ensuring that international institutions

respect food security. Debt relief, development aid, and assistance for family planning programs—the latter representing one of the most cost-effective means to improve incomes, health, and lives of women and their children—are among the international initiatives that can alleviate poverty and reduce hunger. And there remains a need to restructure free trade agreements so that subsidizing food production intended for domestic markets is not viewed as illegal.[63]

In nations where overeating is a problem, policymakers require a different set of policy tools. Technofixes attract great attention: liposuction is now the leading form of cosmetic surgery in the United States, at 400,000 operations per year; designer foods use olestra to offer worry-free consumption of nutritionally marginal foods, and laboratories scurry to find the human "fat gene." But such quick fixes typically fail to confront behavioral patterns that underlie obesity, and fail to promote diverse, balanced diets and active lifestyles. Renewed attention to these nutrition basics will require education—with a focus on prevention—and modification of the food environment.[64]

A high-payoff venue for nutrition education is schools, where children are gathered in one place, meals are often served, and eating habits are still being formed. In Singapore, for example, the Trim and Fit Scheme has reduced obesity among children by 33 to 50 percent, depending on the age group, through changes in school catering and increased nutrition and physical education for teachers and children. Programs like this can have lasting effects. In the United States, the Child and Adolescent Trial for Cardiovascular Health targeted children in grades 3 through 5 with additional nutrition and physical education, and yielded behavioral changes that persisted at least into adolescence: a three-year follow-up, without further intervention, showed dietary fat intake to be substantially lower and daily vigorous activity substantially higher in participating children than in control groups.[65]

"Hands-on" learning, which allows students to cook and taste healthy food, critique food advertisements, and read nutrition labels, is an especially effective approach to nutrition education. One example of this philosophy is found in Berkeley, California, where schools have instituted vegetable gardens to educate students about food and nutrition—and to supply some of the food for the cafeterias, which were required in 1999 to begin serving all-organic lunches.[66]

Nutrition education is especially critical for doctors, nurses, and other health care providers, who are in a key position to educate patients about the links between diet and health. Doctors unfamiliar with the impact of diet on health respond only to the consequences of poor nutrition rather than to early warning signs that suggest the need for a change in diet, increase in physical activity, or other preventive measures. In the United States, only 23 percent of medical schools required that students take a separate course in nutrition in 1994—a benchmark that could be raised.[67]

For society at large, mass education campaigns may be needed to change long-standing nutritional habits. One of the most successful national nutrition education campaigns was carried out by Finland in the 1970s and 1980s to reduce the country's high incidence of coronary heart disease. Government-sponsored media campaigns, national dietary guidelines, and regulations on food labeling were among the diverse educational tools used to hammer home the nutritional message. This broad, high-profile approach—it also advo-

cated an end to smoking, and involved groups as diverse as farmers and the Finnish Heart Association—slashed mortality from coronary heart disease by 65 percent between 1969–71 and 1995. About half of the drop is credited to the lower levels of cholesterol induced by the nutrition education campaign.[68]

Less ambitious but well targeted approaches to nutrition education can also provide a large payoff. A consumers' organization in the United States reports that more than two thirds of all shoppers are "very interested" in increasing their consumption of fresh fruits and vegetables. But the same share of shoppers admitted that they did not know how to prepare or eat these foods—and therefore did not buy them. Supermarkets and others in the food business could help by teaching consumers the skills needed for healthy eating.[69]

A nutritionally literate public will begin to demand different foods, to which the food industry may respond with wider offerings of low-fat, whole grain, and other healthier selections. Governments can help markets to promote good nutrition by allowing nutritional claims—verified by an independent authority—to be displayed on food products. Such a system will have greater credibility, however, if the government also requires equally prominent warning labels on unhealthy foods. In Finland, the government requires "heavily salted" to appear on foods high in sodium, while allowing low-sodium foods to bear the label "reduced salt content."[70]

Similarly, governments may need to regulate harmful nutritional information, especially the advertising of nutritionally poor foods. Sweden and Norway, for instance, do not allow any advertising aimed at children under 12, while the Flemish region of Belgium prohibits advertising five minutes before or after children's programming. And more than a dozen European countries have voluntary, self-regulating codes of conduct for advertising targeted at children.[71]

Some nutrition advocates, including Yale professor Kelly Brownell, have argued that regulating advertising of junk foods is similar to curbing the promotion of cigarettes: both industries target children in an effort to win lifelong customers, both kill in huge numbers, and both drive up health care costs substantially. Does it make sense to allow nutrition-poor foods to be promoted and sold freely while tobacco is increasingly regulated?[72]

As support for healthy eating grows, consumption of nutrition-poor foods can be further reduced using fiscal tools. Brownell advocates adoption of a tax on food based on the nutrient value per calorie. Fatty and sugary foods poor in nutrients and loaded with calories would be taxed the most, while fruits and vegetables might escape taxation entirely. The idea is to discourage consumption of unhealthy foods—and to raise revenue to promote healthier alternatives, nutrition education, or exercise programs. This initiative would best be accompanied by efforts in mass public education, so that low-income people, who make up a large share of the public who consume nutritionally poor foods, are not adversely affected.[73]

A final part of reshaping the food environment is recultivating an appreciation of food as a cultural and nutritional treasure. Groups like the Slow Food Movement, based in Italy, and the Oldways Preservation & Exchange Trust in the United States promote a return to the art of cooking traditional foods and of socializing around food. Their work, which targets chefs as well as consumers, is the kind of cultural intervention that could help more people shift to a healthy diet, similar to the change in consciousness that encouraged a shift away from smoking in the United States. Government encouragement of these groups, perhaps through assistance with marketing and promotional activities, would give broader exposure to this important work.[74]

In an age of unprecedented global prosperity, it is ironic—and wholly unnecessary—that malnutrition should exist on such a massive scale. Indeed, poorly nourished people are a sign of development gone awry: prosperity has either bypassed them and left them hungry, or saturated them to the point of overindulgence. But a nation's development path is the product of human choices, and can therefore be redirected. By providing access to nutritionally sound food for all, governments can help shape a social evolution that is truly worthy of the name development.

Chapter 5

Phasing Out Persistent Organic Pollutants

Anne Platt McGinn

In late spring 1999, Belgium suffered a public health crisis that toppled a government, sent food exports plunging, and captured world press attention. The problems began in January, when two synthetic organic compounds—polychlorinated biphenyls (PCBs) and furans—were mixed with animal oil and fats destined for feed. Investigators do not know what caused the initial contamination, but they suspect used transformer oils. Both compounds have effects similar to dioxins, persistent substances that are one of the most toxic group of chemicals known.[1]

About 10 feed manufacturers sold their products to nearly 1,700 farmers in Belgium, who gave it to countless chickens, pigs, and cows. Although the exposure in each bite of feed was minuscule, these contaminants collect in animal fat, concentrating to much higher levels. By early February, chickens were dying and egg production at many of the farms had declined. In mid-March, the Ministry of Agriculture found PCBs and furans in chicken fat and

eggs at several hundred times the legal and safe limit.[2]

Despite the evidence, national authorities waited another month to warn the public and neighboring nations of the dangers, a decision that came back to haunt them on election day. By the end of May, retailers throughout the country pulled from their shelves all poultry and egg products, including mayonnaise, cakes, and cookies; farmers slaughtered any animals suspected of carrying the poisons; and governments from Australia to the United States banned importation of Belgian animal products. Consequently, Belgian farmers lost nearly $500 million in sales and many politicians were ousted in the June election.[3]

Unfortunately, this widely reported case is but the tip of a much larger scandal that affects people and wildlife in every corner of the globe—the uncontrolled spread of persistent organic pollutants (POPs) like dioxins and PCBs, and their infiltration into our food and water, the environment, and our bodies. In this scandal, we are not only the victims but also the culprits.

POPs are a group of insidious synthetic compounds that share four common properties: they are toxic, they accumulate in the food chain, they persist in the environment, and they have a high potential to travel long distances from their source.[4]

The properties that made some of these chemicals so appealing to industries in the first place have made their control extremely difficult. Rather than being soluble in water, POPs have an affinity for the fat tissue of living organisms. Animals and people accumulate these toxins in their bodies, primarily from the food they eat. As the chemicals move up the food chain, they bioaccumulate, which means that each link or species in the food chain takes up the previous link's exposure, adding it to their own and magnifying the effects.

POPs are also extremely durable. Even where these chemicals have been banned for 20 or more years, as in the United States, they persist in soil, water, and body fat. Many health problems can be traced to chemical releases that ended long before the victims were born. And because they are highly reactive to temperature and pressure, POPs can hitchhike on global wind and water currents, migrating from warmer to colder climates and contaminating people as far away as the Arctic Circle.[5]

As the twenty-first century begins, the legacy of the chemical revolution is clear: from plastics to pesticides, POPs are ubiquitous. Some of these synthetic compounds have helped raise levels of food production, protected human health, and allowed for many of the conveniences of modern life. But these advances have come at a high price. All of us now have about 500 anthropogenic chemicals in our bodies—potential poisons that did not exist before 1920. Many of these are POPs, with PCBs and DDE—a highly persistent breakdown product of the notorious DDT—the most commonly detected.[6]

While scientists have yet to understand fully the long-term health impacts of POPs, they have begun to confirm some of the initial effects. Several POPs are thought to be hormonally active compounds that impersonate natural chemical messengers and throw the body's endocrine and immune systems into disarray. Others are associated with delayed intellectual development, reproductive problems, and cancers. From gulls and fish to seals and eagles, species throughout the food web have shown signs that they are imperiled by POPs. And as Rachel Carson wrote in *Silent Spring*, more than 30 years ago, "Our fate is connected with the animals." Unless actions are taken to reduce and eventually eliminate POPs, these synthetic time bombs will continue to menace life on Earth.[7]

The World of POPs

Today a new chemical substance is discovered about every nine seconds of the working day. Most remain laboratory artifacts. On June 15, 1998, chemists identified the 18 millionth synthetic chemical substance known to science. Of the millions now recognized, fewer than 0.5 percent—possibly 50,000 to 100,000—are actually used in commerce. The vast majority of these are organic substances, meaning they contain carbon, an element essential to life. But in certain molecular combinations, carbon can herald trouble.[8]

Synthetic organic chemicals are largely a twentieth-century creation. They were first manufactured on a large scale during the 1930s, and have been growing in volume ever since. Global production escalated from less than 150,000 tons in 1935 to more than 150 million tons by 1995. In the United States, production increased more than threefold, from 45,000 tons in 1966 to nearly 150,000 tons in 1994. (See Figure 5–1.) Some of these "everyday" synthetic organic chemicals are not of direct environmental concern, such as those in pharmaceuticals, cosmetics, and food additives. More problematic are the numerous compounds introduced into the environment in large quantities, such as dyes, detergents, additives in plastics, organic metals, antifouling paints, and pesticides.[9]

Of some 1,000 recognized environmental contaminants, about half contain chlorine, which tends to impart stability and persistence to the molecule and make it more likely to bioaccumulate. In nature, chlorine is almost never found in its elemental state—it normally binds with a more electron-rich atom such as sodium or carbon. Because it is so reactive, it is the basis for thousands of humanmade chemicals.[10]

Chlorine has been called "the single most important ingredient in modern [industrial] chemistry." About one third of the chlorine made in the United States is used to manu-

facture plastics such as polyvinyl chloride (PVC), which shows up in plastic wrap, shoe soles, automobile components, siding, and medical products such as intravenous bags, tubing, and gloves. Slightly more than half of the chlorine helps synthesize thousands of other commercial chemicals, such as carbon tetrachloride used to make nonstick frying pans, or ends up as chlorinated solvents. The remainder is used to bleach paper or treat drinking water.[11]

Many chlorine-containing chemicals are both innocuous and valuable for commerce and medicine. Some, however, including most of the POPs, are extremely harmful. The challenge is to identify and regulate the most dangerous substances. At the moment, scientists do not even know how many POPs exist. Estimates vary from dozens to hundreds. Many compounds have never been tested for bioaccumulative or persistent properties. More important than the absolute number, however, is their cumulative impact in the environment.[12]

Government officials, members of environmental and health groups, and toxic substance specialists have singled out 12 POPs as the first candidates for global action. These 12 are deemed particularly high priorities as they have serious and well-understood effects. Several of them have been banned in industrial countries for years, making them an easy starting point. Referred to as the "dirty dozen," they include nine agricultural pesticides, one industrial compound (PCBs), and two industrial byproducts (dioxins and furans). (See Table 5–1.)[13]

Ironically, the pesticides that are now blacklisted in some countries were for many years viewed as a chemical triumph over nature. In 1948, for example, Swiss chemist Paul Müller

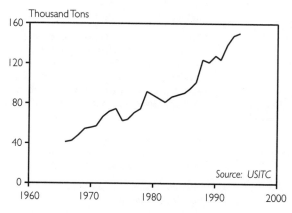

Figure 5–1. U.S. Synthetic Organic Chemical Production, 1966–94

Table 5–1. Production and Use of the "Dirty Dozen" POPs

Persistent Organic Pollutant	Date Product Introduced	Cumulative World Production (tons)	Designed Use
Aldrin	1949	240,000	Insecticide used to control soil pests (primarily termites) on corn, cotton, and potatoes; fumigant
Chlordane	1945	70,000	Insecticide used on variety of crops and for termite control
DDT	1942	2.8–3 million	Domestic and agricultural insecticide used against mosquitoes
Dieldrin	1948	240,000	Insecticide used on fruit, soil, and seed crops including corn, cotton, and potatoes
Endrin	1951	(3,119 tons in 1977)	Rodenticide and insecticide used on cotton, rice, and maize
Heptachlor	1948	(900 tons used in 1974 in the United States)	Insecticide used to kill soil insects and termites. Used for fire ant control. Also for malaria control
Hexachloro-benzene	1945	1–2 million	Used as fungicide. Also a byproduct in pesticide manufacture and contaminant in other pesticide products
Mirex™	1959	no data	Insecticide used on fire ants, leaf cutters, and harvester termites. Also a fire retardant. One of the most stable and persistent pesticides
Toxaphene™	1948	1.33 million	Insecticide, especially used against ticks, mites, and maggots and on cotton. Mixture of 670 chemicals
PCBs	1929	1–2 million	Used primarily in liquid form as dielectric fluid in capacitors and large transformers, hydraulic fluids, and heat transfer systems. Used as nonliquids in weatherproofers, carbonless copy paper, paint, adhesives, and plasticizers in synthetic resins
Dioxins	1920s	(10.5 tons ITEQ[1] of dioxins and furans combined, 1995)	Byproduct of combustion, especially of plastics, and of chlorine product manufacture and paper bleaching
Furans (dibenzofurans)	1920s		Byproduct, especially of PCBs, often with dioxins

[1]International Toxic Equivalency.

SOURCE: See endnote 13.

won a Nobel Prize for creating dichlorodiphenyl-trichloroethane—DDT. During World War II, soldiers and civilians were doused in DDT powder to ward off insect-borne diseases. Farmers sprayed DDT liberally on their crops in the United States through the 1950s and 1960s to deter insects. Cotton farmers in Central Asia may still be using it illegally for this purpose today, while a handful of countries are believed to release DDT directly into the environment illegally through crop spraying and other agricultural applications.[14]

DDT was also initially hailed as a public health miracle for combating vector-borne diseases such as malaria. Indeed, shortages of DDT were once considered a public health hazard by the World Health Organization (WHO). Today, only India, China, and a few other countries still produce DDT, but 30 or so use it to control malaria-transmitting mosquitoes. With about four children dying every minute from malaria, primarily in sub-Saharan Africa, numerous doctors argue that DDT remains the best weapon to combat this disease. Where alternatives are not yet widely available, this may be true. In parts of India and South America, however, mosquitoes are resistant to DDT. In other areas, alternatives such as insecticide-soaked bednets have proved just as effective at reducing exposure. The challenge is to eliminate DDT from agricultural use while preserving its reliability as a malaria-control tool to be used only under specific circumstances until viable alternatives are widely available.[15]

Other pesticides have undergone a similar transition from a heyday of production to a period of more restrictions and less effectiveness. World pesticide use has increased 26-fold in the last 50 years, although growth has slowed recently. (Not all pesticides are POPs.) Since the 1970s, an increasing number of countries have curtailed use of particular pesticides, due to health and environmental concerns.[16]

Meanwhile, about 1,000 species of insects, plant diseases, and weeds now regularly survive chemical applications intended to wipe them out. Improper, incomplete, and sometimes even full application has allowed many pests to evolve stronger, more resistant strains. This has prompted farmers to engage in a vicious circle of more toxic combinations, more frequent applications, and larger quantities. Even though the quantity of pesticides used in industrial countries has declined, the toxicity has increased 10- to 100-fold since 1975, so each application is deadlier to pests as well as to people. Moreover, many of the "dirty dozen" pesticides continue to plague Earth's ecosystems because they are still used in many developing countries. Reducing these persistent and transboundary compounds in the environment therefore requires a global strategy tailored to local agricultural needs.[17]

Polychlorinated biphenyls are among the most notorious POPs. The PCB family consists of about 209 compounds, each of which has between 1 and 10 chlorine atoms in its molecular structure. In industrial countries, PCBs were manufactured mostly between the 1920s and the late 1970s. (They are still manufactured in Russia today.) Because of their stability, low flammability, and low conductivity, PCBs were ideal transformer fluids used to insulate electrical components. They are also found in cables, electrical wiring insulation, and even carbonless copy paper.[18]

During the last 20 years, the levels of PCBs in most environmental media have been declining, particularly in countries where they have been banned. But this trend slowed during the 1990s as PCB use and release continued elsewhere. In developing countries, PCB levels measured in human fat, blood, and milk samples have been increasing recently. Scientists estimate that 70 percent of all PCBs manufactured in the last 70 years are still in use or can be found in landfills. Since PCBs are very stable compounds that do not degrade easily, they are virtually guaranteed to contaminate the environment—and humans—for decades.[19]

In contrast to PCBs and pesticides, dioxins and furans have never been produced intentionally. They are formed primarily as the byproducts of processes such as the bleaching of wood pulp or the incineration of medical and solid wastes with chlorine-containing products. When such wastes are burned in poorly controlled incinerators, they create dioxins and dioxin-like furans. Worldwide, incineration of medical and municipal wastes accounts for 69 percent of dioxin and furan releases to air, some 7,000 kilograms a year.[20]

Japan currently reports the highest total emissions of dioxins and furans in the world. (See Table 5–2.) The government is now trying to contain dioxin releases from more than 3,800 highly polluting municipal incinerators. (In contrast, the United States has fewer than 200 such facilities in operation.) Japan is not the only country facing a dioxin crisis from incineration: on a per capita basis, the Dutch are responsible for almost as many emissions as the Japanese, and Belgians account for more than twice as much. In contrast, Austria, Germany, and Sweden have made significant strides in controlling incinerator releases and now boast the lowest reported per capita emissions in the world.[21]

The true scope of dioxins and furans is hard to estimate: the "global" totals include only 15 countries that responded to a 1995 global inventory by the U.N. Environment Programme (UNEP). Many countries have only recently begun to monitor emissions, let alone try to control them.[22]

The World Bank is now actively promoting medical waste incinerators in 20 developing countries to help reduce their wastes. Instead of advocating cleaner and more environmentally appropriate technologies, the Bank has held on stubbornly to tech-

Table 5–2. Known or Estimated Dioxin and Furan Air Emissions, Selected Countries, Total and Per Capita, 1995

Country	Total Emissions (grams ITEQ/a)[1]	Per Capita Emissions (nanograms ITEQ/a)[1]
Sweden	22	2,500
Austria	29	3,625
Denmark	39	7,464
Slovak Republic	42	7,844
Hungary	112	10,952
Australia	150	8,358
Switzerland	181	25,340
Canada	290	9,792
Germany	334	4,090
The Netherlands	486	31,439
United Kingdom	569	9,758
Belgium	661	65,523
France	873	15,047
United States	2,744	10,276
Japan	3,981	31,728

[1] International Toxic Equivalency emitted into air.
SOURCE: U.N. Environment Programme, Chemicals Division, *Dioxin and Furan Inventories: National and Regional Emissions of PCDD/PCDF* (Geneva: Inter-Organization Programme for the Sound Management of Chemicals, May 1999); U.N. Population Division, *World Population Prospects: The 1998 Revision* (New York: 1998).

nologies that are no longer suitable. As Ravi Agarwal and Bharati Chaturvedi, coordinators for the Delhi-based Indian Campaign Against Medical Waste Incineration, argue, the so-called cure for the medical waste problem—incineration of contaminated syringes, infectious materials, and health care wastes, which results in releases of dioxins and toxic chemicals—is far worse than the disease itself, the piles of waste and garbage.[23]

Rather than serving a useful purpose, dioxins and furans make old problems worse and add new ones to the mix. To a certain extent, the same is true for pesticides and PCBs, which transform systems of food production, disease prevention, and electricity generation into sources of toxic releases. The subset of the chemical universe represented by POPs is thus particularly dangerous because compounds that at first glance seem too good to be true become insidious chemical threats long after they are released in the environment.

Routes of Exposure and Environmental Fate

The usual image of environmental pollution is a massive oil spill or a pipe dumping toxins into the water. POPs, in contrast, are released into the environment through everyday activities—from pesticide applications to waste burning—usually with little thought of the effects. Consequently, humans and animals are exposed to multiple chemicals—mixtures that are difficult to isolate for testing and identification, let alone for predicting how they react together. As Linda Birnbaum, Director of the Experimental Toxicology Division for the U.S. Environmental Protection Agency, says, "We live and breathe in a chemical

soup." Life on Earth has always existed in a chemical soup, but in the industrial age we have added vast quantities of synthetic chemicals to the recipe. Scientists are just beginning to understand the fates of pollutants and the risks they pose to all of us.[24]

The four factors that distinguish POPs from other chemicals—high toxicity, bioaccumulation, persistence, and the potential for long-range transport—also determine how far-reaching their effects are in the environment. The combination of bioaccumulation and persistence, for instance, allows POPs to invade the web of life and to linger on almost indefinitely. Arctic cod and turbot, to cite but one example, have 1,000 times higher concentrations of DDT per gram of fat than the zooplankton they consume. And herring gulls from Lake Ontario harbor nearly 25 million times the concentration of PCBs that are found in surrounding waters.[25]

Although DDT was banned from use more than 25 years ago in both the United States and Canada, enough of the synthetic compound is present in the Great Lakes to reduce substantially the chances that eagles and other top predators will produce viable offspring. Even newcomers fall prey to this residual contamination. Within two years of migrating from pristine inland areas to the Great Lakes shores, bald eagles suffer marked declines in fertility. For this reason, experts refer to the Great Lakes as a "black hole" for bald eagles.[26]

Humans collect POPs from their food just like animals do. People receive about 90 percent of their total intake of bioaccumulative POPs through foods of animal origin. Consequently, monitoring food sources of POPs, particularly meat and fish, is absolutely critical to protecting human health. Foods tested in India, Thailand, and Viet Nam show low to moderate levels of

PCB contamination, a trend that is expected to worsen as these countries become more industrialized. More than 90 percent of human dioxin exposure occurs from eating fatty foods and animal products such as meat, fish, milk, cheese, and even ice cream. In the United States, 46 states have issued advisories against eating local fish because of dioxin contamination.[27]

Because POPs collect in high levels in fatty tissues, they are sometimes inadvertently transmitted, along with vital nutrients, in breast milk. Breast-fed babies in India and Zimbabwe absorb DDE (a highly persistent breakdown product of DDT) at levels six times the acceptable daily intake. By their first birthday, Americans may receive up to 12 percent of their acceptable lifetime exposure to dioxins from their mothers' breast milk. In Europe and the United States, dioxins in breast milk are many times higher than what is allowed for commercial cows' milk. (Nevertheless, experts believe that for most children the benefits of breast-feeding still far outweigh these risks.)[28]

Once they are released, POPs become the chemical equivalent of an invasive species. They work their way into innocent species and distant ecosystems, accumulating to ever higher concentrations along the way. Many migrate through the atmosphere, condense and deposit out in soil or water, and then evaporate again, often repeating this cycle several times, spreading to all corners of the globe. Scientists have tested albatrosses in what would seem a remote refuge: Midway Atoll—aptly named for its position roughly halfway between Japan and California, near the International Dateline. Many of the birds there carry levels of DDT, PCBs, and dioxin-like compounds equivalent to those that are debilitating bald eagles near the North American Great Lakes. Similarly, tree bark sampled from more than 90 sites—including several tropical and temperate developing countries—found that DDT, chlordane, and dieldrin were present no matter how remote the area.[29]

Polar regions are one of the world's main collection sinks for POPs, even though these areas account for just 13.4 percent of Earth's surface area. With less than six months of regular daylight, and with delayed decomposition, evaporation, and metabolism in the bitter cold temperatures, POPs are especially long-lived in these high-latitude areas. Even though only a fraction of the mass of all POPs make the trip to the poles, they readily accumulate to high concentrations in marine species. "The Arctic is not only firmly connected to the global transport system, but it is also especially vulnerable to some of the things being transported," Robie Macdonald of the Institute of Ocean Sciences, Fisheries, and Oceans Canada in British Columbia explains.[30]

Because native peoples in these remote areas eat large quantities of local fish and marine mammals, which are at the top of the food chain and high in fat, their diet poses special risks. A few years ago, scientists looking for control populations with little to no exposure to modern chemicals headed to the Arctic. What they found was quite a surprise. Indigenous people in the remote reaches of Canada's Northwest Territories and Baffin Island, well north of the Arctic Circle, are among the most exposed people in the world. One bite of raw whale blubber, an Inuit delicacy known as *muktuk*, can contain more PCBs than Canadian scientists say should be consumed in a week. Inuit women carry 2 to 10 times as high a concentration of PCBs in their breast milk as women in southern Canada do, and

10 times as much chlordane, even though they are thousands of kilometers from the closest agricultural area.[31]

In addition to the nearly invisible flows on wind and water currents, the wings of birds, and our own dinner plates, some POPs are sold as commodities in international markets. In 1996, total pesticide sales topped $33 billion. The value of world exports of pesticides, some of which are POPs, increased more than eightfold in less than four decades, from $1.3 billion in 1961 to nearly $11 billion in 1997 (in 1997 dollars). (See Figure 5–2.) Most exports originated in industrial countries.[32]

One of the sad ironies of the chemical age is that many countries continue to produce and export chemicals that are deemed too deadly for use domestically. Although forbidden from using them on American soil, U.S. companies are allowed to manufacture more than 4 million tons of chlordane and other pesticides. The majority go to developing countries. And some may come back to haunt Americans in imported food.[33]

Despite growing awareness of the dangers of bioaccumulating chemicals, this trend is getting worse: between 1992 and 1996, U.S. chemical companies increased exports of domestically prohibited pesticides by 18 percent. The rate of exports of pesticides that have never been registered jumped 40 percent during this time. Although domestically restricted and never-registered pesticides represent only a small share of all U.S. pesticide exports, more than 14 percent of the total exported are restricted in some form. Many are transferred without handling instructions, safety data sheets, or any indication of how to properly use and dispose of the chemical.[34]

Tracing the life cycle of POPs from pro-

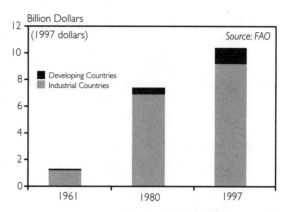

Figure 5–2. Pesticide Exports from the Top 30 Countries, Selected Years

duction to end point is quite difficult: much of the data are proprietary, the science of mapping a compound's environmental fate is still fairly elementary, and the routes to world food supplies remain murky. Although information on the properties of POPs is still emerging, the health effects of living and breathing in a chemical soup are gradually becoming clearer—and harder to ignore.

Health Consequences

Since the 1950s, scientists have linked persistent organic pollutants to a range of health problems in wildlife, including masculinization and feminization, poor egg survival and twisted beaks in birds, deformed limbs, reproductive and developmental disorders, and decreased immunity. A large share of human data come from studying people who have been highly exposed at work or in severe chemical accidents. Combined with a rapidly evolving scientific understanding of POPs and growing public concern of potentially serious and irreversible human health impacts, researchers have begun looking for—and finding—the signature effects of POPs at everyday exposure levels. With data from animals, highly exposed communities, and the general population, scientists have thus started to confirm the health effects of persistent organic pollutants.[35]

Many of the problems can be traced to chemicals that act like, disrupt, or otherwise interfere with the body's hormones.

As early as 1948, just nine years after researchers discovered that DDT could kill insects, scientists measured DDT in human fat tissue. That same year, the World Health Organization convened a panel of experts to examine the toxic effects of pesticides on people and to recommend safe handling procedures. In the 1950s, two American scientists studied chickens exposed to DDT and found that the roosters failed to develop red combs and wattles, typical male characteristics. In addition, their sex organs were almost 20 percent smaller than normal.[36]

In the 1960s, Rachel Carson's writings and mounting scientific evidence of the harm from pesticides helped spark a generation of activists who campaigned against pesticide misuse, toxic pollution, and emissions. Carson's work also united two often disparate communities—environmentalists and biologists with doctors and health specialists, a partnership that continues to honor her legacy by working together on behalf of human and planetary health.

The next wave of attention to POPs came during the 1970s and 1980s, when a series of marine mammal die-offs in the Baltic, Mediterranean, and North Seas prompted researchers to test whether POPs might be responsible. In one experiment, a group of seals were fed highly contaminated fish from the Wadden Sea while a control group ate cleaner fish from Atlantic waters. Scientists found that females who ate the cleaner fish bred normally with males 83 percent of the time. In contrast, only one third of the females who consumed the contaminated fish mated successfully; the rest failed to produce any offspring.[37]

A similar feeding experiment showed that seals carrying higher loads of PCBs experienced declines of 20 and 60 percent in two types of cells that are essential components of a healthy and functioning immune system. With their natural defenses impaired, seals and other species can easily fall prey to normally harmless infections. Given that populations of many marine mammals have declined to historic lows, these are troubling results. Not only are marine mammals more vulnerable to potentially lethal illnesses, but the difficulties they now experience in reproducing successfully greatly diminish the chances that they will repopulate anytime soon—if at all.[38]

The late 1980s and early 1990s saw a sharp increase in our understanding of how these chemicals work and an appreciation that they can cause severe damage at levels dismissed as insignificant only a decade earlier. Pioneering detective work by Dr. Theo Colborn, a wildlife toxicologist, and other scientists helped bring these links to light. Many of the problems can be traced to chemicals that act like, disrupt, or otherwise interfere with the body's network of hormones and receptors, known as the endocrine system. Some endocrine disrupters block hormones from getting through to their receptors by interfering with the levels of different hormones or the process of attachment; others induce hormone-like reactions from receptors or simply increase the number of receptors.

Whichever change occurs, low levels of these unnatural compounds can derail a species' normal development—often with long-lasting and irreversible effects.[39]

In 1993, toxicologists identified 45 pesticides and industrial chemicals as known or suspected endocrine disrupters. By 1998, 15 other compounds were added, bringing the total to 60, including all of the "dirty dozen." Yet one recent estimate, citing the German Federal Environment Agency, identifies 250 such compounds. More are likely to be listed as testing and screening continue.[40]

One of the first signs of the impacts of endocrine disrupters on humans came during the 1950s and 1960s. Nearly 1 million pregnant women in the United States took an artificial hormone, diethylstilbestrol (DES), to prevent spontaneous abortions. The drug had severe side effects. DES daughters suffered from fertility problems, abnormal pregnancies, reproductive organ malfunctions, immune system disorders, and higher rates of a rare vaginal cancer, which is typically only seen in women over the age of 50. DES sons reported cryptorchidism (undescended testicles), abnormal semen, and hypospadias (abnormal urethral opening)—similar to the effects seen in roosters exposed to DDT.[41]

People accidentally exposed to high levels of PCBs develop many telltale signs of acute toxicity and sometimes endocrine disruption. For nine months in 1978, for instance, about 2,000 Taiwanese unknowingly used rice bran cooking oil that had been contaminated by a mixture of PCBs and furans, which came from leaking cooking fuel. The average intake was about one gram of PCBs, nearly a thousandfold the average for Americans, and an estimated 3.8 milligrams of furans, nearly 10,000 times higher than normal exposure.[42]

By 1983, the provincial Health Department had recorded more than 2,000 cases of "Yucheng" or oil disease, which was characterized by skin discoloration, severe stomach problems, hypersecretions from the eyelid glands, and delayed nerve conduction—all characteristic of acute toxicity. In addition, scientists found that rates of chronic liver disease were 2.7-fold higher among the Yucheng victims than among people who had never been exposed. Children who were born several years after their mother's initial exposure suffered from slower cognitive and physical growth through their teens, and more frequent ear and upper respiratory infections.[43]

In the United States, more than 200 children whose mothers ate PCB-contaminated salmon and lake trout from Lake Michigan during their pregnancies developed similar traits of slowed intellectual development. By sixth grade, these children lagged up to two years behind their classmates in reading ability and word comprehension. One hypothesis is that PCBs may have interfered with thyroid hormones, which stimulate nerve cell proliferation and development. Particularly worrying is the fact that these were not especially contaminated fish. In contrast to the women in Taiwan, these mothers consumed levels of PCB that were "similar or only slightly higher than the general population" consumed. Researchers confirm comparable effects among children exposed to PCBs and dioxins in the Netherlands.[44]

Some scientists argue that endocrine disrupters are linked to significant drops in sperm counts over the past 60 years in industrial countries. In 1992, a report in the *British Medical Journal* noted a 50-percent drop in sperm production between 1938 and 1991 among European and American men. In the United States, sperm counts reportedly fell by 1.5 percent a year from 1938 to 1988, whereas in northern Europe, sperm counts declined by roughly the same amount but nearly twice as fast—3.1 percent a year—during the 1970s and 1980s. Other studies have cited a rash of related male reproductive health problems in industrial countries since the 1960s, including a higher incidence of testicular cancer, cryptorchidism, and hypospadias.[45]

But the debate over male reproductive health is highly controversial. Sperm counts collected in seven developing and non-western countries—including Brazil, India, Israel, Nigeria, and Thailand—show no clear trends since monitoring began in 1978, and even an increase in some areas. Moreover, some urban populations in Europe and the United States witnessed no drop in sperm counts. Several long-term projects are now under way to analyze the trends and determine what role—if any—endocrine disrupters may be playing.[46]

A growing body of scientific evidence suggests that people who have been exposed to pesticides and other POPs at work or in accidents may have higher rates of immune system cancers (non-Hodgkin's lymphoma, leukemia, and myeloma), infertility, genital defects among sons, heart disease, and suppression of the immune system, which can lead to higher rates of infections. For example, highly exposed farmers and agricultural workers are at similar risk of infectious ailments as people who have AIDS or genetic immunological deficiencies.[47]

General consumers are not spared from excessive pesticide exposure either. A 1999 study by the U.S.-based Consumers Union found that domestic fruits and vegetables often exceed the safe exposure limit set by the Environmental Protection Agency (EPA) for young children. Peaches, pears, apples, and even frozen green beans could potentially cause severe neurological damage in children because of high levels of an organophosphate, methyl parathion. EPA recently banned this compound from being sprayed on crops. While long overdue, banning a single compound will in the short term have little effect on the amount of active contaminants already in the environment. Like POPs, methyl parathion resists degradation and will continue to affect our food systems for years.[48]

The difficulties of regulating particular compounds are made worse by the fact that most POPs are highly synergistic—two of them mixed together trigger greater effects than the sum of the results when they are used alone. (See also Chapter 2.) Although the health effects of one substance may be known, it can produce unpredictable fireworks when coupled with another.[49]

The Policy Response to POPs

Since the dawn of the chemical era, countless manufacturers have worked hard to spread the gospel about these seemingly miraculous compounds. Today, however, many scientists and political leaders agree that POPs are more dangerous than other synthetic compounds. Moreover, an increasing number of leaders recognize that adequately addressing these compounds requires a unified, global strategy. While the shape of such a strategy is still up in the air, several regional agreements have called for the reduction and eventual elimination of most POPs, based on the precautionary principle—the idea that even in the face of scientific uncertainty, the prudent stance is to take cost-effective steps to restrict or even completely prohibit an activity that has the potential to cause long-term or irreversible harm. There is now significant potential for this principle to become the operative tool for POPs policy.[50]

Traditionally, international agencies have focused their attention on how to use chemicals safely rather than questioning the need for a synthetic compound in the first place. In 1963, for example, WHO and the U.N. Food and Agriculture Organization (FAO) formed a joint committee to evaluate what levels of pesticides are safe in particular foods. The recommendations were channeled through an intergovernmental body, the Codex Alimentarius Commission, which draws up international food safety standards.[51]

For many years, individual countries have taken the lead on chemical control by enacting prohibitions and import bans (see Table 5–3), advocating nontoxic alternatives, and educating the public about the hazards. By the 1970s, several industrial countries had banned the use of several POP pesticides. The number of national bans on particular pesticides worldwide jumped from 53 in 1983 to 689 in 1995. About two thirds of these were in developing countries. Most of the "dirty dozen" pesticides are now banned in a number of countries, but many nations remain addicted to their production and use. And this affects all nations, of course, for POPs do not respect national boundaries.[52]

The roots of an international POPs strat-

Table 5–3. Regulatory Status of the "Dirty Dozen" POPs

Chemical	Countries with Bans or Prohibitions[1]	Countries with Restrictions[2]	Countries with Import Bans
Aldrin	72	10	52
Chlordane	57	17	33
DDT	60	26	46
Dieldrin	67	9	53
Endrin	65	9	7
Heptachlor	59	17	36
Hexachlorobenzene	59	9	4
Mirex™	52	10	n/a
Toxaphene™	58	12	n/a
PCBs	9	4	5
Dioxins	0	23	n/a
Furans	0	22	n/a

[1]Bans includes national laws that ban substance for all uses or from principal use and substances banned by EC directive or regulation.   [2]Restrictions include national laws that severely restrict substance or restrict use of substance.

SOURCE: U.N. Environment Programme (UNEP), *Summary of Existing National Legislation on Persistent Organic Pollutants* (Geneva: 14 June 1999); PCB data from Joint Canada-Philippines Planning Committee, *International Experts Meeting on Persistent Organic Pollutants: Towards Global Action*, meeting background report, Vancouver, Canada, 4–8 June 1995 (revised October 1995); import bans from UNEP and U.N. Food and Agriculture Organization, *Report on Study on International Trade in Widely Prohibited Chemicals* (Rome: 16 February 1996).

egy emerged in regional marine policies of the late 1980s. Looking to reduce sources of upstream pollution, several treaties highlighted the dangers of toxic, bioaccumulative substances in the marine environment and called for their phaseout, including the Third Ministerial Conference on the North Sea in 1989, the International Joint Commission on the Great Lakes in 1991, and the Paris Commission for the Northeast Atlantic in 1992.[53]

At the 1992 Earth Summit in Rio, governments addressed the threat of POPs in the oceans chapter of Agenda 21. They called for eliminating organohalogen compounds (which includes most of the "dirty dozen" POPs) and reducing other synthetic organic compounds in the marine environment. Agenda 21 also laid the groundwork for more specific international commitments concerning POPs and created institutions that could help coordinate such efforts. For instance, the International Forum on Chemical Safety (IFCS), composed of various governments, nongovernmental organizations (NGOs), and intergovernmental bodies, was established to provide policy guidance to national agencies and governments. Although the Agenda 21 chapter on toxics essentially supported business as usual, the oceans section broke new ground for international POPs policy and set such actions in the context of the precautionary principle.[54]

After Rio, public and institutional support for addressing the 12 initial POPs gathered momentum. In late 1995, environment ministers from 103 countries attended a UNEP conference on land-based sources of marine pollution in Washington, D.C. They signed a high-level ministerial declaration calling for a legally binding global treaty to reduce and eventually eliminate POPs. Several months later,

the IFCS concluded that there was enough evidence of the harmful effects of the "dirty dozen" to justify a global treaty to curb their use and production. The World Health Assembly of WHO seconded these findings. Nearly five years after the Earth Summit, the UNEP Governing Council agreed to oversee development of an International Legally Binding Instrument for Implementing International Action on Certain Persistent Organic Pollutants. Formal negotiations began a year later.[55]

At the opening session, in June 1998, UNEP Executive Director Klaus Töpfer declared that "the ultimate goal [for the global treaty] must be the elimination of POPs, not simply their better management." Officials from more than 100 countries and dozens of NGOs have heeded his call and are now working to specify obligations and mandate actions on the "dirty dozen." At the third session, in September 1999, for instance, the parties to the proposed treaty agreed to list 8 of the 10 intentionally produced POPs on Annex A, which calls for eventual elimination.[56]

The most telling test of the delegates' resolve will take place in upcoming sessions, however. Delegates have yet to decide if and where DDT and PCB can still be used. Most likely, these two compounds will be severely restricted, with specific exemptions under certain circumstances. Dioxins and furans will also be the subject of a reduction-versus-eventual-elimination debate. Currently, industry representatives and the United States are opposed to a specific long-term goal of elimination for these ubiquitous byproducts, but Europeans have recognized the need for such a framework.[57]

It also remains unclear whether the accord will focus on the elimination or reduction of all POPs or each individual POP. How these issues are decided will determine the ultimate form—and useful-ness—of the treaty. If it turns out to deal with POPs one by one, it would represent an unfortunate shift away from a broad application of the precautionary approach to a more limited approach favoring busi-ness as usual.

With the nearly universal recognition of the need to expand the scope and coverage of the treaty from the original list of 12, a group of experts has been working on the criteria for adding other POPs once the treaty comes into effect. They have benefit-ed from the recently completed regional agreement on POPs, the Aarhus Protocol to the 1979 Convention on Long-Range Transboundary Air Pollution, which was signed by all European nations, Russia, Canada, and the United States. It specifies threshold criteria for identifying a POP. Establishing such standards in the global accord is critical to ensuring that the final product is truly a POPs treaty and not sim-ply a "dirty dozen" treaty.[58]

Despite the progress to date, several stumbling blocks remain in the final stages of global treaty negotiations. The debate between those who want to maintain the status quo and not ban chemicals and those who advocate adopting a more precaution-ary approach continues. Chemical produc-ers such as those represented by the International Council of Chemical Associa-tions argue that POPs can be managed without bans, through voluntary commit-ments and other means. This disagreement is particularly evident with respect to diox-ins and furans and the specific criteria for new compounds, as representatives disagree on what level of a particular characteristic should lead to global action.[59]

Further complicating the treaty process is a challenge that all nations face: finding the money to pay for the phaseout of POPs.

Technical expertise, financial assistance, and cost-effective alternatives are needed. It remains to be seen whether industrial coun-tries are willing to contribute funds for developing countries to comply with a global treaty. To date, a few have voluntari-ly pledged money for an assistance fund. Funding will most likely come from a POPs convention financial mechanism that has yet to take shape, combined with money from multilateral development banks, bilateral funds from industrial-country gov-ernments, and contributions from manufac-turers of new technologies and alternatives who are eager to tap into new markets.[60]

Complicating the treaty process is a challenge all nations face: finding money to pay for the phaseout of POPs.

Getting a handle on the global disper-sion of POPs also requires addressing trade. Hazardous waste trade is certainly much broader than simply POPs, but several recent international agreements cover these compounds and help set the stage for the global POPs treaty. (See Table 5–4.) Although many are not yet in force, these agreements have the potential to reduce transboundary shipments of POP products and their waste, and to publicize the dan-gers of production, use, and disposal of such hazardous chemicals. The Rotterdam accord, for instance, employs the Prior Informed Consent procedure to empower countries to accept or refuse shipments of dangerous hazardous waste, including many POPs. Significant indicators of sup-port for the goal of reducing shipments are national and regional bans: by 1997, more than 100 countries had enacted hazardous waste import bans covering much of the

Table 5–4. Status, Strengths, and Weaknesses of Selected International Conventions Complementary to the Global POPs Treaty

Aarhus Protocol on Persistent Organic Pollutants to the Long-Range Transboundary Air Pollution Convention, June 1998

Strengths:	• Covers 16 POPs; bans 8, restricts 4, and calls for phaseout of others
	• Includes specific criteria for adding chemicals
	• Includes Western and Eastern Europe, Canada, Russia, and the United States
Weaknesses:	• No developing countries involved
	• Air pollution agreement, does not cover all media
Status:	• 55 countries signed; 1 ratification, 16 needed

Rotterdam Convention on the Prior Informed Consent (PIC) Procedure for Certain Hazardous Chemicals and Pesticides in International Trade, September 1998

Strengths:	• Creates an information exchange procedure on national bans and restrictions to reduce imports of unwanted, harmful chemicals
	• Covers 7 POPs; total of 5 industrial chemicals and 22 pesticides
Weaknesses:	• No enforcement "teeth" until ratified—voluntary procedure until then
	• Applied at the national level, with no global aspects
Status:	• 60 countries and EU signed; 50 ratifications needed

Basel Convention on the Control of Transboundary Movements of Hazardous Wastes and their Disposal, 1989, and 1995 Amendment

Strengths:	• Most POPs qualify as hazardous wastes when destined for final deposition or recycling
	• Singles out wastes contaminated with PCBs, dioxins, and furans
	• Technical cooperation trust fund established to assist developing countries in implementation
	• 1995 amendment will stop exports of hazardous waste from OECD countries to non-OECD countries
	• Legally binding with criminal penalties
Weaknesses:	• Applies to imports only, not production or use
	• Amendment not in force yet
Status:	• Convention: entered into force May 1992
	• 1995 Amendment: 67 countries signed; 15 ratifications, 62 needed

SOURCE: U.N. Environment Programme (UNEP), "Analysis of Selected Conventions Covering the Ten Intentionally Produced Persistent Organic Pollutants" (Geneva: 2 June 1999); Aarhus from Daniel Pruzin, "UN/ECE Draft Protocol on Heavy Metals, Persistent Organic Pollutants Concluded," *International Environment Reporter*, 18 February 1998; U.N. Food and Agriculture Organization, "Rotterdam Convention on Harmful Chemicals and Pesticides Adopted and Signed," press release (Rome: 11 September 1998); Basel Secretariat, UNEP, <www.unep.ch/basel>, viewed 23 October 1999.

developing world, including all of Africa, the South Pacific, and Central America. South and Southeast Asia, however, remain wide open.[61]

Even if all hazardous products and waste trade were halted today, the world has many a "ticking time-bomb," in the words of FAO. More than 100,000 tons of obsolete pesticides are improperly stored in the developing world alone. (Many were shipped there by companies in industrial countries looking for a cheap way to get rid of them, in exchange for nominal payments.) Properly disposing of such wastes is a daunting task. Experts estimate it would cost $80–100 million to clean up the more than 20,000 tons of obsolete pesticides in Africa.[62]

With negotiations on a global POPs treaty well under way, politicians, business leaders, activists, and concerned citizens have an enormous opportunity "to change the way the chemical manufacturers do business," according to Greenpeace toxins expert Charlie Clay. Broad acceptance and application of the precautionary principle can shift the impetus from controlling the release of chemicals to questioning their use and requiring safe alternatives. Treaties and bans alone will not accomplish this goal, but they will help spur the development of cost-effective and viable nontoxic substitutes and ease the difficult transition to the phaseout of POPs. Bolstered by greater citizen involvement and oversight, adequate funding and technical assistance, and the development of viable alternatives, these steps can help consign persistent organic pollutants to history.[63]

Retooling Regulations, Business, and Agriculture

Phasing POPs out requires fundamental changes in regulations, business, agriculture, and society at large. In some cases, this may occur in stages—away from highly persistent chemicals to less persistent ones and then to nonchemical methods or sources. Many tools can help speed the transition, including data inventories and pesticide taxes.

In the wake of the Bhopal, India, disas-

ter in 1986, environmentalists, community activists, and the U.S. EPA successfully lobbied the U.S. Congress to pass the world's first community right-to-know law, over strong protests from industry officials. The Emergency Planning and Community Right-to-Know Act created a national database of toxic emissions and releases by manufacturing plants. Known as the Toxic Release Inventory (TRI), it is an invaluable information tool for industry, government, and the public. TRI data allow citizens and the media to publicize the worst polluters and to bring public attention to the issues of toxic waste management and reduction. Despite the obvious benefits of such accounting, the system does not cover all toxic chemicals or sources. Many POPs are not included on the current list of 600 reportable chemicals. There is, however, an enormous opportunity to expand TRI data to include small releases and to identify sources of POPs.[64]

Today, seven industrial countries and one developing nation—Mexico—have implemented systems similar to the U.S. right-to-know laws. Internationally known as Pollutant Release and Transfer Registers (PRTRs), they contain data on chemicals in most facilities and media. Several other countries are expected to adopt similar systems soon: Egypt and the Czech Republic are in the process of developing PRTRs with assistance from the United Nations Institute for Training and Research, and Argentina, Japan, and five former Soviet bloc nations are in the discussion stages.[65]

Given the mandate from the 1992 Earth Summit and recommendations from industrial-country environment ministers in 1996 that all nations establish such systems, the next few years could witness important milestones in tracking POPs, goading communities to become more involved in the management of toxic chemicals. These registers are also an important tool for developing national POPs inventories, as called for in the draft global treaty. And with the Prior Informed Consent procedure, they help remove a wall of corporate secrecy, improve the policymaking process by encouraging greater public participation, and provide a check against government and corporate abuses.[66]

In June 1999, President Estrada of the Philippines signed the world's first national ban on all waste incineration.

Educating and empowering the public can lead to regulatory changes, as happened recently in the Philippines. Setting a historic precedent, in June 1999 President Joseph Estrada signed the world's first national ban on all waste incineration. The law received strong support from environmentalists, community activists, and the Department of Environment and Natural Resources, which called for a blanket ban on waste incineration. The agency forcefully argued that the country should adopt nonburn technologies such as microwaving to sterilize medical wastes that could not be composted or recycled. In a similar move, all incineration was recently banned in Costa Rica by presidential decree.[67]

Two programs in the United States promise to change the way Americans think about POPs. Washington is the first state to embark on a production-and-release elimination strategy for persistent and bioaccumulative toxics, including dioxins, furans, and several POP pesticides. The initiative is now being publicly debated. Meanwhile, in Passaic County, New Jersey, the nation's first community right-to-*act*

law went into effect in January 1999, granting to groups of 25 or more employees and neighbors in the county the right to form a Neighborhood Hazard Prevention Advisory Committee. Upon approval by the county health officer, the committee has the authority to monitor specific chemical facilities by conducting walk-through investigations with technical experts. Should a company refuse, the citizens could sue for access, a policy that could easily be adopted elsewhere.[68]

Public shame stemming from bad publicity can prompt radical changes in corporate practices, compelling companies to focus more on pollution prevention, thereby safeguarding human health while reducing the costs of pollution abatement. In the mid-1990s, Filipino citizens forced a French company to abandon its plans to build the world's largest waste incinerator just north of Manila. Similarly, after extensive lobbying efforts by environmentalists and subsequent public outcry, multinationals including Nike, Lego, and IKEA have pledged to go PVC-free. And in late September 1999, the world's largest automobile manufacturer, General Motors, announced plans to stop using vinyl in car interiors.[69]

While right-to-know laws are aimed at reducing the supply of POPs, financial tools such as taxes are well suited to reducing demand. Charging a higher price for pesticides and other POPs is expected to discourage and thereby lower their use. In the early 1990s, the Oil, Chemical, and Atomic Workers' International Union proposed a transition fund to clean up the Great Lakes ecosystem, financed by taxes on certain chlorinated compounds. Although it was never implemented, it sets a formula for such a tax. One way to implement the proposal would be to extend existing sales taxes on pesticides—or merely reduce current exemptions. In the United States, a study by Friends of the Earth found that of the $8.8 billion spent in 1997 on pesticides, nearly $276 million in government revenue was lost due to exemptions from state sales taxes. These public revenues could be dedicated to nonchemical crop protection research.[70]

Currently, Sweden has a 7.5-percent pesticide tax per kilogram of active ingredient. Between 1986 and 1993, this levy, combined with farmer training and certification, spray testing, bans, and plant protection programs, led to a 65-percent drop in pesticide usage among Swedish farmers. In a March 1999 study, U.K. economists suggested that a 30-percent tax on pesticides could result in 8–20 percent reductions in their use. Accordingly, the U.K. Department of the Environment, Transport, and the Regions has proposed a tax per kilogram of pesticide sold at a rate determined by the level of hazard, so that more-toxic pesticides would be subject to higher taxes.[71]

The Montreal Protocol on Substances That Deplete the Ozone Layer offers a useful model for developing a financial mechanism to help pay for some of the costs of shifting away from POPs. This 1986 treaty granted developing countries an extended time frame for implementing the provisions of the accord, as well as financial support from industrial countries for the conversion from harmful products and technologies to more benign ones. The protocol called for replacing ozone-depleting chemicals with alternative chemicals that would not harm the ozone layer. In the process of phasing out chlorofluorocarbons, companies actually found cheaper nonchemical solutions, which in many cases saved enormous amounts of money.[72]

In the long run, advocates for a global POPs treaty hope to generate a broad shift away from harmful chemicals to alternative practices and cleaner production methods. Several industries and communities have begun to make notable strides in this direction. (See Table 5–5.)[73]

In response to growing concern about medical waste incineration, hospital administrators and waste specialists have begun to develop viable alternatives. Autoclaves use steam heat and pressure to destroy infectious agents. They are cheaper to maintain than incinerators and already a familiar technology in many hospitals, where labs use them regularly to sterilize equipment. In 1997, an estimated 1,500 nonincinerator medical waste treatment facilities were installed in the United States alone. The World Bank and other multilateral institutions should move quickly to advance these technologies in all countries.[74]

Rooting chlorine out of the life cycle of product formation is by no means easy, but it is ultimately the only way to prevent dioxins from forming and being released. A hospital in Vienna, Austria, has replaced PVC in its products with chlorine-free materials, thereby eliminating PVC entirely from its waste stream. In 1996, the Central Pollution Control Board of India banned the burning of PVC in medical waste incinerators, and the American Public Health Association urged health care facilities and suppliers to reduce their use of PVC and other chlorinated plastics.[75]

For evidence that these steps make economic sense, and can even turn a profit, researchers at the Center for the Biology of Natural Systems at Queens College in New York looked at the employment and financial costs and benefits of replacing all major sources of dioxins in the North American Great Lakes region with alternative processes. They estimated that replacing solid

Table 5–5. Alternatives to the "Dirty Dozen" POPs, by Use or Source

Primary Use or Source	Alternatives
Dioxins/furans	
municipal solid waste incinerators	• Intensive recycling
medical waste incinerators	• Autoclaves (steam and heat to destroy pathogens)
	• Chlorine-free materials
iron sintering plants	• Virgin ore or equivalent, landfill process residues
cement kilns that burn hazardous wastes	• Replace hazardous waste with conventional fuel
pulp mills	• Kraft and soda, totally chlorine free processes based on hydrogen peroxide and/or ozone
	• Recycled paper
incinerators	• Upgrade to optimal combustion technology (850°C or higher, long residence times, sufficient oxygen, thorough admixing of gases, firing control without cold streams)
	• Filter dust scrubbing techniques
	• Solidification techniques to immobilize dioxins and furans
Pesticides	
public health vector control	• Pyrethroid-soaked bednets
	• Screens and mosquito netting
	• Larvae-eating fish
	• Environmental management to reduce mosquito breeding
pest control for crop protection	• Crop rotation, selected natural pests
pest control for buildings	• Alternative chemical (chlorpyriphos)
	• Steel or mesh barriers
PCBs	
hydraulic fluids and capacitor insulators	• Silicone oils, transformer grade mineral oils

SOURCE: See endnote 73.

waste incinerators with intensive recycling would cost about $600 million, to pay off existing debt and additional operation costs, but would save $1.1 billion in waste disposal fees and increased sales of recycled materials, yielding an annual net savings of $536 million for this one sector. It would cost $370 million a year to eliminate dioxins from other sources—medical waste incinerators, iron ore sintering plants, cement kilns that burn hazardous waste, and pulp and paper mills. Overall, industries and communities in the region would save approximately $166 million a year and reap the health and environmental benefits of

getting rid of dioxin.[76]

WHO has recently reconsidered its position on using DDT to control malaria-spreading mosquitoes. Previously, this U.N. agency recommended blanket spraying. Its most recent revisions, in 1993, recommend spraying DDT indoors only, combined with use of safer insecticides. Researchers in Africa have demonstrated that bednets soaked in alternative insecticides such as deltamethrin and other synthetic pyrethroids, which are less toxic to people and have less tendency to bioaccumulate, reduce the transmission of malaria by 30–60 percent and childhood mortality by up to

30 percent. Combined with other prevention and treatment strategies, these bednets can prevent half of all deaths from malaria. In addition, they are easily introduced at the local level and relatively cost-effective: $10 for a bednet plus $1 for a year's supply of insecticide. But alternative insecticides do not eliminate all problems: while the compounds are less persistent than DDT, mosquitoes eventually develop resistance to them, and the potential health effects from chronic exposure to these replacements have not yet been fully examined.[77]

In reality, there is no universal solution for malaria control. Approaches involving nonchemical vector control methods—from larvae-eating fish to the elimination of mosquito breeding sites—combined with rigorous local-level disease surveillance, education, and prevention programs are effective strategies for communities in malaria-prone zones. Research to develop a viable vaccine has been under way for more than a decade, but to no avail as yet. One reason is a shortage of funding. Until industrial countries and major drug companies decide to help pay for tools to fight malaria—which in 1998 killed 1.1 million people in the developing world, primarily children under the age of five, and made another 275 million sick—DDT will remain the weapon of choice.[78]

While the phaseout of pesticides for disease control has been hampered by the slow development of cost-effective alternatives, such options are readily available for agricultural uses. Some farmers are now choosing more selective methods by adopting tools included in a wide variety of techniques that fall under the heading of integrated pest management. This combines judicious use of pesticides with intensive training in nonchemical methods, including crop rotation and mixtures to increase diversity, introduction of natural enemies of pests, and pest barriers such as screens to limit crop losses. In 1991, Indonesian farmers warded off an invasion of white rice stemborers by collecting egg masses, setting traps, and nurturing beneficial insects. In Africa, some farmers use cow urine instead of pesticides on their cotton, an otherwise chemically intensive crop. As of 1996, farmers in nine Asian countries relied on integrated pest management to control pests and promote healthy crops and ecosystems.[79]

With regulators and consumers demanding limited use of chemical fertilizers and pesticides, organic farming is a welcome alternative. With a 20-percent annual growth rate in the United States since 1989, organic farming could be the wave of the future. Organic food sales totaled $4 billion in the United States in 1997 and were expected to reach $6.6 billion in 2000. Some 3 million Japanese consumers regularly buy organic produce. Given the profits and markets at stake, more farmers will likely shun persistent pesticides in the future and instead favor the health of soil, consumers, and the planet. Ultimately, however, consumers may have the most sway in whether or not POPs continue to be used. Education thus has a key role to play.[80]

No longer can we simply expect a stronger pesticide to take care of a problem, or a disposal process to eliminate the health risks and waste problem created by various industrial pollutants. From incineration and industrial processes to consumer goods and byproducts, society would do well to heed to advice of Rachel Carson when she warned of the imminent dangers of the escalation of chemical use in the 1960s. Instead of the current model, society can apply the precautionary approach more widely. If followed globally, it could become the cornerstone for toxics policy, preventing the manufacture and generation of POPs in the first place. But until society rejects the notion of innocent until proven guilty with respect to toxic chemicals, we will continue to live under a dangerous cloud of denial.

Chapter 6

Recovering the Paper Landscape

Janet N. Abramovitz and
Ashley T. Mattoon

"Just imagine a day without paper," an ad in the *Financial Times* for a Finnish paper company reads—"you inevitably see our products every day." And they're right. For the average reader of the *Financial Times*, a day without paper would be almost as impossible as a day without breathing. But in many industrial countries, people are so accustomed to paper—whether it is supplying them with the daily news, drying their hands, holding their groceries, or filling their garbage cans—that its role in their daily lives goes virtually unnoticed.[1]

For most of the 2,000 years since paper was first invented, it was a scarce and valued material, used primarily for important letters and documents. In the last century, however, new technologies, falling costs, and growing economies have allowed the use of paper to skyrocket. Today there are more than 450 different grades of paper destined

for purposes as mundane as wiping noses and as specialized as filtering chemicals.[2]

For all the valuable services it provides, paper entails high costs—many of which go unnoticed by the average consumer. Although the impact of a single envelope, magazine, or box may seem negligible, the process of making it requires many steps that take a heavy toll on the world's land, water, and air. The rapid and excessive growth in paper consumption in the industrial world and the potential for growth in the developing world raise some fundamental questions. How much does paper actually contribute to our quality of life? How much paper do we really need? And how can those needs best be met?

The Paper Landscape

In 1997, the last year for which data are available, the world produced 299 million tons of paper—well over six times as much as in 1950. This much paper could fill the

An expanded version of this chapter appeared as Worldwatch Paper 149, *Paper Cuts: Recovering the Paper Landscape.*

Empire State Building 383 times, or make a pile that could reach the moon and back more than eight times. By 2010, global demand for paper is expected to rise by nearly 31 percent.[3]

Paper consumption is closely correlated with income levels, and the vast majority of the world's paper is produced and used in industrial countries. (See Tables 6–1 and 6–2.) With 22 percent of the world's population, these nations account for more than 71 percent of the world's paper use. As the populations and economies of many developing nations grow, however, their share of world consumption is climbing. (See Figure 6–1.)[4]

On a per capita basis, differences in paper use are even more striking. In 1997, people in the United States used on average 335 kilograms of paper and paperboard, while for industrial countries as a whole, the figure was 164 kilograms. But this is no indication of the paper used each year by the average world citizen. The global figure was 50 kilograms, and for developing nations, it was just 18 kilograms. Africans on average used less than 6 kilograms of paper a year, and in more than 20 African nations the figure was below 1 kilogram. (One kilogram roughly equals 225 sheets of office paper, or two copies of a daily *New York Times*.) Yet it has been estimated that an annual per capita consumption level of 30–40 kilograms is required to meet basic needs for education and communication.[5]

Paper is used for many different purposes. Today packaging claims about 48 percent of all paper use. Printing and writing papers accounts for 30 percent, newsprint another 12 percent, and sanitary and household papers, about 6 percent. While

Table 6–1. Top 10 Producers of Paper and Paperboard and Share of World Production, 1997

Country	Production (thousand tons)	Share (percent)
United States	86,477	29
Japan	31,015	10
China	27,440	9
Canada	18,969	6
Germany	15,939	5
Finland	12,149	4
Sweden	9,779	3
France	9,143	3
South Korea	8,364	3
Italy	7,532	3
Top 10 producers	226,807	76
World	299,092	100

SOURCE: Miller Freeman Inc., *International Fact and Price Book 1999* (San Francisco, CA: 1998).

Table 6–2. Top 10 Consumers of Paper and Paperboard and Share of World Consumption and Population, 1997

Country	Consumption (thousand tons)	Share of Consumption (percent)	Share of Population (percent)
United States	89,900	30	5
China	32,695	11	21
Japan	31,374	11	2
Germany	15,733	5	1
United King.	12,240	4	1
France	10,328	3	1
Italy	9,125	3	1
South Korea	6,836	2	1
Canada	6,652	2	1
Brazil	6,124	2	3
Top 10 consumers	221,007	74	37
World	296,896	100	100

SOURCE: Miller Freeman Inc., *International Fact and Price Book 1999* (San Francisco, CA: 1998).

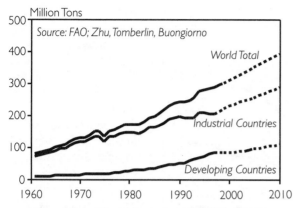

Figure 6–1. World, Industrial, and Developing Country Paper Consumption, 1961–97, with Projections to 2010

the use of all types of paper has increased over time, in recent years consumption of printing and writing paper has grown faster than grades such as packaging paper and newsprint. (See Figure 6–2.) Since 1980, global paper consumption has jumped by 74 percent while that of printing and writing paper has skyrocketed by 110 percent.[6]

Decades ago, at the dawn of the electronic information era, many analysts predicted a "paperless office." Instead, the proliferation of computers and other new technologies such as fax machines and high-speed copiers has gone hand-in-hand with increased use of printing and writing paper. Paper.com, an industry group that examines the correlation between paper use and electronic commerce, e-mail, and the Internet, describes paper as the "currency of the electronic era.... While the Internet and paper will certainly compete in certain areas, the general pattern is one of mutual growth and interdependence."[7]

Their assessment is supported by the trends so far. According to a leading industry analyst, the number of pages consumed in U.S. offices is growing about 20 percent

each year. In 1996, office copiers in the United States churned out more than 800 billion sheets of paper, and laser printers nearly as much. Given the size of the civilian labor force, this means almost 12,000 sheets of office paper per person. Nor has electronic mail replaced traditional letters: in the United States, the number of pieces of mail delivered between 1993 and 1998 increased by 16 percent and the amount of advertising mail rose by 25 percent.[8]

While some technologies have clearly bolstered the use of certain types of paper, others—such as electronic data interchange—are beginning to trim paper use. Still, it is too early to tell what the long-term impact will be on world paper markets. One wild card is the fast-growing markets in some developing countries. As economies grow and technologies spread, the use of printing and writing papers in these countries could skyrocket. But it is also possible that new technologies could allow these countries to avoid assuming the current wasteful habits of industrial nations.[9]

For decades, the world's top producers and consumers of pulp and paper have been fairly constant, with the United States, Europe, Japan, and Canada maintaining leading roles. But in the last 10 years countries such as Brazil, China, and South Korea have emerged as major players. By 2002, some analysts expect that Asia will be the largest producing region in the world.[10]

Another change in the final decades of the twentieth century was a substantial increase in international trade in pulp and paper. Whereas roughly 16 percent of the world's wood pulp and 17 percent of its paper and paperboard were traded internationally in the 1960s, today the figures are

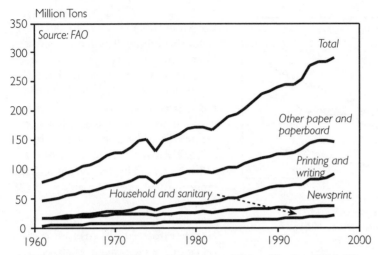

Figure 6–2. World Paper Production, Major Grades, 1961–97

paper imports have gone up nearly fivefold and pulp imports more than threefold. China is now the largest net importer of paper in the world.[13]

These rapid changes in the world's pulp and paper industry suggest that the paper economy in the twenty-first century will be far different than the one we knew in the twentieth. And as investments and production spread to new areas of the world, paper's footprint expands to the forests, air, and water of these far-flung places.

22 and 29 percent respectively. Together these products represent close to 45 percent of the total value of world forest products exports.[11]

China is an extreme example of a country that has an enormous and growing influence in global trade in pulp and paper. Economic growth in the world's most populous country has resulted in soaring pulp and paper consumption. Given its population of 1.25 billion and per capita consumption growing at about 2 kilograms per year, China is an attractive market for the world's pulp and paper companies. Between 1990 and 1997, the consumption of paper and paperboard in China increased by 127 percent while production doubled. By 2010, the country expects to increase production and consumption by more than half again.[12]

Although the Chinese expect tree planting efforts and new mills to meet some of the growth in demand, imports of wood, pulp, and paper will also have to increase substantially. Imports have already risen dramatically in the last decade. Since 1990,

Uncovering the Costs of Paper

The production of a simple piece of paper involves many steps and has many impacts—from soil erosion and species loss when forests are harvested in British Columbia or Chile, to air pollution from pulp mills and waste incinerators in Japan, to the deadly dioxins released by mills along lakes in North America and Russia, to lifeless rivers in China and India. Paper's impacts spread far and wide, and can persist for decades to centuries.

Perhaps one of the most widely recognized costs of paper is the threat it poses to the world's forests. Forests are under a barrage of pressures today, and the insatiable appetite for paper is a major one. The world is currently losing about 14 million hectares of forest cover each year—an area larger than Greece—and even larger areas are being degraded by less obvious threats such as fragmentation, soil degradation, exotic species, and air pollution. The causes of degradation vary greatly among different regions of the world, but logging for pulp, lumber, and fuel, as well as forest clearance for pasture, farmland, and other forms of development, are the leading causes of forest decline.[14]

The virgin wood fiber used to make paper accounts for approximately 19 percent of the world's total wood harvest. Of the wood harvested for "industrial" uses (everything but fuel), fully 42 percent goes to paper production. This share is expected to grow in the coming years since the world's appetite for paper is expanding twice as fast as for any other major wood product. By 2050 it is expected that pulp and paper manufacture will account for over half of the world's industrial wood demand.[15]

Even so, the direct connection between papermaking and forest decline is somewhat difficult to sort out. For one thing, the amount of wood used to make paper is often underestimated due to the lack of accounting for sawmill residues. Of the 42 percent of the world's industrial wood harvest going to paper, almost two thirds comes from wood harvested specifically for pulp, while the rest derives from mill residues such as wood scraps and sawdust. In most global statistics, these residues are not categorized as "pulpwood" and are therefore not accounted for in paper production. The use of mill residues in papermaking will likely decline as engineered wood products use more of the residues and as mill efficiency increases. So in the future more fiber will come from trees harvested specifically for pulp.[16]

Another reason that the link between paper and forests is not always easy to see is that today, just 55 percent of the fibers used to make paper come from virgin wood. Recycled or "recovered" fibers account for about 38 percent of today's total fiber supply for paper, while nonwood fibers such as wheat straw and bamboo contribute about 7 percent. (See Figure 6–3.)[17]

The sources of virgin wood fiber for paper were fairly constant for most of the time since wood pulping technologies were invented in the mid-1800s. But in recent decades the industry has entered a period of rapid flux. The regions, species, and forest types have begun to change.

As noted earlier, pulp production has begun to shift from traditional suppliers in the North to new producers in the South. Countries such as Brazil, Indonesia, and Chile have become global players in the world's pulp market by making the most of a strategic advantage—climates conducive to fast growth rates. In Chile and New Zealand, the widely planted radiata pine grows at about 25 cubic meters per hectare a year, whereas loblolly pine in the southern United States grows at about half that rate. Hardwoods such as eucalyptus—the plantation tree of choice in the tropics—grow at about 40 cubic meters per hectare in Brazil and 26 in Chile, while a comparable hardwood species in Sweden grows at a sluggish 5 cubic meters per year. Faster growth and lower production costs mean that some southern hemisphere producers can provide wood pulp at about half the cost of traditional suppliers in the North.[18]

In addition to the new geography of pulp, there has also been a significant shift in how trees are grown. Rather than managing natural mixed-species, mixed-age forests, the industry has shifted toward a more agricultural model, in which genetic strains are carefully bred and selected, and seedlings are planted and developed into

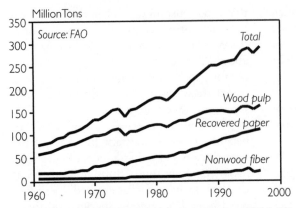

Figure 6–3. World Fiber Supply for Paper, 1961–97

well-organized, single-species, single-aged stands and treated with fertilizers, herbicides, and pesticides. These pulpwood plantations are generally harvested in 6–10 year rotations in the tropics and 20–30 year rotations in temperate regions.

Pulpwood plantations currently account for about 16 percent of the world's total fiber supply for paper. Second-growth forests provide 30 percent, and old-growth forests, another 9 percent. (The remaining 45 percent is recycled and nonwood fiber.) Most of the old-growth forests that are still being logged for pulp are in boreal regions of Canada and the Russian Federation. A smaller share comes from original temperate and tropical forests in countries such as Australia, Indonesia, and Malaysia.[19]

While the current contribution from plantations may seem fairly small, large investments in plantation resources in recent decades ensure that the upward trend in these sources will continue. For many countries, the prospect of gaining a larger share of the wood products market has led to heavily subsidized plantation programs and a rush of foreign investment. In Japan, for example, the shortage of domestic timber has led the industry to invest heavily in pulpwood plantations in Australia, Chile, China, New Zealand, Papua New Guinea, Viet Nam, and elsewhere.[20]

Today there are approximately 13 million hectares of fast-growing tree plantations, yielding more than 15 cubic meters of wood per hectare each year. Nearly all such plantations are used specifically for pulp production, and about 80 percent are in South America and the Asia-Pacific region.[21]

Proponents argue that intensively managed plantations will create jobs, rehabilitate degraded areas, combat climate change by absorbing carbon, and help "save" forests by providing most of the world's wood needs from a much smaller parcel of land than do forests themselves. A 1999 report from the World Commission on Forests and Sustainable Development suggested that it might be possible to meet the world's demand for pulpwood in 2050 with 100 million hectares of fast-growing plantations—an area equal to more than 70 percent of the amount of the world's cropland planted in corn. The implication is that the need to log natural forests will be reduced and that more forests could be protected.[22]

This sounds like an attractive scenario, and some types of plantations could certainly play an important role in improving the industry's impact on forests. But plantation development as it is currently unfolding within the pulp and paper industry is not without drawbacks. Plantations may provide more ecosystem services such as wildlife habitat and soil protection than degraded farmland does, but compared with a mature, native forest, they simply don't measure up. Like virtually all large-scale monocultures, plantations are susceptible to disease and pest outbreaks, so they commonly require regular applications of insecticides and fungicides. Herbicides are also used to prevent invasion of competing vegetation. The frequent harvests and site preparation procedures can result in soil degradation that reduces the long-term viability of the land. A mature pulpwood plantation might look like a natural forest to an undiscerning eye, but it actually has about as much in common with a natural forest as a cornfield does with a native prairie.

Another concern is that in some parts of the world, natural forests are being cleared to make way for plantations. In Indonesia, where pulp production has more than quadrupled in the last decade, more than 1.4 million hectares of natural forest have been replaced by plantations. Plantation expansion and the timber industry have been heavily subsidized by the government for years. Satellite data showed that 80 percent of the fires that burned over 2 million hectares of Indonesian forest in 1997–98 were set largely to clear land for palm oil and pulpwood plantations.[23]

The loss of natural forest in favor of plantations is not only a developing-country phenomenon. In the United States, the expansion of pine plantations for pulp and sawnwood has also come at the expense of natural forests. According to U.S. Forest Service data, pine plantation cover in the Southeast grew by nearly 8 million hectares between 1952 and 1985, while natural pine forest cover declined by 12 million hectares, an area nearly equal in size to the state of Mississippi or three times as big as Switzerland. The Forest Service predicts that this trend will continue into the future, anticipating that by 2030 there will be about twice as much area in plantation pine as in natural pine stands.[24]

In addition to the environmental impacts, pulpwood plantations have also had adverse effects on local people. In Indonesia, some plantations have displaced indigenous Dayak communities. Companies have failed to negotiate land acquisition agreements with villagers, have broken promises to provide facilities, and at times have harassed and intimidated local people. Similar problems have occurred in many other parts of the world. In Brazil, the Tupinikim and Guarani have been fighting for decades to have their traditional territories restored. These lands were lost to the Brazilian paper company Aracruz Celulose when it appropriated thousands of hectares of "uninhabited" land in the 1960s.[25]

Producing pulp and paper casts a long ecological shadow beyond its impact on the world's forests. To make a ton of paper, at least 2–3.5 tons of trees are brought to the mill. Converting them into paper requires large amounts of energy, water, and chemicals, and it generates vast amounts of air and water pollution and solid waste.[26]

Worldwide, pulp and paper is the fifth largest industrial consumer of energy, accounting for 4 percent of all the world's energy use. In Canada, this industry is the largest consumer of energy, and in the United States, the second largest. Although there have been some improvements in reducing energy use, pulp and paper is still one of the most energy-intensive industries in the world. In developing countries, energy use in pulp and paper is often double that in industrial countries, because of greater reliance on outmoded technologies.[27]

In most papermaking nations, energy efficiency has improved in recent decades due to regulations and cost concerns. In Japan, the amount of energy used per ton of paper decreased by half since the mid-1970s. And in the United States, the figure declined by 22 percent between 1972 and

1996. But these efficiency gains were overshadowed by the fact that total energy use increased by 49 percent, a result of production levels rising by 89 percent.[28]

The pulp and paper industry uses more water to produce a ton of product than any other industry—in the United States, some 44,000–83,000 liters per ton of virgin fiber paper, depending on the grade. Writing paper, the fastest-growing grade, uses more than others because of extensive bleaching and washing. Yet, as with energy, some improvements have also been made in efficiency. In the United States, water use has fallen by 50–90 percent per ton since 1950. And in Japan, the amount of water needed per ton of product has dropped by two thirds in just the last 10 years.[29]

Converting wood into paper is a complex process that begins by stripping trees of their bark and then chipping them into small pieces. Chemicals or mechanical grinders are then used to separate, or "pulp," the cellulose fibers. The pulp is rinsed and washed several times and often bleached to make it white. Finally, it is formed into paper.

There are a number of ways to accomplish this transformation. Chemical pulping, especially the "kraft" process, is the most common method and is used to produce strong papers such as printing and writing paper, grocery bags, and corrugated boxes. Chemical pulping is not very efficient in converting wood into pulp, however, and only about half of the wood ends up in paper. The sludgy processing "waste"—the other half of the wood, mixed with leftover chemicals—is usually burned to help fuel the mill and to recover some of the chemicals. Mechanical pulping uses about twice as much energy as other processes but is about 95 percent efficient in converting the wood into pulp. The paper made from

mechanical pulp tends to be weaker and yellows easily, so it is generally used to make newsprint and telephone books.[30]

Pulp and paper mills have long been considered "bad neighbors" because of the foul-smelling air and sickening water they produce. Even with improvements, U.S. pulp and paper factories have one of the highest pollution intensities, or emissions per value of output, of the 74 industrial sectors monitored by the government's Toxics Release Inventory. In many developing countries, mills can be even more polluting, especially if they rely on outdated technology or if pollution is less well regulated. China, India, and other Asian nations, for instance, have thousands of small mills that have no chemical recovery systems, and that pour untreated black "pulping liquor" directly into waterways.[31]

A host of air pollutants are released when paper is made, including volatile organic compounds, nitrous oxides, sulfur oxides, acetone, methanol, chlorine compounds, hydrochloric and sulfuric acids, irritating particulate matter, and carbon monoxide. It is the sulfur compounds that give kraft pulp mills their characteristic "rotten egg" smell. In addition to having well-documented human and ecosystem health effects, some of these air pollutants contribute to climate change and others are ozone-depleting substances.[32]

Thousands of substances are released into water bodies from pulp and paper mills, including dissolved wood, chemicals, and other compounds—many unidentified—that result from interactions between wood and the pulping and bleaching chemicals. This mix can reduce oxygen levels in the receiving water system and thus kill aquatic organisms, cloud and acidify the water, and release toxic chemicals. While some aquatic life is killed immediately, other effects are long-term and persistent as chemicals accumulate and work their way up the food chain to people. (See Chapter 5.)[33]

Many of these highly toxic compounds are produced during conventional bleaching processes that use chlorine. To make whiter paper, wood pulp is bleached, most commonly with elemental chlorine (usually in the form of chlorine gas). Chlorine bleaching has come under intense scrutiny since the discovery of dioxins in mill effluent in the mid-1980s. Some mills have switched to other types of chlorine, such as chlorine dioxide, which can cut measurable discharges by 90 percent. Unfortunately, many of the chlorine-based compounds are toxic at levels that are too low to detect using standard measures.[34]

Most improvements in pollution levels in industrial countries have come in response to government laws and regulations. The earliest tended to deal with end-of-the-pipe (or smokestack) pollution and to prescribe specific technologies. They succeeded in reducing the gross levels of pollution that characterized the pulp and paper industry of 30–50 years ago, when dead and discolored rivers and lakes were commonplace.

Today, there is greater emphasis on pollution prevention or source reduction. This newer approach tends to set longer-range emissions targets and allows industry to figure out how best to meet these goals by examining and modifying production processes. In Sweden, strict standards with more flexible rules for compliance have helped the country achieve higher environmental performance and greater competitiveness. The recent U.S. Environmental Protection Agency "Cluster Rule" also takes a more integrated approach to air and water regulations. As the public, regulators, and the industry have become more sophis-

ticated, there is a growing awareness that the most effective way to achieve environmental and economic goals is to prevent pollution in the first place; wasted resources are wasted profits.[35]

Reducing the Burden of Production

Although significant progress has been made in reducing the environmental costs of producing paper in many parts of the world, there is still considerable room for improvement. From expanding the use of alternatives to virgin wood fiber (such as recycled and nonwood fibers) to improving forest management and cleaning up mills, ample opportunity exists to reduce the hidden costs of paper production.

Increasing the collection and reuse of old paper is one of the most promising ways to lessen many of the problems associated with paper. Producing new paper from old is an efficient process. For each ton of used paper, nearly a ton of new paper can be produced—far more efficient than the 2–3.5 tons of trees used to make 1 ton of virgin paper. And because recycled paper has already been processed, just 10–40 percent as much energy is required during reprocessing as in virgin pulping. And little, if any, bleaching is needed because white papers had already been bleached during their original production and many grades (like corrugated boxes) do not need bleaching.[36]

Recycling can make use of a huge, barely tapped supply of materials—the urban forest. This term has been used to describe cities not because trees grow there, but because cities generate enormous amounts of wood and paper waste that is all too often thrown away. A fast-growing industry is making use of this vast resource to produce useful products, reduce waste, and create jobs. A new paper mill is under construction at an old industrial site in the Bronx area of New York City that will use 100-percent recycled newspaper and get its water from treated sewage. It will produce far less pollution and use half the energy of virgin production. By locating near the source of its raw material and the buyers for its products, the company's transportation energy costs will be slashed by 94 percent. The paper is expected to cost 28 percent less than virgin newsprint.[37]

In many countries, the primary motivation for increasing recovery rates has been the need to reduce the flow of waste to landfills and incinerators. The volume of waste generated in many industrial countries has grown substantially in recent decades—more than doubling in the United States alone since 1960. Paper accounts for the largest share of municipal solid waste in many industrial countries. In the United States, for example, it accounts for 39 percent (by weight) of this waste. And even though almost half of used paper is now diverted for recycling, some 44 million tons are still discarded in the United States each year—more than all the paper consumed in China.[38]

As the benefits of recycling have been recognized, recovered paper's contribution to the global fiber supply for paper has nearly doubled—from 20 percent in 1961 to 38 percent in 1997. Between 1975 and 1997, the volume of paper recovered worldwide more than tripled and the share of paper that was recovered increased to more than 43 percent.[39]

Yet wastepaper recovery rates vary dramatically among countries. Legislation to aggressively reduce solid waste in Germany has resulted in paper recovery rates of nearly 72 percent. And in Japan, the world's second largest paper producer, limited domestic resources and a shortage of waste disposal options have encouraged the heavy use of recovered paper. On the other hand, the recovery rate in China is only 27 percent.[40]

In the 1970s and early 1980s, only about one quarter of wastepaper was recovered in the United States. Thanks to a variety of laws and private initiatives, such as banning paper in landfills, establishing curbside recycling programs, and issuing mandates for recycled content paper, by 1997 the U.S. recovery rate had risen to 46 percent. In Europe, where there are also high levels of wastepaper recovery, a 1994 European Union directive targeted a recovery rate of 50–65 percent for packaging waste by 2001, and a new directive calls for a nearly two-thirds reduction in the amount of biodegradable material (such as paper) sent to landfills.[41]

Recycling has served more as a supplement than a substitute for virgin fiber. Global paper and paperboard consumption is increasing so rapidly that it has overwhelmed gains made by recycling. Indeed, while the amount of material recovered has increased severalfold and its contribution to the fiber supply has nearly doubled, the total volume of virgin wood pulp and paper consumed and waste generated continues to rise, overtaking these important successes.[42]

The potential for using old paper to make new paper has yet to be fully exploited. Today's 43-percent recovery rate is far below the 70 percent or more that is possible. Some grades of waste paper, such as old corrugated boxes and newspapers, are more widely recycled than others, and there are well-developed markets for pulping them into similar new products. Other grades, such as mixed office paper, have lower recovery rates, and very little of what is collected is used in making new office paper. Instead it is "downgraded" for other uses such as card-board, because of the variety of inks used and the demand for ultra-bright white office paper. In fact, more than 90 percent of the printing and writing paper made in the United States is from virgin fiber.[43]

For recovery and recycling to reach its full potential, a number of barriers need to be overcome—some technical or economic, others social. One barrier has been price volatility in recovered paper. For a long time the capacity to pulp wastepaper was limited, and the volatility of supply and price made it difficult for mills to invest in new facilities to handle the paper, creating a vicious circle. With the more widespread adoption of municipal and business recycling programs and the expansion of recycled mill capacity, supply and prices are now more predictable. Unfortunately, wide-spread subsidies and overcapacity for virgin fiber still put recovery and recycling at an economic disadvantage.[44]

Despite the expansion and success of recycling programs, their economic benefits are usually underestimated, and thus municipalities tend to underinvest in recycling. In Massachusetts, these economic benefits are two to four times the waste disposal fees. The additional investment needed to expand recycling fully would amount to just 1–2 percent of the economic benefits that would accrue to municipalities.[45]

In recent years there have been dramatic advances in the quality of recycled papers, thanks to innovations made in processing. And as the volume of recovered paper use rises each year, even more gains are being made. These technological advances need to be more widely adopted. The most common standards for judging writing papers—opacity and brightness—are easily met by today's recycled papers. The strength of recycled paper is on a par with virgin paper as well (a

concern for printers, because breaks in large paper rolls can be very costly).[46]

Nonwood fibers provide another option for improving paper's fiber mix. There are two main types of nonwood fibers for paper: agricultural byproducts from crops such as wheat, rice, and sugar, and crops that can be grown specifically for pulp, such as kenaf and industrial hemp. Wild plants, such as reeds and grasses, and recycled textiles are also used as a fiber source in some parts of the world.[47]

> **Though almost half of used paper is diverted for recycling, some 44 million tons are still discarded in the United States each year.**

Nonwood fibers were once the sole source of raw material for pulp; wood pulping techniques were not even invented until the middle of the nineteenth century. The first piece of paper, produced in China in A.D. 105, was made of tree bark, old rags, hemp, and used fishing nets. Today, nonwoods provide only about 7 percent of the world's total fiber supply for paper. But there is a strong case for increasing that share to 20 percent or more. Doing so could make use of a resource that is currently burned in many parts of the world, provide farmers with an additional source of income, reduce chemical use in pulping, and cut the demand for wood pulp.[48]

Developing countries account for 97 percent of the world's nonwood pulp use. China is responsible for the vast majority of this—83 percent. In the United States, nonwood fibers provide less than 1 percent of the paper industry feedstock, whereas in China, the figure is nearly 60 percent.[49]

Agricultural byproducts currently provide about three quarters of the world's nonwood pulp supply. By one estimate, one hectare of cereal grain can yield up to one ton of straw—and this includes an allowance for half of the straw to be tilled back into the soil. More than 2 billion tons of agricultural byproducts could be available each year, a far cry from the roughly 35 million tons currently used. The majority of agricultural residues should be composted and recycled on farmlands, and in some countries, residues represent an important source of fuel. Yet in some parts of the world, large amounts of these residues are burned, resulting in polluted air and a wasted resource.[50]

Crops planted specifically for pulp, such as kenaf and hemp, also have potential for expanded use. These sources currently account for about 9 percent of the world's nonwood pulping capacity. They can cost more than agricultural byproducts, but many have properties that can yield high-quality pulps as well as environmental and social benefits. One of the primary advantages they have over wood is their low lignin content. Lignin binds cellulose together, and the removal of lignin is part of what makes the pulping process so energy- and chemical-intensive.[51]

Whether kenaf and hemp are preferable to wood fiber depends to a large extent on local social, economic, and ecological conditions. In some places, the climate may be more suitable for nonwood crops; in others, farmers may be looking to diversify and could get a much faster return from an annual fiber crop than from a pulpwood plantation that could take 15–20 years to mature.

Even with expanded use of recycled and nonwood fibers, 25–30 percent of the world's fiber supply for paper will likely continue to come from virgin wood. Tree plantations and forests are integral to the modern paper production system. But in a world with a rapidly expanding population and a rapidly declining forest endowment, reforming forest management is essential. A continuation of the practices of the twentieth century and a growing demand for wood will leave us with severely degraded, fragmented systems 50 years hence. Large expanses of healthy, intact forest ecosystems that are not in protected or extremely remote areas would become a distant memory, and vital ecosystem services would be severely compromised.

There are some encouraging signs that forest management can be improved. In recent years, there has been a growing interest in sustainable forestry practices on the part of local communities, foresters, the industry, policymakers, and concerned citizens. These practices involve changing harvesting techniques; managing for multiple species, age classes, and uses; and protecting wildlife habitat and watersheds while also providing products and livelihoods.

Some types of plantations can play a role in reducing the environmental impacts associated with producing pulp for paper. Farming trees in a sustainable way is clearly preferable to harvesting the world's last remaining old-growth stands. But plantations can be managed much better than they are now. First and foremost, they should be established on lands that truly are "degraded"—that are not currently forested, farmed, or inhabited, and do not have high potential to regenerate naturally. Instead of providing subsidies for plantation establishment in recently cleared or highly productive areas, governments could offer financial incentives to plant in degraded areas. The subsidies could make up for some of the profits lost from slower growth rates.

Other ways to improve plantation management include long-term landscape planning, watershed protection, local stakeholder involvement, reduced chemical use, and the use of native trees. Some companies have already taken steps to reduce their environmental impact. For example, a Brazilian company intersperses its plantations with plantings of native rainforest species. They do this in part to improve their environmental image, but also to provide a natural means of pest control.[52]

Consumer demand has played a role in encouraging the trend toward sustainable forest management and cleaner paper production. In particular, the late 1990s were characterized by increased consumer awareness in some parts of the world. The European market has been especially influential in catching the attention of major forest products industries. And companies ranging from Dutch publishers to McDonald's have recently been demanding products free of old-growth wood.[53]

The growing concern about forest management practices has contributed to an expansion of certification initiatives. The Forest Stewardship Council (FSC) is the best known and most credible third-party certifier. FSC accredits certifiers who audit forest management practices, at the request of companies wishing to use the FSC logo, and who certify products for the entire chain of custody—from forest to transport to processing. FSC-certified forests must follow strict standards set forth in regionally specific principles and criteria for sustainable forest management. Many companies are seeking FSC certification, and by the beginning of 1999 more than 15 million hectares of forest—almost three times the size of Costa Rica—had been approved by FSC certifiers. In late 1998, the first U.S.-produced paper containing FSC-certified tree pulp arrived on the market.[54]

Regardless of the fiber source, there is ample room to improve the way that paper is made. In recent years, new methods have been developed for producing paper with less energy, less pollution, and fewer virgin raw materials. Some businesses have begun to embrace these changes. Many companies were initially pushed by regulations, but more and more are finding that it can be more profitable (and publicly acceptable) to generate less waste and more environmentally benign products. On the whole, however, the pulp and paper industry has been rather slow to change, and too few companies have been willing to adopt innovations that do not pay for themselves in the next quarter.

Plantations should be established on lands that truly are "degraded"—not currently forested, farmed, or inhabited.

Reusing water, chemicals, and other materials in a manufacturing operation is a time-tested way of increasing profitability and improving environmental performance. The ultimate goal is a closed-loop system that completely eliminates releases to the water, air, and land. There are already several zero-effluent mills in the world. One is a mini-mill located in the New Mexico desert that uses old corrugated containers to make 100-percent recycled linerboard (for items such as boxes). It takes steam and water purchased from a nearby electric plant and recycles it over and over in a closed loop. And in Wisconsin, a recycled paper mill has completely eliminated discharge.[55]

As noted, progress has been made in reducing the volume of waste produced in pulp and paper mills, and the effluent levels of some of the most toxic chemicals like dioxin. Using better methods to remove lignin that reduce chemical and energy use and give higher fiber yields, switching to safer bleaching chemicals, and better waste treatment are all major steps toward cleaner paper. Modern mills that have adopted improved processes have cut effluents by 80 percent, which in turn helps cut energy and chemical demands, saves money, and makes it easier to comply with environmental permits.[56]

A number of other new technologies have the potential to improve pulp and paper processing significantly. "Biopulping," for example, takes advantage of naturally occurring local fungi to break down lignin in the wood chips, thereby reducing the amount of energy needed in processing and improving the water quality of effluents while also increasing the strength of the resulting paper. Researchers at the Forest Products Lab of the U.S. Department of Agriculture (USDA) are developing "fiber loading" as a way to produce lighter-weight printing and writing papers from recycled or virgin pulp with the same qualities as heavier paper. Because fiber-loading technology is quite inexpensive, mills that retrofit their equipment could pay for their investment in one year on the material and energy savings alone.[57]

A major focus of efforts in most industrial nations continues to be cleaning up the bleaching process. Reducing or totally eliminating chlorine from processing has a number of advantages, both ecological and economic. Chlorine is corrosive to processing equipment (which limits its recyclability within a plant and requires more expensive equipment), dangerous to workers, and extremely harmful when released to the environment. Luckily, there are ways to eliminate its use.

"Totally chlorine-free" (TCF) processes replace chlorine-based chemicals with oxygen-based chemicals such as hydrogen per-

oxide. Dioxin releases are virtually eliminated, almost no hazardous air pollutants are released, less water is needed, and the water can be reused many times (an essential step toward developing closed-loop systems). The water that is discharged needs less treatment, and energy use is far lower—increasing profitability. Scandinavia has already shifted 27 percent of its production to TCF. The United States and Canada lag far behind, with less than 1 percent. TCF mills are cheaper to build because they need less equipment and can use less expensive metals than mills using chlorine. Worldwide, about 6 percent of the world's bleached pulp is now totally chlorine-free.[58]

So far, most of the industry (especially in North America) has chosen to adopt the "elemental chlorine free" (ECF) approach rather than TCF, because it requires fewer modifications to existing plants. ECF uses chlorine derivatives (such as chlorine dioxide) rather than elemental chlorine, which reduces some—but not all—of the most harmful compounds in the effluent. Worldwide, 54 percent of the bleached pulp produced in 1998 was ECF, a big jump from 17 percent the preceding year.[59]

Adoption of these newer bleaching methods has resulted in significant reductions in toxic discharges. For comparison, an average paper mill using standard chlorine bleaching releases about 35 tons of organochlorines a day, while ECF mills release just 7–10 tons. TCF mills release none.[60]

The reluctance to move to the full range of cleaner technologies is surprising, given the proven cost benefits. A 1995 study of 50 mills in six countries found that the earlier a mill had invested in improved technologies (such as extended delignification, ECF, or TCF), the more profitable and competitive it became. These findings held true even in nations without strong pollution regulations. As the study published in the industry journal *Pulp and Paper International* noted: "Business people have been brainwashed by classes in traditional economics to believe that investing to reduce pollution is a waste of money. The problem with this view is that it makes assumptions that do not hold true in the real world."[61]

Trimming the Costs of Consumption

Although the paper industry has a key role to play in reducing the environmental impacts of paper and shifting toward sustainability, consumers also have a pivotal role. Their purchases and preferences send signals to the industry. And their discards help create mountains of waste, much of which consists of paper. Strategies such as reduce, redesign, reuse, and recycle can easily be practiced by businesses, government, and individual consumers alike.

Business consumers can have a particularly strong influence on the industry. Office paper is the fastest-growing use of paper, and the cost of printing, copying, mailing, storage, and disposal can exceed the initial purchase price of paper by as much as 10-fold. Luckily, there are easy ways for businesses to audit—and reduce—their paper use and costs. Reductions of 20 percent are possible in most offices, just through "good housekeeping" practices such as eliminating "extra" paper purchases and needless copies. Photocopying and printing on both sides can save a substantial amount of paper, and office machines that print "duplex" (double-sided) are widely available. Greater use of electronic communication can also save paper and money.[62]

Many companies have already begun using such strategies. Bank of America, now the largest commercial bank in the United States, decided in 1994 to reduce its paper use by 25 percent in two years. It reported exceeding that goal by 1997 by using online reports and forms, e-mail and voice mail instead of paper forms and memos, duplex copying, and lighter-weight papers. Just trimming the weight of the paper in automated teller machines by 25 percent saved 228 tons of paper a year. Bank of America now recycles 61 percent of its paper, saving about a half-million dollars a year in waste hauling fees. And 75 percent of all its paper purchases have some recycled content.[63]

Many printers and publishers have managed to reduce the amount of paper waste generated at the printing plant and to recycle the waste they create. Advances in computer-assisted layouts and other design tools allow printers to maximize the amount of printing on a sheet. Some printers are using lower-weight papers. The weight of the paper in American newspapers, for example, was gradually reduced by 9 percent between the late 1960s and the mid-1980s. Some newspapers and magazines are saving even more money and paper by also reducing the size of their publications by a fraction. It is still common for magazine publishers, however, to print and ship many more copies than they expect to send to subscribers or to sell at the newsstand: at least half the magazines printed in the United States are never sold.[64]

The rapidly growing overnight and express shipping industry is another large consumer of paper. United Parcel Service (UPS), the largest of these companies, ships more than 3 billion packages a year. Several companies (including Airborne, UPS, and the U.S. Postal Service) now use more than 80 percent post-consumer wastepaper for boxes and paperboard, have eliminated bleached paper, offer reusable materials, and are using lighter packaging.[65]

Smaller businesses can make a difference as well. An increasing number of Japanese offices are coming together to collect paper in sufficient quantities to make it economical for pickup by a private recycling company, thanks to an initiative started in 1990 by a concerned office worker. One such "office town council," a group of 280 offices, now diverts more than 8,000 tons of high-grade office papers from the waste stream each year, saving nearly 87 million yen (approximately $712,000) in disposal costs. The group has also promoted purchases of recycled-content paper products. Its example has been replicated elsewhere in Japan.[66]

Each household may not use as much paper as a business, but collectively household paper use is considerable, as is the influence that individuals have on businesses that provide them with goods and services. Guides are now available to help consumers audit and reduce their own paper use and to identify and support more environmentally benign purchases. They can eliminate excessive "junk mail" and catalogues, choose items with less packaging, and support community recycling efforts. An essential step is buying recycled products, for without a market, paper that is collected for recycling may end up in a landfill or incinerator.[67]

Some manufacturers are using creative ways to produce the same desired product while using less raw material. In the 1980s, for example, standards for corrugated shipping containers shifted from weight-based standards to performance-based ones. By "lightweighting," a substantial amount of raw material was saved, and stronger containers resulted. Since nearly half of the world's paper is turned into packaging and shipping products, such a shift is significant.[68]

Products and services can be redesigned to use far less material or more environmentally benign materials. Consumer products giant Procter and Gamble shaved the amount of paper packaging per unit of product by 24 percent in just four years. In Europe, some padded shipping envelopes are a "bag-in-a bag" that can easily be reused and recycled. The plastic bubble wrap bag slips out of the paper envelope, and both can be recycled.[69]

Entire products and functions are being redesigned to virtually eliminate paper use. Documents such as phone directories, parts catalogues, and technical reference manuals, can now be accessed on-line or on CD-ROMs, saving millions of dollars and tons of paper. The potential is illustrated by the fact that the contents of all the phone books in the United States could be put onto a few CD-ROM disks for just a few cents and are already available on the Internet.[70]

While the "paperless office" predicted by some in the 1970s may never materialize, today's new generation of information and communication technologies is showing how businesses can function with far less paper. Electronic mail, electronic data interchange, document scanners, intranets, and the Internet can radically reduce paper use, while also saving time and money. Many companies are now producing and processing core business documents such as invoices, purchase orders, and reports electronically, greatly increasing their efficiency.[71]

Paper products in the office—from copy paper to corrugated containers, shipping envelopes, and magazines—can be reused before recycling. Manufacturers of a range of products like furniture and computers are also switching to returnable and reusable packaging, which can save money and resources.[72]

Municipalities and businesses have found

that expanding recycling has reduced the volume of waste. Retailers that receive large shipments have already discovered that recycling old corrugated containers can be profitable, and their efforts help account for the fact that corrugated paper has the highest recovery rates of all paper types. In the United States, more than 70 percent of old corrugated containers have been recycled since 1995; grocery stores are now the single largest source of this valuable material.[73]

As very large consumers of paper, publishers have a key role to play by making their operations more efficient and moving the market to supply better papers. Dutch paper buyers, for example, are demanding old-growth-free paper from Finland, and German publishers are making sure that their suppliers practice good forest management. Printers are finding that customers are asking for these papers and that firms that supply them have a competitive edge. Members of like industries, such as publishers or environmental groups, have come together to form buyers groups, so that collectively their purchases can help spur change.[74]

Governments are also big paper purchasers. The U.S. government, which accounts for 2 percent of all paper bought in the country, has mandated that its paper purchases must have a minimum of 30-percent post-consumer recycled content. Japanese government offices now buy recycled-content copy paper with whiteness of 70 percent (instead of the old standard, 80 percent), thanks to a campaign by concerned businesses and nongovernmental organizations. The campaign also taught consumers that whiteness standards for paper were unnecessarily high.[75]

A number of creative alliances are bringing changes. Five years ago the U.S. Environmental Protection Agency set up the WasteWise program to encourage public and private entities to reduce waste by sharing strategies and highlighting achievements. Last year, more than 900 partners cut waste by 8 million tons and saved about $40 million in avoided purchases and disposal fees. The overnight shipping industry and the fast-food giant McDonald's have worked with the Environmental Defense Fund to reduce materials use. The U.S. Postal Service is collaborating with USDA's Forest Products Lab to develop better stamp adhesives that will make paper more recyclable.[76]

In recent years there have been many advances in the quality and availability of more "environmentally friendly" paper. The perception of recycled paper, for example, as weak or coarse and flecked is no longer true, as a range of high-quality recycled paper that meets the demands of all types of printing is available at competitive prices. Likewise, the range and availability of "tree-free" papers and those made without chlorine bleaching has expanded considerably, and guides are available to help consumers find suppliers. Labels that indicate recycled content, chlorine bleaching, and so forth are becoming widespread in many parts of the world. Consumers are using this information to buy better products and to encourage the industry in a sustainable direction.[77]

Designing a Sustainable Paper Economy

As we enter the new century, the need for a healthier paper diet is more urgent than ever. In its most recent projections, the U.N. Food and Agriculture Organization (FAO) predicts that global paper consumption will reach nearly 391 million tons by

2010, more than a 30-percent increase over today's level. With such growth would come the felling of more trees, and more pollution and waste. But ever-rising consumption and all the costs that it entails are not a foregone conclusion. Accelerating the use of promising technologies and promoting leaner consumer habits can bring a healthier paper diet within reach.[78]

Paper use has traditionally been closely correlated with income levels, and industry analysts generally treat rising consumption as a sign of a healthy economy and improving quality of life. But using more paper is no more a prerequisite for rising economies and standards of living than is increased automobile traffic, television viewing, pollution, or heart disease—other trends that have historically grown with income. In Japan, total paper use rose by 75 percent between 1980 and 1997, yet it is questionable that well-being rose by a comparable level. Paper use can be "decoupled" from economic growth thanks to improvements in efficiency and conservation, much as energy use has been.[79]

There are many parts of the world where access to paper does need to increase for education, communication, and sanitary purposes. About 80 percent of the world uses less than the 30–40 kilograms per person a year recommended as the minimum for basic literacy and communication needs. But consumption levels in industrial nations can be substantially reduced without losing the benefits of the services that paper now provides. Such changes are both possible and practical, as described earlier, and they promise many economic and environmental benefits.[80]

For individuals and companies alike, there are ways to provide the benefits and fill the functions of paper with far less material, and without racking up the costs that come with today's high level of paper use. Businesses, for example, can become more "eco-efficient" by adding value to products and services, while reducing material and energy use, pollution, and waste. Today's box manufacturers could sell shipping services, as some are already doing by selling or leasing reusable containers, and companies could sell information transmittal or communications services rather than reams of paper and copy machines.[81]

These strategies, combined with the many others described earlier, can yield substantial savings by "capturing the magic of compounding arithmetic," as Paul Hawken, Amory Lovins, and Hunter Lovins describe in their book *Natural Capitalism*. If, for example, a process has 10 steps, and you can save 20 percent at each step, the net savings is a whopping 89 percent. The highest gains are made by reducing consumption, as the impact is multiplied back through the entire chain of production, from the mills to the forests.[82]

Applying this principle to paper consumption and production, dramatic savings could be achieved using some fairly conservative estimates of improvements along the chain. For example, if per capita consumption in industrial countries were trimmed by one third, an amount largely possible just through "good housekeeping," global paper consumption would fall by 5 percent, and developing-country consumption could rise to 30 kilograms per person a year, improving the chances that basic needs are met. And if at the same time production efficiency increased by 5 percent, while recycled paper as a fiber source expanded to 60 percent and nonwoods as a fiber source doubled, then we could save 56 percent of the wood fiber now used for paper making.[83]

Policymakers have a key role to play in making the transition to a sustainable paper economy. Through policies and regulations, governments can spur cleaner production, encourage recycling, promote sound forest management, shift consumption patterns, and save taxpayers money. Reforming a host of policies that are environmentally destructive and fiscally wasteful is also essential.

Fiscal reform is a top priority, starting with reducing the extensive subsidies for raw materials use (such as wood, water, and energy), tax incentives for forest conversion, and low concession and stumpage fees. Many countries rely on such policies with the mistaken assumption that forest exploitation, trade, and increased consumption are engines for economic growth. These policies are not only wasteful in their own right, they also put relatively sustainable alternatives and activities—such as reuse-and-recycle and nonwood fiber sources—at an economic disadvantage. The overall effect is to make paper produced from virgin wood fiber inexpensive, encouraging overproduction and wasteful consumption.

Eliminating numerous widespread subsidies would reap financial and environmental benefits. In Canada—the world's largest timber exporter—British Columbia subsidizes the timber industry to the tune of $7 billion a year. Indonesia's subsidies for forest exploitation range between $1–3 billion a year. Some subsidies are direct, such as the $811 million in tax breaks the U.S. government gives the forest industry each year. Others are indirect, such as tax-exempt bonds for landfills and incinerators, or the subsidized energy, water, and transportation infrastructure that make virgin extraction and processing more profitable. For example, the pulp and paper industry is the biggest beneficiary of the U.S. taxpayer-funded $2-billion Tennessee-Tombigee waterway, which gave companies access to land-locked forests and made the explosive growth in wood chip exports possible.[84]

Taxes can be an effective tool for shifting the industry in a more sustainable direction. Higher taxes for exceeding pollution levels, excess packaging, waste incineration, and landfilling have the double benefit of discouraging things people want less of (such as pollution) while lowering other taxes, and thus encouraging things people want more of (such as jobs and investment). The Netherlands, for example, has gradually increased taxes on water pollution in recent decades, which spurred major improvements in water quality as industry adopted cleaner technologies. The United Kingdom's new landfill tax is expected to stimulate paper recovery and help trim paper imports.[85]

There are ways to provide the benefits and fill the functions of paper with far less material.

Most governments have been pursuing a policy of trade promotion and liberalization that encourages forest degradation. Current global trade negotiations, for instance, are set to further reduce tariffs on forest products—a potential stimulus to production. And some positive steps, such as national laws covering packaging, recycling, certification, and eco-labelling, are viewed by the World Trade Organization as nontariff barriers to trade, and are also thus targeted for elimination under expanding global free trade rules. Also under attack are national sanitary standards intended to prevent the accidental introduction of exotic pests and diseases that can hide in wood products and decimate forests in the importing nation.[86]

In addition to reforming unsustainable policies, scaling up policies with positive effects—such as recycling promotion, pollution prevention, mandates for recycled content in procurement, and so forth—would make a difference. Laws mandating waste reduction and recycling have helped expand the supply of old paper available for paper manufacturing. In many countries, "pay-as-you-throw" programs provide an incentive to generate less trash and to recycle more by charging households only for the amount they throw away, much the way many businesses already pay trash haulers based on volume. Helping recycling mills to open in urban areas close to fiber sources and markets, or agricultural fiber mills to set up in areas where such fibers are little used, can simultaneously make use of wasted resources and provide regional economic benefits. Regulators and taxes can encourage wider adoption of integrated pollution prevention methods that slash pollution and save money.[87]

Sometimes sound policies and regulations fail to translate into action because of poor implementation or enforcement. The efforts of many developing countries to enforce pollution control measures or halt illegal logging and land conversion are often hampered by understaffed and underfunded environmental agencies. Good laws may also be stymied by corruption at various levels of government. And regulators may bow to pressure by industry, as is happening with British Columbia's pollution policy that mandates zero discharge of deadly dioxins.[88]

Relatively small investments in research by government and industry can yield big dividends. Given the economic importance of paper, surprisingly little is spent by government and industry on research and development into new technologies. In fact, the pulp and paper industry spends less on research than any other major industry—about 1 percent of sales, compared with 4–5 percent in manufacturing industries as a whole. A hopeful sign is that several developing nations, notably India, Indonesia, and China, have established research institutes that focus on the environmental problems of their pulp and paper industries.[89]

Continuing the current course of paper production and use carries costs that are too high to pay—from impoverished landscapes, to pollution and waste, to unmet needs. These costs will become increasingly burdensome as the world's population and consumption continue to expand. A high standard of living can be enjoyed in all countries without incurring such costs, for the tools to bring about a sustainable paper economy are available.

Chapter 7

Harnessing Information Technologies for the Environment

Molly O'Meara

The Silicon Valley Toxics Coalition (SVTC) is a grassroots organization in California that monitors the pollution generated by computer production. In 1999, the group discovered a new way to use computers themselves to uncover some of the toxic consequences of computer manufacturing. The organization commissioned experts in geographic information system (GIS) software to create computer-based maps showing pollution from semiconductor manufacturing in northern California.

The maps, posted on the coalition's Web site, show a patchwork of diamonds and squares rimming the San Francisco Bay that indicate sources of toxic chemical releases and hazardous waste sites monitored by the federal government. Internet users can point and click on a town's name to zoom in for a closer view. A further click on a symbol will bring up a text box with details about the hazardous site.[1]

This story touches on three broad areas in which information technologies intersect with efforts to build an environmentally sustainable society. The first is the environmental impact of the production, use, and disposal of information technologies. The toxic chemicals used to make semiconductors, circuit boards, and electronic monitors can cause pollution, as SVTC's maps attest. The disposal of information technologies also creates environmental hazards: obsolete computers burden landfills, just as old satellites add to the celestial junkyard.

Yet the net environmental effect of the use of information technologies is far from clear. On the downside, computers require electricity and use paper, while radios, televisions, and the Internet broadcast advertising and programs that may drive people to buy resource-intensive products. But there are myriad ways in which the use of information tools may benefit the environment—for instance, by substituting data for materials and energy, or communication for transportation.

The second area is monitoring and modeling the environment. The computer-generated maps with pollution data on SVTC's

Web site are just one instance of information technology helping people to monitor the environment. Other examples abound. Satellite sensors are giving us clearer pictures of environmental change than ever before: spreading fires in the tropical forests of Southeast Asia, ozone loss above the Antarctic, and the shrinking of the Aral Sea, among many others. In addition, researchers are using computers to study various environmental scenarios, from urban transportation alternatives to the burning of fossil fuels worldwide.[2]

In 1996, the worldwide growth in cellular phones surpassed the growth in fixed-line telephones.

Networking for sustainable development is the third area of overlap. By placing the maps of toxic sites on the Internet, SVTC gives anyone with a computer and a modem access to the data. New communications systems, such as the Internet and cellular telephones, are speeding the exchange of all types of information, including environmental data. By linking far-flung people, the network helps researchers and activists work together to solve environmental problems. The expanding communications network also transmits information to remote areas, where it can be used to boost human development—helping teachers to extend educational programs, doctors to provide information and emergency aid, and rural farmers and entrepreneurs to reach urban markets.

This chapter considers the environmental implications of a large family of electronic technologies that help people gather, store, analyze, share, and distribute information. Given the tremendous growth and power of these technologies, people everywhere stand to benefit from thinking about how to harness these tools to promote sustainable development—and how to avoid the potential pitfalls.

An Expanding Global Network

In June 1999, reporters from around the world flocked to Bhutan to watch the tiny Himalayan kingdom officially enter the information age. Bhutan's rulers had long tried to protect the nation's traditional Buddhist culture from outside influence with limits on tourist traffic and bans on satellite television receivers. But now King Wangchuck was celebrating the twenty-fifth anniversary of his reign by inaugurating Bhutan's first Internet hookup and addressing his subjects during the nation's first television broadcast. A new state-of-the art digital telephone system would serve both computer modems and television.[3]

Bhutan has joined a diverse and rapidly expanding global network. Radio remains the most common form of electronic communication in the world, with more than 3 billion sets in use. Televisions are the next most prevalent device, followed by fixed-line telephones, computers, and cellular phones. (See Figure 7–1.) Between 1990 and 1996, international telephone traffic more than doubled, from 33 billion to 70 billion minutes per year.[4]

The newest technologies are catching up with the most established. In 1996, for the first time, the worldwide growth in cellular phones surpassed the growth in fixed-line telephones: more than 55 million new cellular subscribers compared with some 49 million main-line connections. The number of personal computers in use worldwide more than tripled in the 1990s, to top 370

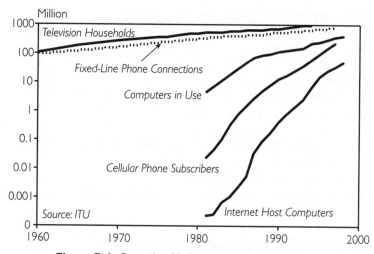

Figure 7–1. Growth of Information Technologies

million by the end of 1998. During the same period, the number of Internet "hosts"—computers that are linked directly to the Internet—grew 115-fold to 43 million, allowing an estimated 147 million people to go online.[5]

There is also growth in the satellite remote sensing and computer software technologies that obtain, store, and analyze information about Earth. Altogether, more than 45 Earth observation missions now operate, with more than 70 planned during the next 15 years by civil space agencies and private companies. And the number of people using geographic information system computer software is swelling by roughly 20 percent each year. The leading GIS software company, ESRI, grew from fewer than 50,000 clients in 1990 to more than 220,000 in 1999.[6]

The contribution of these technologies to the global economy is growing rapidly as well. Industry analysts estimated the worldwide market for information and communications technologies at $1.83 trillion in 1997 and commerce over the Internet at $7–15 billion in 1998. The boom

in Internet businesses helped push the net worth of the world's 200 wealthiest people from $463 billion in 1989 to beyond $1 trillion in 1999.[7]

The digital revolution now under way may mark the third wave of technological change in communications, each of which has transformed societies. (See Table 7–1.) Humans were writing for about 4,000 years before Johann Gutenberg invented the printing press in the fifteenth century, but only a tiny elite could obtain the texts. The printing press brought information to the masses, spurring the growth of Christianity as Bibles became widely available, the codification of law as the same legal tomes could be shelved at all courts, and the development of intellectual property rights as writers laid claim to their publications.[8]

The second wave came in the nineteenth century, as people began to exploit electrical current and the electromagnetic spectrum to transmit information quickly over great distances. Samuel Morse launched the electrical telegraph in 1844 by tapping out a message from Washington, D.C., to Baltimore, asking "What hath God wrought." And by 1875, telegraph pioneers in Victorian England had built a network of cables linking the United Kingdom, the United States, Australia, India, and other points in Asia. The telegraph shrunk the world as never before, shortening the time it took to send messages from London to Bombay and back from 10 weeks to a mere four

Table 7–1. Key Events in Communications History

Year	Event
around 3000 B.C.	Written language arises
A.D. 1448	Johann Gutenberg invents the printing press—bringing information to the masses
1489	The Thurn und Taxis family formalizes an extensive postal system in the Holy Roman Empire
1844	Samuel Morse launches the first electrical telegraph system—connecting people in a communications network
1876	Alexander Graham Bell invents the telephone
1896	Guglielmo Marconi patents a device for generating and detecting radio waves (radio telegraph)
1929	First public television transmission
1945	ENIAC, the first electronic digital computer, is completed
1948	Transistor is invented
1962	First communications satellite launched
1969	First link in what would become the Internet is installed—marrying the computer to the communications network
1971	Microprocessor is created First e-mail program written by Ray Tomlinson
1990	Tim Berners-Lee at CERN (European Laboratory for Particle Physics) creates the first World Wide Web server
1993	Development of first graphical Internet browser

SOURCE: See endnote 8.

minutes. Subsequent innovations—telephones, radio, television, and communications satellites—built on the progress of telegraphs to overcome the barrier of distance in communications.[9]

The current wave of change stems from the marriage of the computer to these communications links. Telephone and television networks have long conveyed sound and pictures as analog waves. Now, however, many types of information—text, sound, picture, or video—can be transmitted digitally, as compressed bits in the binary language of computers. Digitization is sweeping through the telephone, photography, remote sensing, broadcasting, film, and music industries. As a result, the lines separating telecommunications from computers, information processing, publishing, recording, and entertainment are increasingly blurred.[10]

Although this digital movement is bringing information to more people, it has a long way to go to reach the majority of the world's population. Just 23 industrial countries account for 62 percent of all phone lines, even though they are home to less than 15 percent of the world's people. (See Figure 7–2.) Finland, Norway, and Sweden actually have more phone links—fixed-lines and cellular combined—than people. And some 84 percent of cellular phone subscribers and more than 90 percent of Internet users live in industrial countries. (See Table 7–2.)[11]

In the developing world, telecommunications investment is concentrated in 30 emerging economies. So although the number of phone connections per 100 people in developing countries jumped from two to six between 1985 and 1997, the gains occurred mainly in parts of Latin America and East Asia. Buenos Aires, with a population of more than 6 million, has twice as many links to the fixed-line telephone network as does all of Eastern Africa, which is home to nearly 250 million. Variations within nations are often just as great, with urbanites better connected than rural dwellers. In the United States, for instance, people in urban areas are more than twice as likely to have Internet access as those earning the same income in rural areas.[12]

Technology changes are beginning to narrow this gap. One technological driver is growth in the capacity to handle information, in both computing and communication. The basic components of a computer have shrunk dramatically, from the bulky vacuum tubes of the first machines to microscopic transistors so tiny that researchers expect silicon chips will reach the physical limits of miniaturization early

Table 7–2. Internet Hosts by Region, January 1999

Region	Share of Total (percent)
Canada and United States	64.0
Europe	24.3
Australia, Japan, and New Zealand	6.3
Developing Asia-Pacific	3.4
Latin America and the Caribbean	1.6
Africa	0.4

SOURCE: Network Wizards, "Internet Domain Surveys, 1981–1999," <www.nw.com>, updated January 1999.

in this century. In the last three decades, the computing power of a single computer chip has increased by a factor of 64,000.[13]

Similarly, communications capacity has risen as copper wires have been replaced by ever more powerful optical fibers, thin glass strands that transmit light signals. At a given instant, all of North America's long-distance telephone traffic could theoretically be carried on a single pair of optical fibers, each the thickness of a human hair.

As the Internet explodes in size, transmission of data is expected to surpass voice calls within the next few years. To meet this demand, companies are boosting installed capacity—laying undersea fiber-optic cables and launching satellites at a record pace.[14]

High-capacity digital connections, or "broadband" services, allow communications service to be provided in new and potentially cheaper ways—for instance, telephone bundled together with television or Internet service. The cost of installing one new link can be defrayed by payments for more than one service.[15]

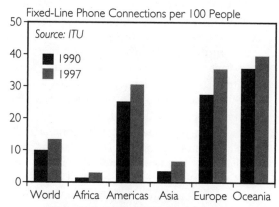

Figure 7–2. Density of Telephone Penetration, by Region

Wireless technologies are well suited to reach the most remote locations. In several developing countries where the fixed-line network is concentrated in urban areas and where wireless services offer competitive prices, cellular phones are substituting for traditional service. Mountain farmers in yak caravans in Laos and Myanmar now use cell phones to find the best route to market during the rainy season. Wireless phones are also being adopted in places where copper wires are often stolen for scrap value and in countries such as Cambodia, Lebanon, and the Democratic Republic of Congo (formerly Zaire) where phone lines have been damaged by war.[16]

Mountain farmers in yak caravans in Laos and Myanmar now use cell phones to find the best route to market during the rainy season.

Communications satellites are beginning to cover underserved markets as well. Since 1965, Intelsat, a treaty-based cooperative that now has 140 member nations, has dominated commercial satellite communications with a fleet of satellites that are "geostationary," orbiting the equator high enough to remain over the same point on Earth. The first satellite could provide 240 telephone circuits (or one television channel) between Europe and the United States; within two decades, Intelsat satellites were providing the bulk of the world's intercontinental telephone and television links.[17]

A new generation of satellites is designed for low orbit (700–1,400 kilometers from the surface). These are cheaper to launch, can cover the most remote parts of the planet, and are better suited than geostationary satellites for interactive voice and video connections because they are close enough to Earth's surface to eliminate the time lag in transmission.[18]

Teledesic, the largest constellation of low-orbiting satellites, aims to provide high-speed Internet access from anywhere on Earth. It is scheduled to begin service in 2002 and plans to reserve some of its capacity to extend communications links to rural parts of the developing world. Iridium, a system of 66 such satellites completed in 1998, is the first global satellite phone network. At $3,000 per handset and $1–3 per minute, its use is limited to wealthy business travelers. But more modest one-satellite systems, which allow Internet links for just two hours a day, are already being used to transmit medical information in Africa.[19]

Spurred by private companies, communications satellites rocketed from 14 percent of all satellite launches in 1967 to 69 percent in 1997, the first year in which commercial launches exceeded government ones. In 1998, a private venture, World-Space, launched the first of three communications satellites intended to provide digital radio service to underserved populations in Africa, Asia, and Latin America.[20]

Along with new technologies, new market liberalization and competition policies are expanding the communications network. In this respect, the tumultuous changes in today's communications industry are reminiscent of the telephone explosion of a century ago. When Alexander Graham Bell's patent on the telephone expired in the 1890s, competition in the United States surged for a short period, resulting in lowered costs that made the telephone affordable to more people. The rise of a giant telephone monopoly slowed the drop in prices. Although state-controlled monopolies have dominated the phone business for most of its history, the World Trade Organization ushered in a new era in February

1997 when 69 nations, accounting for more than 90 percent of world telecommunication revenues, agreed to open their markets to foreign competitors.[21]

From Asia to Africa, governments in the 1990s have been dismantling state-run monopolies and opening their telecommunications markets to competition. Since 1996, Côte d'Ivoire, Ghana, and South Africa have sold parts of their phone networks to private companies that agreed, as part of their contracts, to expand service substantially. Similar policies have succeeded in extending service to underserved populations in Mexico and Argentina, which were among the first Latin American countries to privatize in the 1990s.[22]

Squandering or Saving Natural Resources?

Throughout their "lives," computers, satellites, televisions, and other telecommunications instruments take a toll on Earth's resources. Their effects on the environment at birth and death—production and disposal—are fairly easy to estimate. But the net environmental effect during their useful lives is much harder to gauge.

Making computers requires energy and water. The production of the silicon semiconductors that form computer "chips" is particularly energy- and water-intensive. A single large semiconductor manufacturing plant, producing 5,000 eight-inch wafers a week, could use as much electrical power and water as a small city.[23]

Manufacturing computers and televisions also generates waste, much of it hazardous. Toxic solvents, acids, and heavy metals are used in the manufacture of semiconductors, printed wiring boards, and cathode ray tubes for computer monitors and television screens. In 1993, the Microelectronics and Computer Technology Corporation analyzed the waste created in manufacturing a typical computer workstation. Its study suggests that 63 kilograms of waste, 22 of them toxic, are generated in producing a 25-kilogram computer.[24]

California's Silicon Valley provides a case study in the environmental hazards of computer production. Before the Santa Clara Valley was transformed into a computer-producing hub in the 1970s, pristine underwater aquifers supported agriculture. Three decades later, 29 Superfund hazardous waste sites—including 23 former chip-making factories—blight the valley, and the county of Santa Clara must now import much more water. In some cases, these wastes have harmed either the workers assembling the devices or the local environment surrounding the factory.[25]

Computers and mobile phones also present a tremendous disposal problem—in part because they become obsolete so quickly. A recent study by the U.S. National Safety Council estimated that 20.6 million personal computers became obsolete in the United States in 1998—and of these, only 11 percent were recycled and 3 percent were resold or donated. Because computers become outdated rapidly, repair is costly compared with the price of new goods. When computers are trashed, the lead in monitors, the mercury and chromium in central processing units, and the arsenic and halogenated organic substances inside the devices all become health hazards.[26]

Recycling of computers and telephones is difficult because most are not designed to be recycled. Thus it is usually not cost-effective for recycling businesses to pay for transportation and labor to dismantle and sort 10–25 kilograms of used computer into three types of plastics and four types of metals, which may be worth as little as $1.80. Small electronic devices, such as mobile phones, are particularly difficult to recycle.[27]

Spurred by activists and threats of government action, high-tech companies have begun to deal with the pollution and waste that arise from manufacturing and disposal. In the United States, the Silicon Valley Toxics Coalition has taken the lead in publicizing the toxic side of electronics manufacturing, and has organized other nongovernmental organizations (NGOs) worldwide in an International Campaign for Responsible Technology. U.S.-based companies such as Motorola, IBM, and Intel now publish annual environmental health and safety reports that show progress toward reducing the energy, water, and hazardous materials used in manufacturing.[28]

In 1998, Taiwan started a take-back system for computers, televisions, and large home appliances.

In Europe, the idea of making producers responsible for their products at both the cradle and the grave is becoming more accepted—and may ultimately transform the production of information technologies. The best example so far of "extended producer responsibility" is Germany's packaging law, which went into effect in 1991 and has required manufacturers and distributors to recover their packaging and reuse or recycle it. If producers are to take computers and cell phones back cost-effectively, the products' design will have to change to make them easier to recycle. Proponents argue that this will actually save companies money because it favors simple designs with fewer parts.[29]

Since 1995, the European Union (EU)

has been discussing a draft directive on take-backs for electronics. It prohibits the use of certain toxins—mercury, cadmium, hexavalent chromium, and brominated flame retardants—in electrical goods by 2004. And it requires producers to pay for collection systems, and distributors to take back an electronic device when supplying a new one. Between 70 and 90 percent of all material recovered by weight must be reused or recycled.[30]

Austria, France, Germany, Japan, the Netherlands, Sweden, Switzerland, Taiwan, and the United Kingdom are the countries where the strongest support is found for take-back plans for electronics. In 1999, Germany began to move the draft decree through its parliament. The Netherlands is aiming for 100-percent recovery by 2000, with different recycling targets for large appliances, consumer electronics, metal, and plastics. Industry is allowed to impose a surcharge to fund the take-back schemes.[31]

The prospect of an EU directive has spurred some companies into action. The five cell phone producers who account for the bulk of the European market—Motorola, Ericsson, Nokia, Alacatel, and Panasonic—started pilot programs in Sweden and the United Kingdom to take phones back for recycling. At Sony Europe, new TV designs now have snap-together parts and fewer screws to make them easier to take apart. And Siemens in Germany has built a personal computer that is much easier to take apart—an added benefit is lower production costs, as the fewer parts mean that the personal computers can be assembled more quickly as well.[32]

Outside Europe, Taiwan started a take-back system for computers, televisions, and large home appliances in 1998 that requires retailers to accept used electronics, regardless of where they were sold. And in 1999,

Australia announced a national recycling scheme for mobile phone batteries.[33]

Like consumer electronics, used satellites also create a waste disposal problem. Humanmade space debris—an orbiting graveyard of dead satellites, camera lens caps, spent rocket boosters, and other bits of wreckage—now poses a greater collision risk to spacecraft than do natural meteoroids. The U.S. Space Command catalogues these objects, which now number more than 8,000. In recent years, companies that produce satellites for geostationary orbit have agreed to design their satellites to retain enough fuel at the end of their "life" to boost themselves into a graveyard orbit. And satellites closer to Earth will use their last fuel supply to move downward into the atmosphere, where they will burn up. Researchers are also trying to develop reusable rockets, smaller satellites, and ways to reel in old space objects.[34]

Cleaner production processes and fledgling efforts to reuse and recycle information technologies may alleviate many of the environmental hazards of their manufacturing and disposal, but the most profound environmental effects of information technologies are likely to occur from their use. Enough data exist to estimate some of the environmental effects of operating information technology—for instance, the paper and electricity requirements of computers. Much less is known, however, about the extent to which materials and fuels may be conserved when people substitute information exchange for activities that require more natural resources.

Paper is one resource whose consumption is clearly linked to computer use. Between 1988 and 1998, average per capita consumption of printing and writing paper in industrial countries shot up by 24 percent. (See Chapter 6.) Computers have not sated the appetite of paper-hungry industrial countries—they have merely altered their tastes. While office paper use has risen steadily, newsprint consumption has stagnated in the United States. Electronic commerce is expected to lower the demand for paper used for advertisements, but increase the need for packaging paper.[35]

Among the many reasons that computers have yet to supplant paper is that computer screens are hard on the eyes and their horizontal orientation is harder to scan than vertically oriented office paper. Better electronic reading devices may change this. The first "e-book"—a portable machine to read electronic documents—went on sale in November 1998. The size and weight of a paperback book, the e-book could be used to store and display any number of heavy textbooks. Several teams of researchers are working on versions of an "electronic paper" technology that would be about twice the thickness of an ordinary piece of paper, but just as readable and portable, and capable of being reused millions of times. In 1999, Xerox and 3M announced a partnership to market an e-paper technology.[36]

Another requirement of electronic information tools—although a relatively modest one—is electricity. A typical desktop computer with a monitor uses 120–180 watts of electricity, about the same as two lightbulbs. Efforts to make computers more efficient include designing sleep modes, in which the computer uses 30 watts or less, or having a machine shut off entirely when not in use. The Energy Star campaign of the U.S. Environmental Protection Agency (EPA) has helped boost production of energy-efficient computers and other office equipment—on an annual basis, this efficient equipment uses as little as half the electricity of conventional equipment. A 1998 report by the World Resources Institute highlights electronic products by Sharp and "Instantly Available PCs" from Intel that bring electricity use below even Energy Star levels.[37]

Moreover, growth in energy use has been slower than economic growth in many industrial countries. And this decoupling appears to be led by information technologies, which are transforming manufacturing and commerce. Computers and other office equipment represent the fastest-growing demand for electricity. At the same time, most computer processing units are embedded in devices other than computers, and these electronic controls have increased energy efficiency in a tremendous number of industries. For instance, semiconductors permit 3-D data visualization of energy use in buildings for the building operator to see—and this can improve energy efficiency by 15–20 percent. In the 1999 *Annual Energy Outlook*, the U.S. Department of Energy found that electricity demand was only slightly higher than the previous year because the increase in commercial demand from rapid growth in computers and other office equipment was offset by a decrease in industrial demand from efficiency improvements.[38]

Another question related to energy is whether communications technologies will become so effective that they will decrease the need for transportation. Speaking at the Governing Council meeting of the U.N. Environment Programme (UNEP) in February 1999, a representative of a large U.S. telecommunications company laid out a vision "that the transmission of bits and bytes may increasingly substitute for the burning of hydrocarbons—that search engines will eventually prove more important and prevalent than internal combustion engines." Telecommuting and videoconferencing are on the rise in indus-trial countries—but it is far from clear that they are causing an overall reduction in transportation.[39]

Although telecommunications may substitute for actual travel in many instances, it can also stimulate transportation in a variety of ways—by allowing people to live farther apart; by making people aware of conferences, events, and stores that can be reached by travel; and by making travel time more productive with mobile phones and pagers. No communications technology in history has ever been associated with a net reduction in travel. Between 1880 and 1910, telephones were paralleled by commuter railways and metropolitan subways; between 1920 and 1940, radios by automobiles and airplanes; and between 1950 and 1970, the television by highways and commercial jets. New information technologies that allow people to work from home more easily may play an important role in alleviating peak traffic congestion, but they will not eliminate the need for integrated transportation and land use planning. Indeed, such planning will be critical to prevent "telesprawl."[40]

Similarly, both environmental opportunities and dangers accompany electronic commerce, as more people and companies buy and sell products over the Internet. Environmental analyst Nevin Cohen believes that, like the rise of shopping malls and the globalization of production, "e-commerce will have significant environmental consequences." For instance, online companies can avoid waste by holding limited inventory. Products such as books, music, photos, and videos can be bought in electronic form and sent directly over the Internet, saving materials and energy. And the popularity of online auction houses such as E-bay indicates that the Internet can help companies and individuals who have used products to match up with those who can use them, turning wastes into resources.[41]

Such environmental benefits, however, could easily be offset if the spread of information technologies causes a surge in the consumption of products that harm the environment or human health. Television, the world's most ubiquitous information technology, is also a potent medium for advertising. Automobiles, cigarettes, and fast food constitute the top advertisers in the world. Television programs that depict lavish consumer spending further promote consumerism. As satellites beam television to remote areas, and as commerce and entertainment expand on the Internet, these technologies have the potential to spur greater consumer excess.[42]

Monitoring and Modeling

In October 1999, astronauts from some 15 countries met in Romania to discuss space-based research to protect the environment. Since the first manned flights to the moon in the 1960s, space travelers have been struck by the sight of a fragile blue planet suspended in the vast blackness of space. Noting the large swaths of clearcut forests and the drab spread of urban settlements visible from 258 kilometers above Earth's surface, in the early 1990s one U.S. astronaut explained the power of the global perspective: "Most people don't get to see how widespread some of the environmental destruction is. From up there, you look around and see that it's a worldwide rampage."[43]

Today, a growing number of satellites capture such images of human activities on Earth. Computers and GIS software allow images obtained by satellites to be stored, analyzed, and manipulated. Together with on-the-ground monitoring and other data,

this information can help researchers examine pollution and other environmental hazards, identify areas rich in particular resources, and model changes to the environment. And it can help decisionmakers and planners to manage the environment better. (See Table 7–3.)

> **No communications technology in history has ever been associated with a net reduction in travel.**

Although other types of environmental monitoring are equally valuable, satellites are unique in being able to collect detailed information about parts of Earth that are otherwise difficult to get to—the far reaches of the atmosphere, the depths of the oceans, the icy polar regions, and forest interiors. Moreover, remote-sensing instruments aboard Earth-orbiting satellites can frequently record changes over large areas and long periods of time.

The first effort to monitor Earth systematically from space began in 1972, with the launch of a U.S. government satellite that was eventually named Landsat. Hand-held cameras carried by the earliest astronauts were limited to the visible range of the electromagnetic spectrum, but the Landsat sensors could also record heat and reflected energy invisible to the human eye. Five subsequent Landsats have created a continuous data series from 1972 to the present. The French space agency has a similar series, System Probatoire d'Observation de la Terre (SPOT), which began in 1986.[44]

Although these systems have global coverage, they also have fairly high spatial resolution, with a 10- to 30-meter-square piece of land on Earth corresponding to one picture element—or "pixel"—of data. Compiling such a detailed picture of the

Table 7–3. Mapping the Environment: Global Digital Data Sets on the Internet

Data	Sources
Elevation and Drainage Basins	United States Geological Survey/EROS Data Center <edcwww.cr.usgs.gov/landdaac/gtopo30/gtopo30.html>
Population Distribution	CIESIN <www.ciesin.org/data.html>
Land Use	Global Land Cover Characteristics Data Base, U.S. Geological Survey <edcwww.cr.usgs.gov/landdaac/glcc/globe_int.html>
Soils	FAO Soils Map <www.fao.org/catalog/new/products/v8600% 2De.htm>
	Global Assessment of Human-Induced Soil Degradation <grid2.cr.usgs.gov/data/glasod.html>
Oceans and Sea Surface Temperatures	Marine Geology & Geophysics Images, <web.ngdc.noaa.gov/mgg/image/images.html>
	University of Wisconsin, Space Science and Engineering Center <www.ssec.wisc.edu/data/sst.html>
Water	Global Hydrology and Climate Center <wwwghcc.msfc.nasa.gov>
Stratospheric Ozone	CIESIN <sedac.ciesin.org/ozone>

entire planet would take a great deal of time, so these data are better for close-up local analysis: to categorize the types of land use in a country or to assess vegetation health, water resources, or coastal boundary changes. For instance, scientists have recently compared declassified images from covert military satellites pointed at Antarctica in 1963 to more recent SPOT data to gain insight into changes in the continent's ice cover.[45]

Another class of satellites, designed primarily for meteorology, has much coarser resolution, with each data point covering at least one square kilometer. This imagery is well suited to frequent regional or global monitoring and analysis. For instance, the U.S. National Oceanic and Atmospheric Administration uses the Advanced Very High Resolution Radiometer (AVHRR) to look at the changes in global sea surface temperature that characterize El Niño. This picture shows how warming off the coast of Peru is linked to droughts in Australia and tropical fish landings in California. Imagery from this type of sensor is also appropriate for mapping large areas of vegetation. Boston University Professor Ranga Myneni and colleagues at the National Aeronautics and Space Administration (NASA) analyzed AVHRR data to detect an increase in "greenness" from plant growth in northern latitudes in summer months between 1981 and 1991; this lengthening growing season was linked to warmer temperatures.[46]

A different set of meteorological satellites forms the backbone of the most effective global environmental monitoring program to date: the World Weather Watch. Operated by the U.N. World Meteorological Organization, this network combines satellite observations with ground, sea, and air monitoring stations, telecommunication links, and computer analysis centers.[47]

A key element of the weather program's success is the multitude of observations from both sky and land. In 1985, such a combination proved essential to revealing the damage to the ozone shield that protects Earth from ultraviolet radiation. Satellite instruments that calculate atmospheric ozone were initially filtered to discount abnormally low data as system error. Thus an instrument on the ground in Antarctica was the first to detect the ozone "hole." The satellite data were then recovered to provide a more complete picture of how widespread the ozone decreases were.[48]

Most satellite sensors measure either reflected light or heat, but a few relatively new systems use radar—transmitting short bursts of microwave energy to Earth and recording the strength of the reflected energy that comes back. Microwaves can penetrate the atmosphere in all conditions, so radar can "see" in the dark and through haze, clouds, or smoke. Radar sensors launched by European, Japanese, and Canadian space agencies in the 1990s have mainly been used to detect changes in the freezing of sea ice in dark, northern latitudes. But experiments suggest a wide range of applications exist. For instance, satellite-borne radar has been used to help map the topography of the ocean floor, providing insight into ocean currents, tides, and the upwelling of nutrient-rich water that sustains fisheries.[49]

Governments have launched most of the current global fleet of Earth observation satellites, but private companies are now beginning to enter the picture. One of the first was OrbImage, a U.S. company that launched a satellite called SeaWiFS in 1997. Originally designed to measure ocean color and temperature, SeaWiFS has monitored fires in Indonesia, floods in China, and dust storms in the Sahara and Gobi Deserts.

Through a public-private partnership, NASA is purchasing the data for its science program, while OrbImage retains rights for commercial purposes. In a similar vein, three private companies are teaming with government-run space agencies in the European Union and Russia to start a Global Environmental Service that will provide data and maps for nature preservation, regional planning, forestry, agriculture, and emergency management.[50]

Other commercial satellites are beginning to deliver images as sharp as those obtained by government-owned spy satellites. With one-meter resolution, such instruments will be able to identify individual houses and cars and provide more detailed information about crop health. Companies expect orders from not only intelligence agencies but also urban planners and farmers. The first of a new generation of commercially owned satellites was the IKONOS instrument, launched by Colorado-based Space Imaging in September 1999. Under ideal conditions, a customer will be able to get a one-meter image as quickly as 30 minutes after it is taken. At least two other U.S. companies are planning similar systems.[51]

A key reason for the commercial interest in remote sensing is the growing demand for geographic information, as advances in computer software now allow satellite images to be combined with other data in computerized maps. Human eyes can often identify patterns more quickly on maps than in written text or numbers. The programs that create these maps are geographic information systems.

In much the same way that old medical encyclopedias depict human anatomy, with transparencies of the skeleton, circulatory system, nervous system, and organs that can be laid over a picture of the body, a GIS stores multiple layers of geographically referenced information. The data layers might include satellite images, topography, political boundaries, rivers, highways, utility lines, sources of pollution, and wildlife habitat. Researchers can now take air or water samples at a given location, and feed the data directly into a GIS with latitude and longitude coordinates supplied by a network of 24 navigation satellites operated by the U.S. Department of Defense.[52]

Maps that are stored in a GIS allow people to exploit the advantages of computers, which can store huge amounts of data and do repetitive and complex calculations. Thus when geographically referenced data are entered into a GIS, people can use the computer to look at changes over time, to identify relationships between different data layers, to change variables in order to ask "what if" questions, and to explore various alternatives for future action.

When GIS software was first developed in the 1960s, it required huge computers. Some environmental modeling and analysis still require tremendous computing power. Among these applications are the programs that use the basic laws of physics that govern the climate, taking into account Earth's topography, along with land, oceanic, and atmospheric processes. These climate models can simulate an accretion of various gases: both the heat-trapping gases such as carbon dioxide that warm the atmosphere and the sulfate aerosols, another byproduct of fossil fuel burning, that cool it. Scientists who have studied the models think they see a telltale human "fingerprint" on the climate: observed temperature patterns closely match the patterns that models generate.[53]

Computer software has also aided in European efforts to halt transboundary air pollution, which has killed fish in Scandinavian streams and trees in Germany's Black Forest. Meeting in Oslo in 1994, European negotiators agreed to reduce their emissions of sulfur dioxide, a byproduct of fossil fuel burning that forms acid rain. In setting their emissions goals, they relied heavily on a computer model. Devised by scientists at a nongovernmental research institute in Austria, this model analyzed the environmental effects of a range of sulfur emissions scenarios. The computer showed the policymakers how they could target their cuts to prevent ecosystems from exceeding "critical loads" of acid deposition, beyond which long-term damage was likely.[54]

With advances in computing power, some geographic information system software packages can now be run on desktop computers. As a result, GIS is beginning to reach more people. For instance, this tool is helping environmental and community activists identify local sources of pollution, allowing energy agencies in developing countries to determine the best sites for renewable energy installations such as wind turbines, and helping conservation groups craft strategies for natural resources management and the protection of biological diversity.

The New York Public Interest Research Group has shown how maps can empower local activists with the Community Mapping Project it launched in 1997. With relatively simple GIS tools, community activists in polluted neighborhoods of New York City have been able to create maps that overlay facilities such as garbage transfer stations, oil refineries, and sewage treatment plants with locations that report high levels of asthma and cancer cases.[55]

With a more complex program requiring greater analytical expertise, researchers at the U.S. National Renewable Energy Laboratory have mapped the available wind resources of parts of Asia, Latin America, and North America. The computer program combines elevation data with meteorological information to identify areas where strong winds might be harnessed to generate clean, renewable electricity. In the Dominican Republic and the Philippines, for instance, detailed wind maps have stimulated public and private interest in wind energy development.[56]

Several large conservation organizations have also embraced GIS. Washington-based Conservation International was among the first to bring the technology to developing countries. The group has developed a relatively low-cost GIS in English, Spanish, Portuguese, and French, and has invested heavily in training local people to create databases and maps to manage national parks and other natural resources better. The software is now used by more than 200 institutions in at least 30 countries.[57]

Another environmental group, the World Wildlife Fund (WWF), has been using GIS for the past six years to support ecoregion conservation in projects ranging from the local to the global. By overlaying satellite imagery with many other types of data, such as road networks and national parks, the group can help local and national governments identify priority areas for biodiversity conservation. WWF has used GIS to develop a global map of the world's terrestrial, freshwater, and marine ecosystems. This project was the basis for WWF's Global 200, a list of outstanding examples of Earth's diverse biodiversity and habitats.[58]

In 1996, the World Resources Institute invested in GIS to analyze threats to natural resources. Researchers combined ground and satellite data on forests with information about wilderness areas and roads to map the world's remaining large, intact "frontier" forests and identify "hot spots"

of deforestation. And an investigation of the threats to coral reefs worldwide pulled together information from 14 global data sets, local studies of 800 sites, and scientific expertise to conclude that 58 percent of the world's reefs are at risk from development, overfishing, and pollution.[59]

Some geographic information system software packages can now be run on desktop computers.

Remotely sensed data and GIS software can help researchers study the global environment, but their use is limited by several problems. While these systems can complement existing environmental protection efforts, only relatively few people, mainly in industrial countries, know how to use them. Developing countries often need computer hardware and software, but what is absolutely essential at the same time is training in how to use them.

Another problem is that launches of Earth observation satellites are outpacing the growth of ground- or ocean-based environmental observation programs. Remotely sensed images must be compared with data obtained on the ground if they are to be interpreted correctly. But ground measurements of basic environmental indicators are not available for much of the world. For instance, the Global Environmental Monitoring System for Urban Air Pollution (GEMS/AIR), a joint project of the World Health Organization (WHO) and UNEP, ran from 1975 through 1996, introducing pollution monitoring technology to a number of developing countries. Although more than 80 cities in 50 countries contributed to the GEMS network, many were sporadic or short-term participants, so data from 1975 to 1995 only exist for a few cities, data before 1989 may not be reliable, and differences in monitoring methods make city-to-city comparisons difficult. Among the other indicators not tracked globally are transboundary air pollution, groundwater quality, land degradation, deforestation, and ocean productivity.[60]

Other types of data problems thwart projects. Although volumes of digital, geographically referenced data are now available, much of the material is out of date. Moreover, data often must be converted into a format that a particular software program can use—and this process can be time-consuming. In addition, aside from the World Weather Watch, there is no process for coordinating a worldwide, long-term time series of comparable data from Earth observations. Rather, individual scientists collect data to answer specific questions for their own projects. In recent years, national space administrations have teamed up with research funding agencies and two international research programs to support an Integrated Global Observing Strategy that would create a framework for uniting environmental observations. If it succeeds, this project would be an important contribution to human understanding of the global environment.[61]

Networking for Sustainable Development

The networking power of communications technologies has wide-ranging implications for sustainable development. Scientists have easier access to environmental data for their research. Environmental NGOs can respond to emergencies quicker and mount campaigns. Both governments and NGOs are more able to publicize information about polluters. And communications links can aid

in education, health care, and job creation—all essential to raising the quality of life.

The most prevalent communication devices—radio and television—can be used to educate and inform. Population Services International is one group that has used broadcast technologies in countries as varied as Bangladesh, Haiti, and South Africa to spread information about family planning, child survival, and AIDS prevention. For instance, the group collaborated with the National AIDS Committee of Côte d'Ivoire on a public health campaign that included a weekly television soap opera that explored AIDS transmission, detection, and prevention. Well-know Ivorian entertainers donated their time to act in the program, which became quite popular.[62]

The Internet combines broadcasting with two-way information exchange, which makes it easier for scientists and researchers exploring questions about the state of the environment to share ideas and collaborate. In 1999, for instance, the World Resources Institute initiated a Global Forest Watch program by linking up with more than 25 groups around the world who were already collecting forest data. These partners will maintain a set of interlinked Web sites to share data. Initial participants included researchers in Cameroon, Canada, Gabon, and Indonesia; additional contributors are being sought in Central Africa, Southeast Asia, South America, North America, and Russia.[63]

Another environmental project made feasible by communications technologies is the proposed Global Biological Information Facility. Funded by the Organisation for Economic Co-operation and Development through its Megascience Forum, this project would put more than 350 years of information on biological resources currently housed in major museums around the world on the Internet. Each of the world's named species would be listed on a Web site, with links to a description, references in the literature, and information on which museums house specimens.[64]

As the global communications network has grown, it has become an increasingly important tool for NGOs pursuing humanitarian and environmental agendas. A directory of NGOs operating in three or more countries maintained by the Union of International Organizations documents a surge from 985 groups in 1956 to more than 20,000 in 1996. With the aid of new technologies, these citizens' organizations are affecting government policies.[65]

For example, a loose international coalition of NGOs launched a campaign to ban land mines, making use of a Web site, e-mail, and newsletters. They succeeded in getting a land mine treaty, an accomplishment that earned the organizers the 1997 Nobel Peace Prize. The Rainforest Action Network is another pioneer in Internet campaigns, using its Web site to disseminate information on logging and mining in tropical forests from its partners in more than 60 countries. The group enlists the help of online visitors, who are urged to write a letter to government officials or company executives and to fax it directly from the Web site.[66]

The largest audience for online activists is in the United States. In December 1997, when the U.S. Department of Agriculture proposed relaxing the standards by which it defined "organic" produce, a nonprofit group used the Internet to mobilize a Save Organic Standards campaign. These activists pointed out that the proposed standards would allow genetically modified food, irradiated food, and food grown with municipal sludge to be labeled "organic." After receiving a public response that included more

than 237,000 e-mail messages, the government dropped its proposal.[67]

According to Richard Civille, cofounder of the U.S.-based Center for Civic Networking, Internet campaigns are most effective as part of a broader strategy. An ideal campaign would use traditional radio and print ads to direct people to a Web site dedicated solely to the campaign, which would include a page with an underlying database to collect names and addresses of visitors. Using these tenets, the nonprofit Technology Project (now e-group) designed <www.ourforests.org> for the Heritage Forest Campaign, which prompted some 175,000 people in May and June 1999 to send e-mails to U.S. Vice President Gore, pushing him to protect national forests from road construction. Rob Stuart of e-group dubbed the effort a "word of mouse" campaign.[68]

In countries with limited communications infrastructure, environmental networking can aid in emergencies. One example comes from Karachi, Pakistan, where representatives of the World Conservation Union–IUCN saw an alarming news story in May 1993. Karachi police, finding a warehouse owner and his driver dead after inhaling the fumes from a 2.5-ton canister of chemicals in their truck, had dumped the material into the already polluted Lyari River. Together with local NGOs, IUCN helped the government remove the canister from the river and identify it as meta-dinitrobenzene, but they had no idea how to handle it or dispose of it safely. They did not have access to the Internet, so they phoned colleagues in Lahore, who sent an appeal for information to two computer-based conferences—<en.toxics> and <en.alerts>—on the Association for Progressive Communications (APC) network. Within days, the NGOs in Karachi received detailed advice and offers of help by fax from environmental experts around the world.[69]

The activist network that came to the rescue, APC, is one of the oldest NGO associations on the Internet. It is an international group of more than 60 networks and Internet providers that has been serving NGOs and activists since 1990. Its partner in the United States, the Institute for Global Communications, was formed in 1987. In the 1990s, a number of other global environmental networks joined APC in cyberspace—some sponsored by NGOs, others by U.N. agencies. (See Table 7–4).[70]

In fact, the NGOs in Karachi were able to reach the APC network with the help of a U.N.-sponsored effort, the Sustainable Development Networking Programme (SDNP). Following the 1992 Earth Summit, the U.N. Development Programme launched this program in 12 pilot countries, including Pakistan. The idea was to provide $150,000–200,000 over two to three years to help developing countries make effective use of a communications network. The funds generally pay for a manager, a technical specialist, hardware, software, training, and Internet connectivity. In many countries that lacked Internet access, SDNP effectively served as an Internet service provider. Between 1992 and 1998, it spread from 12 to more than 80 countries.[71]

By publicizing pollution, a government agency or NGO can often force a company to clean up its act. This approach—dubbed "regulation by revelation" by Ann Florini, an analyst at the Carnegie Endowment for International Peace—was pioneered in the United States. A 1986 law, the Emergency Planning and Community Right-To-Know Act, required EPA to prepare an annual Toxics Release Inventory (TRI) beginning in 1987. (See Chapter 5.) The publication

of these data provides an incentive for companies to reduce pollution; indeed, toxic releases in the United States fell by 43 percent between 1988 and 1997.[72]

Two nonprofit groups, operating the Right-To-Know Network, were the first to offer online, public access to the TRI database. (EPA now provides the data on its Web site.) Recently, the Environmental Defense Fund used TRI data to create an online scorecard—<www.scorecard.org>—that ranks facilities to reveal the biggest threats. There are maps that show schools, roads, and facilities that release toxic materials. People visiting the site can send faxes to polluting facilities.[73]

Networking has a number of implications for poor and rural communities—including health care, education, and job creation. "Telemedicine" refers to the use of communications technologies to link doctors to far-flung colleagues and patients. A prominent example is SatelLife, a nonprofit in Cambridge, Massachusetts, started in 1989 to promote the exchange of health information through a network called HealthNet. A small, relatively cheap, low-orbiting satellite picks up and delivers e-mail as it passes over ground stations.[74]

This system began with ground stations in seven African countries and now links 19,500 health care workers in more than 150 countries worldwide. Doctors use the network to collaborate on patient care, collect data, and do research. A weekly e-mail newsletter features excerpts from leading medical journals. An e-mail conference, the Program for Monitoring Emerging Diseases (ProMED) is used to post medical alerts. In July 1996, a doctor in Switzerland posted a message about the death of a yellow fever patient, alerting the Pan American Health Organization, which found that the patient had traveled through Manaus,

Table 7–4. Selected Global Environmental Networks

Network	Date Launched	Description
Sponsored by NGOs:		
Association for Progressive Communications (APC) <www.apc.org>	1990	Links NGOs promoting human rights and environmental justice.
OneWorld Online <www.oneworld.net>	1995	A "supersite" that links to hundreds of Web sites to provide information on development.
Global Forest Watch <www.wri.org/gfw>	1999	Linking WRI to NGOs in five countries to monitor the world's large, intact "frontier forests"
Sponsored by U.N. Agencies:		
UNEPNet (UNEP) <www.unep.net>	1997	Links eight UNEP offices and at least nine other partner institutions by satellite to improve the flow of global environmental information.
Global Urban Observatory (Habitat) <www.urbanobservatory.org>	1998	Links researchers worldwide to compile statistics and examples of best practices in urban management.
HORIZON Solutions Web site (UNDP, UNEP, UNFPA, UNICEF, IDRC, Harvard, Yale) <www.solutions-site.org>	1999	Provides case studies on solutions to problems of water, waste, energy, transportation, toxic chemicals, public health, industry, desertification, biodiversity, air pollution, and agriculture.

Brazil; the agency contacted authorities in Manaus, who launched an immunization campaign to stave off an epidemic.[75]

Some efforts are under way to use the Internet to open markets to rural businesses, but it is probably too soon to evaluate their success. In 1995, for instance, the Center for Civic Networking received a grant from the U.S. Department of Agriculture to start a Public WebMarket project. The goal is to stimulate Internet businesses that will reduce dependence on natural resource extraction in areas that are losing industries, such as sugar cane plantations, coal mining, or logging.[76]

To connect people to the information network more quickly, it will be important to provide access to communications centers. The "telecenter" or "telecottage" movement started in rural parts of Scandinavia in the 1980s. A community telecenter might include public telephones, fax machines, computers, and access to the Internet.

In Estonia, the nonprofit Estonian Association of Rural Telecottages grew from 3 in 1993 to 32 in 1997, connecting a larger share of the country's rural population. And in Ghana, the postal service is being bolstered by Message Link, a network of e-mail hubs, where messages are sent and then delivered to recipients by bicycle. Since the mid-1990s, telecenter projects have been supported by a number of development agencies, including UNESCO, the Netherlands-based International Institute for Communications and Development, the U.S. Agency for International Development, and Canada's International Development Resource Centre.[77]

The European experience with telecenters suggests that for such projects to be sustained over the long term, local people must be able to make money by running them as a business. In Bangladesh, the world's largest wireless pay phone project allows villagers to purchase cellular phones on a lease program and then sell calls to their neighbors. Farmers use the phone to check on the price of crops in Dhaka in order to avoid being cheated by intermediaries.[78]

A complementary technology is small-scale power generation. Neither phone lines nor power lines reach some parts of the developing world, so portable, wireless power systems—for instance, solar electric panels—are a perfect match for remote computers, televisions, and radios. (See Chapter 8.) The Solar Electric Light Fund, a nonprofit group that has demonstrated the feasibility of solar power in remote villages, is now beginning to link solar-powered computers in South African schools to the Internet via satellite. Other groups are just starting to promote solar-powered telecenters in Palestine, west central India, and Uganda.[79]

Information Tools for a Healthy Planet

The fusion of computing and communications has quickened the pace of growth and innovation in information technologies. Governments and citizens face new challenges and opportunities to use these powerful tools to conserve natural resources,

educate people, and diminish inequities.

As cellular phones and computers proliferate, it is increasingly important that they are manufactured safely and designed so that the materials in them can be easily reused. The EU's proposed directive on producer responsibility for the waste from electrical and electronic equipment could serve as an important model.

New, high-resolution satellite imagery and computer software can be used to augment environmental laws and treaties. For instance, one of the leading fishing nations, Peru, has begun to monitor its own waters to prevent overfishing that leads to collapse of the fisheries. And in Italy, the city of Ancona is now planning to buy satellite photographs to better detect illegal waste dumps.[80]

As information technologies permit new patterns of living, working, and organizing industries, environmental policymakers will need new approaches to take advantage of the opportunities for substituting information for energy and materials. David Rejeski, a policy advisor at EPA, warns: "Ten years from now, the environmental policy community may wake up and realize that they missed the Information Revolution."[81]

Even more important than alert policymakers will be an environmentally literate public who are able to steer information technologies to benefit the environment. For instance, advertising over the Internet threatens to unleash more consumption of resource-intensive products. Yet an environmentally conscious consumer could perhaps make use of software robots, which hold the promise of extensive online comparison shopping, to find products that were produced in an environmentally friendly manner and designed for recycling.[82]

To use information tools effectively, the global community will need much higher levels of literacy and education. Roughly 22 percent of the world's population, some 1.3 billion people, are illiterate. Although communications technologies permit the cross-fertilization of ideas, they are currently dominated by the English language and by American perspectives. Nearly 60 percent of the world's online population uses English to gain access to the Internet. As governments focus on bringing communication tools to more people, the percentage of non-English content is likely to rise. [83]

Education will be essential to helping people make sense of the growing glut of data. Whereas the cost of a printing press or television antenna once limited the numbers of people broadcasting information, the Internet gives this power to many. The World Wide Web grows by roughly a million electronic pages per day, adding to the hundreds of millions that already exist.[84]

Although remote sensing, GIS, and other technologies can help us understand how we are changing the planet, they cannot substitute for first-hand knowledge of the environment. Even while information technologies help swell databases of environmental knowledge and link people, they may disconnect people from the world around them.

Time spent in front of a computer collaborating with faraway colleagues by e-mail or talking on a cell phone is time not spent in face-to-face dialogue or interacting with nature. A recent study found a link between heavy Internet use and depression, suggesting that electronic communication does not make up for time spent nurturing personal relationships. In the early 1990s, environmental writer Bill McKibben surveyed the content of a day's worth of programming on more than 90 U.S. television channels and compared it with a day spent in the mountains. Television, he concluded, immersed him in an artificial, people-centric world, whereas his time in the wilderness taught him that human beings are not all-important.[85]

Information technologies not only shape our world view, they also give us greater power to change our world. We have a responsibility to harness these tools to build a healthier, greener, and more equitable future.

Chapter 8

Sizing Up Micropower

Seth Dunn and Christopher Flavin

When the first Olympic Games of the new millennium open in Sydney, Australia, in September 2000, one highlight of the environmentally conscious "Green Games" will be the Olympic Village, whose buildings will include rooftop solar cells that generate all the electricity they need. When the Games conclude, the village will become a 1,500-residence "solar suburb" that its designers calculate will eliminate 7,000 tons of carbon pollution annually that would have been produced by the coal-fired power plants that now provide most of Sydney's electricity.[1]

In a variation on the Olympic motto of "swifter, higher, stronger," the village's power system might be described as "cheaper, cleaner, more reliable." These tiny new household-scale generators represent a dramatic reversal of the bigger-is-better philosophy of electricity generation that has prevailed throughout most of the power industry's first century—culminating in the multibillion-dollar coal and nuclear behemoths that now provide most of the world's electricity. Today, a confluence of technological, political, and environmental forces is breathing new life into an old concept: decentralized, small-scale power.

During the next few years, manufacturers plan to introduce several new generating systems at scales that were unimaginable just a few years ago. Stirling engines, microturbines, fuel cells, and other devices can all be deployed in sizes that make them suitable for power generation in hotels, schools, hospitals, small businesses, and even homes. The smallest of these systems produces just 2 kilowatts of power—one five-hundred-thousandth of the power generated by an individual nuclear plant. With the average U.S. commercial customer using just 10 kilowatts of power and the average residential user just 1.5 kilowatts, the new generation of technologies is well matched to the market.[2]

E.F. Schumacher may not have had micropower specifically in mind when he argued that "small is beautiful," but today's small-scale electricity generators are without doubt clean, efficient, and economical. The new technologies emit levels of air pollution, including climate-altering carbon dioxide, that are 70–100 percent lower than conventional systems, in part because they are fueled by natural gas or renewable energy and in part because they are more efficient. And since they are deployed where the power is needed, the waste heat from micro generators can be captured for use— leading to total thermal efficiencies of 80–90 percent, compared with the 30 percent that is typical of today's centralized power system.[3]

The shift to small-scale power generation has other advantages. Power produced locally puts fewer demands on electric transmission systems, and provides added reliability when local power lines are cut due to severe weather or other problems. Most industries as well as vital institutions such as banks and hospitals are now highly automated, and the loss of power for even a few-hundredths of a second often leads to major economic losses. Local power also gives consumers who want it an added measure of independence from distant economic institutions.[4]

The advent of small-scale electricity may be most consequential in the developing world, where central power grids are notoriously vulnerable to frequent blackouts, where power plant and grid construction has saddled government treasuries with multibillion-dollar debts, and where 2 billion people still have no electricity at all. Micropower can reduce substantially the cost of providing basic services to a rapidly growing urban consumer base. And in rural areas, the new technologies may allow the development of stand-alone village power systems, doing away with the need for costly grid expansion.[5]

In a special August 1999 issue of *Business Week* entitled "21 Ideas for the 21st Century," local, "personal" power plants top the list. Indeed, the electric power industry is beginning the new century in a period of ferment not seen since its early days 100 years ago. Already, the flow of entrepreneurial enthusiasm and venture capital has attracted the interest of giant companies like BP Amoco and General Electric.[6]

Exactly where all of this is headed is uncertain. In the short term, the liberalization, restructuring, and reform of the electric power industry now under way in many nations will set the pace of change. The new rules of the road will determine how open the markets are to the new, small-scale competitors. The vested interests of the incumbent monopolies can be expected to resist change, but as with the restructuring of telecommunications, their resistance will almost certainly slacken over time.[7]

It is impossible to know whether two decades from now we will be getting 30, 50, or 80 percent of our electricity from small-scale power—or what the full implications of a new era of "energy independence" are. But one thing is certain: electricity's future is unlikely to resemble its recent past.

Miniaturized Machines

At the dawn of the electrical age in the late nineteenth century, Thomas Edison envisioned a decentralized system, with numerous small companies competing to install generators near the point of use. Since power distribution wires were still rare, the Edison Electric Illuminating Company found most of its early business installing generators in small, independent factories, department stores, hotels, and residences.

By 1888, Edison had installed 1,700 small-scale plants. And by the early twentieth century, more than half of U.S. electricity was generated by industrial facilities that produced their own power, harnessing the waste heat and selling any excess electricity to nearby customers.[8]

The first small-scale generators were noisy, smoky, and unreliable, however. These problems, combined with advances in steam turbines, transformers, and alternating current, led to the development of ever-larger, more economical central power stations, located further and further from the end-user. As demand soared and prices fell, the power business was recast as a large-scale "natural monopoly"—to avoid the wasteful duplication of equipment—and industries turned from on-site generation to purchasing power from electric utilities, which by 1970 supplied more than 90 percent of the world's electricity.[9]

Plant size "hit the wall" in the 1970s, however, as environmental concerns, energy crises, and multibillion-dollar meltdowns at massive nuclear power projects led many to question the wisdom of the centrally planned electricity paradigm. By the 1980s, the trend toward larger power stations reversed, as smaller, mass-produced gas turbines entered the market. The average size of a new utility power station in the United States fell from 600 megawatts in the mid-1980s to 100 megawatts in 1992 and 21 megawatts in 1998.[10]

Yet recent evidence suggests that the downsizing may have only just begun. The past five years have witnessed the emergence of a new generation of pint-sized power technologies that are a tiny fraction of the size of the large generators that are now the mainstay of the electric power industry. Reflecting advances in metallurgy, synthetic materials, electronics, and other fields, these devices span a wide array of technologies, ranging from improved internal combustion (IC) engines to generators relying on electrochemical and photoelectric processes. Although micropower is often still relatively costly compared with conventional generators, prices are likely to fall rapidly as the technologies mature and as mass production lowers costs further.[11]

Micropower technologies are defined here as less than 10 megawatts (or 10,000 kilowatts) in size. This is small enough to be connected to low-voltage local distribution systems, rather than requiring direct connection to the high-voltage transmission system, and to be installed in most commercial and residential buildings. Power systems of this scale are small enough to be factory-built in modular units and transported in one piece by truck or rail to the site where they are installed. Installation often takes a few hours or less. These characteristics contrast sharply with the massive power plants of the 1980s, unable to benefit from mass production and built on site over periods of up to a decade. (See Table 8–1.)[12]

The leading edge of the commercial micropower industry is the small "genset," using a diesel or spark-ignition reciprocating engine, that is closely related to the engines found in trucks and buses and that for decades has provided power for off-grid applications. Even as these traditional markets expand, a growing number of these systems are now being installed as "packaged" natural-gas-burning generators for commercial and even residential buildings. (See Table 8–2.)[13]

Since these engines are already mass-produced by scores of manufacturers, including well-known firms such as Caterpillar, Detroit Diesel, and MAN, they cost as little as $600 per kilowatt, and are backed by an extensive sales and maintenance infrastruc-

Table 8–1. Power Plant Downsizing

Plant Type	Typical Size (kilowatts)
Nuclear power station	1,000,000
Combined-cycle power station	250,000
Industrial cogeneration plant	50,000
Commercial gas turbine generator	10,000
Grid-connected wind turbine	1,000
Commercial microturbine or fuel cell	200
Residential fuel cell	10
Household solar panel	2

SOURCE: See endnote 12.

ture in some countries. The global diesel generator market more than doubled between 1990 and 1997, and by one estimate is as large as 35 gigawatts annually.[14]

The efficiency of these generators in producing electricity ranges from 20 to 45 percent, depending largely on the size of the plants, which range from 5 kilowatts to 10,000 kilowatts. In most of these packaged engine generator sets, attached heat exchangers capture the waste heat from the engines (known as cogeneration), which can be used for water or space heating or for industrial process heat, increasing the efficiency of the total system to 80 percent or more. Waukesha and Caterpillar, for example, offer 25-kilowatt generators that are suitable for small commercial applications such as fast-food restaurants. Some manufacturers have assembled units that use both electricity and waste heat to run cooling systems.[15]

On-site power systems relying on internal combustion engines and natural gas do produce local air pollution, principally emissions of nitrogen oxides, one of the main contributors to urban smog. To lower these to the stringent levels currently required in many urban areas, engineers have adapted "lean burn" engines and catalytic converters—widely used on automotive engines—to stationary use. While this increases the cost of the gensets, they are still competitive with traditional power plants. Noise, which is generally not a problem in commercial settings but remains a concern for some residential applications, can be suppressed by mufflers and soundproofing, as the Canadian firm CDH District Heating has done with two 2.5-megawatt engines.[16]

Servicing thousands of small, distributed power systems also raises issues not presented by large power plants, which can afford full-time maintenance personnel. Like car engines, generator engines must have their oil changed, their oil filters, air filters, and spark plugs replaced, and their valves adjusted as often as every 2,000 hours of operation. Most genset owners have maintenance contracts, and their systems are regularly serviced by professionals, just as central air conditioning systems are. Complete overhauls are required less frequently. Mid-sized commercial diesel engine gensets are estimated to cost 1.0–1.5¢ per kilowatt-hour for all maintenance and overhauls, representing as much as one third of the total cost of the electricity produced.[17]

The most immediate small-scale challenger to these conventional engines is the microturbine—a tiny jet engine that uses heat released by combustion to spin a single shaft that in turn spins a high-speed generator. Microturbines are derived from commercial jet engines and the gas turbines now dominating the power market. Although the most popular turbines for

Table 8–2. Combustion-Based Micropower Options

Characteristic	Reciprocating Engine	Microturbine Engine	Stirling Engine
Current Size Range (kilowatts)	5–10,000	30–200	0.1–100
Electrical Efficiency (percent)	20–45	30–38	20–36
Current Installation Cost (per kilowatt)	$600–1,000	$600–1,100	$1,500
Expected Installation Cost with Mass Production (per kilowatt)	<$500	$200–250	$200–300

SOURCE: See endnote 13.

utility use are 50 megawatts and above, mid-sized turbines for industrial use are now widely available in the 1- to 20-megawatt scale, from companies such as Solar Turbines. Proponents believe that even smaller microturbines could supply power for portable applications like laptop computers.[18]

Microturbines that generate 15–300 kilowatts are expected to be slightly more efficient than IC engines at comparable scales, and considerably more if waste heat is reused. Their chief advantage is their low cost: with just two moving parts, they are in principle easy to manufacture. They are also long-lived (perhaps 40,000 hours of operation), and unlike conventional engines do not require liquid lubricants or coolants, which simplifies operations and maintenance. They produce less nitrogen oxide pollution than IC engines, and have similar, manageable noise control problems. And they are adaptable to a wider array of fuels than conventional engines: natural gas, diesel fuel, kerosene, propane, and biogas can all be used in microturbines, making them viable even where piped gas is not available.[19]

Though the commercial market is most-

ly untested, Capstone Turbine has 200 orders for its 30-kilowatt unit and began offering a 75-kilowatt version—after testing at restaurants, factories, bakeries, and banks—in late 1999. Microturbines are especially well suited for small businesses, which use anywhere between 25 and 300 kilowatts of power. Capstone President Ake Almgren, who predicts a $1-billion microturbine industry in five years, estimates that at a volume of 100,000 units per year, 30-kilowatt turbines could cost $400 per kilowatt. Turbines generating 100 kilowatts would cost just over $200 per kilowatt—less than half as much as the most economical power plants being built today.[20]

Another "newcomer" to the small power market is Scottish engineer Robert Stirling's engine. Invented in 1816, and used extensively in the late nineteenth century, the Stirling is now being revived by new, efficient piston designs that reduce friction and wear. These pistons are driven by a gas that is heated by "external combustion"—a cycle that allows the engine to be very small and to accommodate most combustible materials, including agricultural and forestry residues. In one configuration, Stirling engines can be powered by heat generated by parabolic dishes that concentrate the sun's radiation. Stirling generators have achieved efficiencies of over 30 percent, and are relatively simple, quiet, and durable, with operating lifetimes projected at 30,000–60,000 hours.[21]

Because they are highly efficient at small scales, the early Stirling engines on the market are sized from 30 kilowatts to as small as several hundred watts. A number have been installed in remote regions, and several companies are beginning to market packaged systems suitable for home use, with competitive prices projected at $1,500. A Norwegian firm has demonstrated a 3-kilowatt Stirling cogeneration system with 95-percent efficiency. The Stirling Technology Company is developing a 1-kilowatt model, as well as off-grid systems ranging from 3 to 350 watts.[22]

Cool Power

The most revolutionary new micropower devices require no combustion and have no moving parts. The fuel cell, invented by the British physicist William R. Grove in 1839, is an electrochemical device that splits hydrogen into ions that either run along an electrode or combine with oxygen, producing electricity and water. Though deployed extensively in the U.S. space program, fuel cells have generally been considered far too expensive for terrestrial use.[23]

Most of today's fuel cells are hand-built by electrochemists, and require expensive, high-tech membranes and significant quantities of platinum to catalyze the reactions. But advances over the last 10 years have yielded designs with the potential for far lower costs at a wider range of scales and applications. Those attracting the most attention are phosphoric acid fuel cells (PAFCs), already available commercially, and proton exchange membrane (PEM) fuel cells, which several companies plan to market in the next few years.[24]

Fuel cells have several advantages over combustion generators. They are virtually soundless, making them ideal for use in buildings where noise is a problem—libraries, office buildings, hospitals. Where they use hydrogen fuel, the only byproduct is water. Most commercial fuel cells will initially derive their hydrogen from natural gas using a fuel processor, which produces some nitrogen oxides and carbon dioxide but at lower quantities than most engines. Some fuel cells can be built at sufficiently small sizes to power electronic devices such as laptop computers or cell phones. With no moving parts, they are more reliable than conventional power plants and require little maintenance.[25]

The central challenge with fuel cells is to lower their current high cost. The 200-kilowatt PAFCs on the market today are used in several hundred commercial buildings and as power boosters at electric utility "substations." At up to $3,500 per kilowatt, fuel cells are still in a demonstration phase, and can be justified economically only on the basis of a need for ultra-reliable power. (See Table 8–3.) Companies around the world are striving to reduce the cost of fuel cells by developing lower-cost materials and designs, and by figuring out ways to mass-produce the electrochemical devices. According to one estimate, some 85 organizations are now doing research on PEM fuel cells.[26]

The commercial prospect for fuel cells is being accelerated by some of the world's leading automakers. DaimlerChrysler, Toyota, Ford, and Volkswagen are among those that have proclaimed fuel cells as the probable successor to the internal combustion

Table 8–3. Non-Combustion-Based Micropower Options

Characteristic	Fuel Cell	Wind Turbine	Solar Cell
Current Size Range (kilowatts)	<1–10,000	<1–3,000	<1–1,000
Electrical Efficiency (percent)	35–50	—	—
Current Installation Cost (per kilowatt)	$2,000–3,500	$900–1,000	$5,000–10,000
Expected Installation Cost with Mass Production (per kilowatt)	$100–300	$500	$1,000–2,000

SOURCE: See endnote 26.

engine; at least four plan to have fuel cell cars in showrooms by 2004. Most automotive research focuses on the PEM fuel cell, led by Ballard, a small company in Vancouver, British Columbia. DaimlerChrysler, for example, has a $500-million joint venture with Ballard to develop fuel cell "engines" near its factories in Stuttgart. Larger systems are being developed for use in hydrogen-fueled city buses, which are being test-driven in Chicago and other cities.[27]

PEM fuel cells at the scale now being developed for transportation can be readily adapted to a range of commercial and residential uses. In fact, such fuel cells will become competitive for many power sector applications well before they get to the mass production target.[28]

Although stationary fuel cells will have to be more durable, and will require a fuel processor to derive hydrogen from natural gas, costs of less than $2,000 per kilowatt appear achievable within the next several years. Major power companies are actively pursuing the fuel cell power market, including General Electric in the United States, Alsthom in France, Ebara of Japan, and Siemens of Germany, some in partnership

with small firms like Ballard and Plug Power. Their current focus is the 100–300 kilowatt commercial market—particularly facilities that need electricity and either heating or cooling during a large part of the day. Larger facilities may simply use several of these modular units wired together in a utility room.[29]

The residential market for fuel cells is likely to emerge in small niches at first, and will open wider once prices fall below $500 per kilowatt. An experimental fuel cell the size of a dishwasher was installed in a model home in upstate New York in 1998.[30]

Equally radical are small-scale power technologies that are powered by sunshine and wind. Technical improvements over the past two decades have dramatically lowered the costs of wind turbines and solar cells, making them the world's fastest-growing energy sources in the 1990s.[31]

Wind turbines, a technology that emerged in modern form in the 1980s, generally consist of three-bladed mechanical devices that point into the wind. Although the early commercial market for wind power started with machines of less than 50 kilowatts, the scale of the machines has steadily increased, even as the rest of the power industry has downsized. The most popular commercial models are now 600–1,000 kilowatts, and several 2–3 megawatt models are on or nearing the market.[32]

Many grid-connected wind projects consist of large collections of turbines, called "wind farms," but in Germany and Denmark, the first and third largest users of wind power, most of the turbines are sited individually or in small clusters, connected directly to local distribution systems. The cost of such installations—which generally range in size from 600 kilowatts to 3 megawatts—is under $1,000 per kilowatt, making them competitive with traditional power plants in areas where winds are strong. Unlike thermal generators, distributed wind turbines are generally located on farms, and are often owned by a farmer or group of farmers. In addition, smaller wind turbines are widely used in rural areas to provide electricity, including in China, where more than 150,000 wind turbines are as small as 200 watts each.[33]

Although some regions in Denmark, Germany, and Spain already get 10–25 percent of their electricity from the wind, and although installations have grown by 22 percent annually since 1990, the global potential has barely begun to be tapped. Inland regions like the U.S. Great Plains and China's Inner Mongolia have wind resources sufficient to meet the entire electricity demand of their countries. Decentralized wind development can provide an important power source and revenue generator for rural areas. The amount of wind power developed in a given area will be determined by the capacity of the local distribution system and the availability of other generators to provide power when the wind is not blowing.[34]

The most versatile small-scale power option is the photovoltaic (PV) cell, based on semiconductor chips that convert sunlight directly into electricity. Surpassing even the fuel cell in modularity, solar cells have for nearly two decades been used in off-grid applications such as satellites, telecommunications, and handheld calculators. Solar cells are now entering the grid-connected micropower market, in the form of residential and commercial rooftop installations. These are typically 2–5 kilowatt systems, enough to provide half or more of the annual power of a residential building, with the rest coming from the grid.[35]

To drive down costs, several governments, including those of Germany, Japan, and the United States, have ambitious rooftop solar programs that provide financial and technical support for interested individuals and businesses. The most successful program so far is in Japan, where 25,000 homes now have solar cells on their rooftops. Japanese solar system owners sell power back to the utility at the same high price they pay for it, and also receive an upfront subsidy from the government.[36]

Government subsidies, market growth, and technological gains promise to deliver sharp cost reductions in coming years. The vibrant solar home market may motivate this transition, as improved solar-powered roofing shingles and window glass can be substituted for sometimes expensive building materials. Solar cells can also be deployed along highways, over parking lots, and atop municipal buildings and transit stations.[37]

Is Smaller Better?

For many people in the United States, the summer of 1999 tested their faith in an electric power system that most had taken for granted. In Chicago, rising demand for power from soaring temperatures and equipment failures cut off 2,300 businesses and the Chicago Board of Trade on a midweek afternoon. In New York City's Washington Heights, 200,000 homes were blacked out for 18 hours, and cancer and AIDS research experiments were ruined by the loss of refrigeration at Columbia Presbyterian Medical Center. New York City Mayor Rudy Giuliani denounced the city's power supplier and filed a lawsuit to recover the city's losses.[38]

The 1999 power failures exposed decades of underinvestment in local power distribution by utilities, a neglect that may be exacerbated by the cost-cutting now under way as the industry becomes more competitive. "The outages during the summer revealed very dramatically that we do indeed have substantial problems," according to David Helwig, a senior vice president at Commonwealth Edison in Chicago. "But the issue isn't about generating enough power, it's about not being able to deliver it."[39]

The weaknesses of local distribution systems point to the need to spend more on upgrading power lines and transformers, but also to the potential value of small-scale, distributed generators. By producing some power within the local system, small generators can reduce loads on distribution equipment. And in downtown centers where power lines are underground, where replacing them requires tearing up streets, adding small generators sometimes costs a fraction as much as upgrading the system.[40]

Just as aging equipment and strained transmission networks are casting doubt on the ability of large, centralized power systems to handle rising electricity demands and more extreme weather conditions, growing dependence on digital, computerized processes is heightening interest in high-quality, "non-interruptible" power. And as the information age creates a premium demand for power quality, new communications technologies are speeding the way to a more reliable and decentralized system.[41]

In many areas of the world, the main threat to the reliability of the electricity system is the disruption of local power supplies caused by weather-related damage to distribution lines or the overloading of lines due to excessive demand. Distribution system failures cause 95 percent of the electricity outages in the United States. In the developing world, these systems—if they exist at all—are even more brittle. While heat waves can cause power demand from air conditioning to overwhelm electricity distribution systems, weather-related disasters—such as floods, ice storms, or hurricanes—can cause widespread outages by knocking down lines. U.S. transmission and distribution expenditures have exceeded those for generation since 1994; $800 million to $2.5 billion of these expenditures could be profitably diverted to small-scale generators and improved energy efficiency, according to a report prepared for the Energy Foundation.[42]

Businesses affected by extreme weather events commonly cite electricity as the most important "lifeline" service—more crucial than telephones, natural gas, or water. Disrupted power can account for as much as 40 percent of the total insured losses claimed after a disaster. Increasingly, small-scale power is seen as a way to mitigate against these losses. While nearly 3 million people were left without power during a 1998 ice storm that hit New England, New York, and Quebec, those with wind power systems stayed up and running. One New York homeowner, using a wind turbine and battery, provided his community with hot showers and laundry services; the utility took 25 days to restore power to the county.[43]

The growing role of computers and the Internet throughout the economy and the rise of computerized manufacturing make users more vulnerable to even momentary voltage fluctuations or outages. In the past, such "glitches" were less important, causing lights and motors to dim or slow but not fail. But growing reliance on computers means a growing need for voltage stability: computer networks cannot withstand disruptions longer than eight-thousandths of a second, a period utilities do not even consider a failure. U.S. businesses lose some $26 billion annually from computer failures.[44]

If computers at the heart of the financial system shut down just for a brief moment, data can be lost and millions of dollars of transactions involving loans, credit cards, and automatic teller machines prevented. In 1997, the First National Bank of Omaha—which estimates a one-hour outage costs it $6 million—responded to a computer system crash that followed a power "flicker" by investing in four fuel cells, backed by IC engines and flywheels. The cells supply the majority of the data center's power, and the system's computer can cut itself from the grid at any sign of instability. The fuel cell power system has the side benefit of reducing carbon emissions by 45 percent and other air pollutants by 95 percent.[45]

In addition, supermarkets, restaurants, insurance companies, hospitals, and postal service offices are all beginning to look to micropower to avoid costly interruptions. The world's largest fuel cell system will provide backup power and heating for the U.S. Postal Service facility in Anchorage, Alaska; Harvard Medical School is exploring the use of fuel cells to power its teaching and laboratory sites. High-tech industries—pharmaceuticals, chemicals, semiconductors, and biotechnology—that increasingly rely on computerized manufacturing processes are another near-term application. Computer chip manufacturing plants, for example, can use fuel cells for reliable electricity as well as the byproduct of hot distilled water.[46]

Distribution system failures cause 95 percent of the electricity outages in the United States.

The other major advantage of small-scale power is its much smaller environmental footprint. Today, 62 percent of the world's electricity comes from fossil fuels—38 percent from coal alone—concentrated energy sources that can be economically burned in large-scale power plants. Although many of these plants have been fitted with costly emissions controls in recent decades, they still belch vast quantities of pollutants into the atmosphere. In the United States, electric generators are responsible for 28 percent of nitrogen oxides, 67 percent of sulfur dioxide, 36 percent of carbon dioxide, and 33 percent of mercury emitted nationwide each year. Globally, electricity generation is the largest single source of carbon emissions, accounting for more than one third of the total.[47]

The combination of higher efficiency—effectively using up to 90 percent of the energy in a fuel, as opposed to 30 percent in today's system—and the use of cleaner fuels allows micropower to vastly reduce the environmental burdens imposed by today's electric power system. Replacing all U.S. coal-fired power plants with microturbines, fuel cells, and renewables could cut national carbon emissions by more than 20 percent.[48]

Recent research by Amory Lovins and Andre Lehmann of the Rocky Mountain Institute (RMI) shows that deploying small-scale generators closer in size and location to customers has important advantages over giant plants, greatly increasing their value to the power system as a whole. This cost gap is large enough to make even today's pricey solar cells cost-effective in many applications. The RMI research identifies more than 70 ways in which small-scale, distributed power provides additional economic and environmental value. (See Table 8–4 for a synthesis of the major benefits.)[49]

For decades, electric utilities have insisted that the reliability of the electric grid depends on maintaining centralized control of the system. Allowing thousands of generators to be controlled by consumers was, they argued, a prescription for chaos. A growing number of analysts, however, now argue the contrary: a more decentralized, dispersed control may prove far more resilient than a centralized, hierarchical system. Such a system could evolve along the lines of resilient biological systems—such as ecosystems, or the human body—that decentralize control among numerous feedback loops rather than rely on a centralized hierarchy. Just as the brain does not need to track every bodily process for the system to function, power networks need not have a central point through which all information flows.[50]

Table 8–4. Seven Hidden Benefits of Micropower

Benefit	Description
Modularity	Micropower system size can be adjusted, by adding or removing units, to match demand
Short Lead Time	Small-scale power can be planned, sited, and built more quickly than larger systems, reducing the risks of overshooting demand, longer construction periods, and technological obsolescence
Fuel Diversity and Reduced Price Volatility	Micropower's more diverse, renewables-based mix of energy sources lessens exposure to fossil fuel price fluctuations
Reliability and Resilience	Numerous smaller plants are less likely to fail simultaneously, have shorter outages, are easier to repair, and are more geographically dispersed
Avoided Plant and Grid Construction, Losses, and Connections	Small-scale power can displace construction of new central plants, reduce grid losses, and delay or avoid adding new grid capacity or connections
Local and Community Control	Micropower provides local choice and control, and the option of relying on local fuels and spurring community economic development
Avoided Emissions and Environmental Impacts	Small-scale power generally emits lower amounts of particulates, sulfur and nitrogen oxides, and carbon dioxide, and has a lower cumulative environmental impact on land and water supply and quality

SOURCE: Amory B. Lovins and Andre Lehmann, *Small is Profitable: The Hidden Economic Benefits of Making Electrical Resources the Right Size* (Boulder, CO: Rocky Mountain Institute, forthcoming).

Innovations in power electronics, information technologies, and storage devices are making it easier to connect small-scale systems to the grid (or to allow them to remain off it), while improving the ability of the grid to adjust to "peak" demand and power fluctuations. The outcome may be a more "omni-directional" grid, in stark contrast to the traditional one-way street between central plants and end users. Like the Internet, such a power system would be "controlled" by thousands or even millions of businesses and individuals who follow a common set of protocols. (See Figures 8–1 and 8–2.)[51]

Advanced power electronics, including miniaturized chips, wires, and sensors, are creating new ways to connect, invert, and control power flows: microprocessor sensors, for example, can collect data on the electricity demands of automated building and industrial operations. Advances in inversion and conditioning equipment make it possible to invert direct current to alternating current (or vice versa) at a reasonable cost, synchronize small generators with the grid, and isolate them if the grid fails—keeping the system reliable and allowing utility workers to repair power lines and transformers safely.[52]

In the longer run, the confluence of communications and power technologies may lead to an "intelligent" distributed digital grid that will allow all parts of the electricity system to respond to the needs of the system as a whole—from central station and micropower plants to transformers, power lines, and even individual electric appliances. Controlled by computers, such a system could respond instantaneously to any problems and maximize its overall efficiency. For example, the owner of a refrigerator with appropriate communications and control equipment could, for a small reduction in his or her monthly bill, allow a utility to

shut down the refrigerator's electric motor for brief moments when demand on the overall network is high. This would save consumers money and improve reliability. Similarly, a price signal sent from a utility's computers to a consumer's fuel cell could trigger it to turn on when needed to supply the neighborhood's electricity.[53]

Distributed storage technologies may also aid micropower's entry into the grid, allowing intermittent sources to provide more-dependable electricity. Flywheels, batteries, and supercapacitors are among the devices being developed for storage purposes. A particularly promising option is to produce hydrogen from rooftop solar cells when the sun is shining, and then use that hydrogen in a fuel cell to provide electricity when electricity and heat are needed. Just as the internal combustion engine spurred demand for gasoline, the advent of automotive fuel cells may speed the emergence of a network for storing and transporting hydrogen, an idea that is attracting the interest of companies such as Royal Dutch Shell and DaimlerChrysler.[54]

In tandem with rising demand for reliability, these grid innovations are pushing power toward a more dispersed system—and along a historical continuum. During the twentieth century, the closest analogue of the electric power system was the Soviet structure of hierarchical central planning. At the beginning of the twenty-first, electricity is heading more in the direction of the Internet model: decentralized, nonhierarchical, and market-oriented.

Remaking Market Rules

Most of today's power markets still carry the legacy of the state-granted monopoly invented in the early twentieth century by

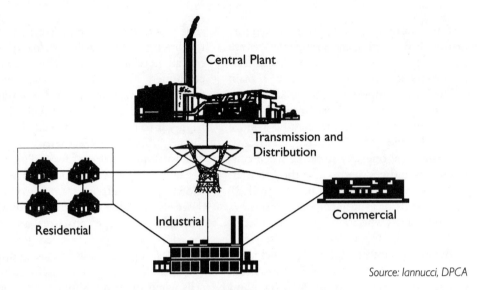

Source: Iannucci, DPCA

Figure 8–1. Example of a Centralized Power System

the American speculator Samuel Insull, who convinced lawmakers that consolidation was essential for the rapid spread of electricity to the public. Insull's sprawling empire of Chicago-based utility holdings—and his personal wealth that derived from it—collapsed under the Great Depression, but a slew of regulations reinforcing large, central-station power remain. As Walt Patterson of the Royal Institute for International Affairs puts it, "All too often… inherently decentralized technologies find themselves 'playing away,' on the home terrain of the centralized system and according to its rules."[55]

If micropower is to realize its potential, electric power regulatory systems will need to be reformed so as to remove the thicket of market barriers that currently limit its penetration. (See Table 8–5.) One step is for electricity policymakers to ensure that micropower generators are paid a price that reflects the full value of the electricity they provide. A 1998 European Commission study found that distributed solar power in

Italy has a total value of just over 10¢ per kilowatt-hour, half of which is the generation value, and half of which represents the added reliability and distribution system support that are unique to small-scale, locally based generators.[56]

Denmark, Germany, and Spain have introduced "in-feed" laws that currently require electric utilities to purchase renewable electricity at prices ranging from 7–10¢ per kilowatt-hour. These laws have led to the development of more than 5,000 megawatts of wind power since the mid-1990s. Similarly, Japan and 29 states in the United States now have "net metering" laws for solar power, allowing the solar system owners to sell excess power back to the grid at the retail price—as high as 23¢ per kilowatt-hour in Japan. They were enacted mainly for environmental reasons, rather than to elicit the broader benefits of distributed power. Nonetheless, they clearly show the potential effectiveness of simple, well-designed laws that give distributed generators access to the grid at a steady, fair price.

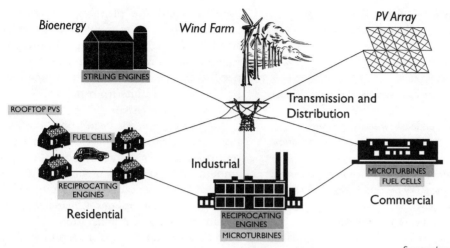

Bioenergy

Wind Farm

PV Array

STIRLING ENGINES

ROOFTOP PVS

FUEL CELLS

RECIPROCATING ENGINES

Residential

Transmission and Distribution

Industrial

RECIPROCATING ENGINES

MICROTURBINES

MICROTURBINES
FUEL CELLS

Commercial

Source: Iannucci, DPCA

Figure 8–2. Example of a Distributed Power System

Micropower tariffs on the order of 7¢ per kilowatt-hour could stimulate rapid growth in most countries.[57]

A June 1999 report prepared by the RAND Corporation for the Pew Center on Global Climate Change recommends that developing countries go beyond the traditional analysis of the "least cost" means of generating electricity, an approach that neglects the economics of the transmission and distribution system. The study recommends using a broader analysis of the system as a whole, an assessment that might well lead to increased investment in distribution infrastructure—and in micropower generators that can support that system. In competitive markets, assessments of delivery costs could be required in all bids for new power generation. According to the study, this strategy—together with the use of low-emissions technologies and improved efficiencies of electricity supply and demand—could meet growing power needs while cutting local air pollution and lowering carbon emissions by up to 42 per-

cent relative to business-as-usual trends.[58]

Another critical reform is to standardize requirements for safely connecting micropower systems to the grid. In many regions, electric utilities accustomed to large-scale generators impose a melange of complicated and unnecessary requirements that typically add thousands of dollars to the cost of systems priced only at $3,000–10,000 to begin with. These requirements stem from antiquated utility engineering standards, as well as the understandable desire of electric utilities to staunch the loss of customers for their centralized power. Efforts to streamline such requirements are under way in the United States, where the Underwriters Laboratory and the Institute of Electrical and Electronics Engineers are developing standards for safely connecting small solar systems to the electric grid.[59]

Additional policies are required to prevent utilities from unfairly blocking micropower development. Rules preventing access to the network must be rewritten, and utilities should be required to offer

Table 8–5. Eight Barriers to Micropower

- Higher initial capital costs
- Ownership rules
- Customers not rewarded for relieving peak load
- Impacts on local reliability ignored
- Unfair standby charges, exit fees, transition costs
- Burdensome interconnection requirements
- Discriminatory permitting, fire, building, and other codes
- Inequitable emissions policies

SOURCE: Joseph Iannucci, Distributed Utility Associates, "Distributed Generation: Barriers to Market Entry," Presentation to Board on Energy and Environmental Systems, National Research Council, Washington, DC, 6 May 1999.

straightforward "power purchase" contracts for micropower systems, rather than discouraging them with unnecessarily dense legal documents. Additional fees, used by utilities to penalize customers who reduce their purchases of grid power, need to be minimized. The state of Massachusetts, for example, has reduced "stranded cost" charges, which fund the retirement of uneconomic plants, for customers who use on-site systems.[60]

Other obstacles to micropower stem from siting, permitting, and emissions regulations that were designed before micropower became an option. Small-scale electricity is not accounted for in most building, electrical, and safety regulations, nor do local code and zoning officials tend to be familiar with the technology. U.S. homeowner associations concerned about lower property values often retain restrictions on modifications—such as solar roofing—well after developments have been finished. Land use planning and zoning laws favor the right to build over the "solar access" of neighboring property owners. Environmental regulations in many nations need to be revamped to credit the reduced pollution that results from deploying efficient small-scale systems.[61]

Monopoly utilities need some motivation to encourage small-scale power. Current rules compensate utilities based on the amount of capital equipment bought and installed or the amount of electricity delivered, giving these utilities a financial interest in blocking self-generation. Thomas Starrs and Howard Wenger recommend in a Renewable Energy Policy Project report that regulators create performance-based regulations that remove these disincentives, and offer local utilities an incentive to encourage distributed generation.[62]

Tom Casten, president of Trigen Energy Corporation, a leading U.S. builder of mid-sized industrial cogeneration plants using gas turbines, has identified a plethora of laws, regulations, policies, and practices that are based on the assumption that power plants are large and do not sell associated heat, penalizing small plants that combine the two. The tax depreciation life of a cogenerating gas turbine is 15–20 years, for example, even though a small, on-site plant lasts only 5–7 years. Casten recommends that such rules be removed or reformed, and supplemented with per-megawatt pollution and efficiency standards.[63]

Efforts to restructure power laws are generally dominated by incumbent monopolies keen on protecting past investments in large, centralized power plants. Progress in reforming those laws to make way for a new class of small-scale generators that barely exists in the marketplace today has been predictably slow. However, as the number of companies investing in these technologies grows and is augmented by major firms, and by millions of consumers who would like to own their own power systems, the tide will turn.[64]

Potential models are emerging in the United States, where the California Alliance for Distributed Energy Resources and the Distributed Power Coalition of America advocate a range of reforms in state and national regulations. These groups seek to create better conditions for micropower in a manner opposite to that of Samuel Insull—rather than one individual fighting for consolidation and monopoly, a political network pushing for decentralization and fair competition.[65]

How Far, How Fast?

Towering above New York City's Times Square, a new 48-story "green skyscraper" wears a skin of solar cells along its south and east sides. In the building's basement sit two 200-kilowatt fuel cells that supply both electricity and hot water. Incorporating additional measures such as efficient lighting and insulating windows, the "cool power" design is estimated to result in greenhouse gas emissions that are 40 percent lower than in a traditionally powered building.[66]

Micropower's high-profile appearance in one of the world's leading financial centers—just blocks from Thomas Edison's tiny 1882 power station at Pearl Street—is a homecoming of sorts. Still, many analysts remain skeptical that mass-produced, small-scale generators can provide a major alternative to large, central power plants. Similar doubts were observed among railroad owners confronted with the early automobiles, and among mainframe computer manufacturers facing the first personal computers. It

remains to be seen whether industry leaders and policymakers will learn from these lessons of economic history.[67]

One argument leveled against small-scale power is that such tiny devices cannot possibly make a major contribution to the electricity supply of a modern economy. The weakness of this argument is revealed by a simple calculation. If the average capacity of an automobile power system is taken to be roughly 100 kilowatts, the U.S. auto industry's annual production of 13 million new cars and light trucks is equivalent to 1,300 gigawatts of generating capacity. The capacity of all the U.S. power plants in operation in 1998, by comparison, was just 778 gigawatts. In its total car and truck fleet, the United States already has more than 200 million reliable, self-generating power plants—each with a per-kilowatt capital cost one tenth that of a large, central generator.[68]

Recent market assessments suggest that a boom in small-scale power is on the way. A July 1999 study by the Business Communications Company concludes that small-scale power will account for "a significant portion" of the 200 gigawatts of new capacity added by 2003 worldwide. Estimating U.S. sales of power equipment between 1 kilowatt and 5 megawatts at $4.2 billion in 1998, the study projects the U.S. market to grow at an annual rate of 32 percent in the next several years, surpassing $16 billion in 2003. The fuel cell market is projected to grow from $305 million in 1998 to $1.1 billion in 2003, with the microturbine market exploding from virtually zero in 1998 to $8.5 billion in 2003—reaching almost half the U.S. small-scale power sector.[69]

Looking further ahead, projections of micropower's share of new U.S. generation in 2010 range from 5 to 40 percent; one study predicts renewably fueled fuel cells will be a $10-billion global market by then.

But the true potential may be much greater. Small-scale systems could displace significant amounts of existing central generation in the industrial world, and avoid the same in developing countries. With the right policies, micropower could well dominate power markets worldwide within the next 5–10 years. In the long run, it is not inconceivable that over half of societies' power would come from small-scale local systems, with the rest coming from larger renewable energy plants such as wind and solar farms—making central thermal plants no longer necessary.[70]

Joseph Iannucci of Distributed Utility Associates—who has logged nearly 300 small-scale power consultations in the United States since 1990—identifies 10 "market accelerators" for micropower. (See Table 8–6.) He concludes that if electric utilities do not take the lead in promoting distributed power, then customers—supported by aggressive new companies—will. The near-term growth of micropower will be most affected by the direction of electric industry restructuring, the pace of decommissioning of nuclear and coal-fired power plants, and new environmental regulations. Some rules in the United States and Europe are being rewritten with small systems in mind, but much remains to be done.[71]

As the grid market for micropower unfolds, an even more important role for the technologies is likely to open in rural areas of the developing world for the 2 billion people lacking electrical services. In many cases, small-scale renewables already compare favorably—in terms of technical performance, economic competitiveness, and reliability—with extending transmission lines to unserved areas. Microturbines and Stirling engines burning locally available biofuels could be used to provide power for isolated villages, as could wind

Table 8–6. Ten Micropower "Market Accelerators"

- Simplified interconnection standards
- Modest or unpredictable growth in electricity demand
- Aggressive new market entrants, such as natural gas and energy service suppliers
- More-efficient electricity pricing schemes
- Saturation of electric transmission and distribution systems
- Siting difficulties for new central generation plants and transmission and distribution capacity
- Streamlined, standardized permitting procedures
- Dissatisfied electricity customers
- Technology advances
- Demand for green energy

SOURCE: Joseph Iannucci, "Overview on Distributed Generation," Presentation to Windpower '99 Conference, American Wind Energy Association, Burlington, VT, 22 June 1999.

turbines or solar cells, backed by fuel cells that run on hydrogen produced by the renewable generators. Trends in some countries suggest developing nations may set the pace in adopting decentralized, small-scale electricity systems in off-grid areas, just as some have "leapfrogged" stationary phones to pocket beepers and mobile phones. (See Table 8–7.)[72]

Small-scale power for remote areas is ideal for critical tasks, such as ice-making, water desalination, purification, and pumping, and for operating rural schools, health clinics, and microenterprises. Although small diesel gensets are widely used to provide off-grid power today, their need for maintenance and fuel that is often difficult to obtain suggests that the new micropower technologies would be far more appealing.[73]

Institutional support is also key. Daniel

Table 8–7. Small-Scale Power Applications, Developing Countries

Country	Application
China	More than 150,000 small-scale wind turbines (200–1,000 watts) operating; over 500,000 people served (one third of non-grid-connected population)
Dominican Republic	More than 9,000 solar home systems installed to date
India	An estimated 1,025 megawatts of wind power capacity installed
Indonesia	Over 20,000 solar home systems installed to date
Kenya	More than 80,000 PV systems installed (over 200,000 people served)
Mexico	More than 40,000 solar home systems installed to date
South Africa	An estimated 50,000 PV systems in place, including 1,300 rural schools, 400 rural health clinics, and 2,000–4,000 water pumping systems
Zimbabwe	Approximately 9,000 PV systems installed in rural homes, schools, and clinics; 70 local design and installation companies in business.

SOURCE: See endnote 72.

Kammen of the University of California at Berkeley argues that small-scale, decentralized power systems could make it far easier for developing countries to expand the availability of power without further environmental deterioration. In the past, however, small-scale power has suffered from institutional neglect, and in many cases, active discrimination. Kammen concludes that even modest increases in investments can yield disproportionately large returns in developing countries—bringing cleaner energy at a lower cost.[74]

Both on and off the grid, broader understanding of micropower's benefits will be critical for realizing its potential. List-serves on the Internet, Web sites, and conferences related to small-scale power are multiplying, as is coverage in the mainstream media. The U.S. National Renewable Energy Laboratory has an extensive database of past projects and an on-line discussion group, and cosponsors Village Power conferences with the World Bank. Also growing is willingness among consumers in Australia, Europe, and the United States to pay small "green power" premiums to support clean energy investments or installations. In California, the Sacramento Municipal Utility District—an early promoter of distributed resources—has mounted more than 450 solar panels on the roofs of "PV Pioneers," financed by modest increases in the customers' monthly electricity bills.[75]

The financing of on-grid micropower systems is not likely to pose a major problem once the technologies mature. The loan for a building-based power plant can simply be built into the commercial loan or mortgage used to finance the building. In fact, developing countries could substantially reduce the financial burden entailed in building up their electricity systems by passing some of the burden to the cash-flush developers of new hotels, office buildings, and factories.[76]

Off-grid micropower presents more difficult financing hurdles. People in rural areas need affordable, small-scale financing to help cover the high upfront expenses that make purchasing or leasing a system prohibitive for most. Innovative "microfinance" strategies are emerging in Bolivia, the Dominican Republic, India, South Africa, and Bangladesh—where the Grameen Shakti, an affiliate of the Grameen Bank, for more than 20 years a leading provider of small amounts of credit to the poor, is bringing solar and wind systems to rural customers via small loans. Funders such as the World Bank, the Inter-American Development Bank, and U.S. foundations are working to set up microenterprises that distribute, fund, and install solar systems in developing nations. Insurers like Swiss Re—for whom micropower is a "risk management" strategy against power losses and weather-related damages—are exploring low-interest, long-term loans for home solar and wind systems. Governments and nongovernmental groups in the United States and Europe publish guides for homeowners and development workers on buying or financing solar and other small-scale systems.[77]

The move to micropower may accelerate the evolution to a carbon-free hydrogen economy.

Until the late 1980s, solar electrification in Kenya was confined to a handful of donor projects; the national Rural Electrification Program had connected less than 2 percent of rural households to the power grid. But falling PV costs, and the efforts of private and volunteer organizations to provide information and training, have fostered a vibrant commercial market with dozens of homegrown assembly, sales, installation, and maintenance companies. Some 80,000 systems have been sold—the largest PV penetration rate in the world—providing power to more than 200,000 Kenyans; sales are increasing 10–18 percent annually.[78]

From Kenya's experience, observers have drawn several broader lessons on how to spread small-scale power. Training and support are crucial, but can be met by a small group of organizations and individuals. Subsidies can help, but may not be as necessary as other steps to promote technology transfer, such as logistics and performance standards. Attracting a range of commercial interests is of vital importance in sustaining the emerging technology's market. And government and international policies can be valuable, particularly when they support independent power producers, lower or remove taxes and tariffs on cleaner technologies, and offer affordable credit.[79]

Greater international cooperation could help spur small-scale electricity. Climate change commitments might accelerate the adoption of micropower in a growing number of countries. Other objectives—technological leadership, reduced oil import dependence, lower air pollution, lessened nuclear safety risks, new export markets—may also motivate support for micropower. Citing a projected $10-trillion market for worldwide energy supply in the next 20 years, a 1999 White House report urges the government and industry to step up investment and forge international alliances in clean energy. The California-based Nautilus Institute has already installed three small wind turbines in power-deprived, famine-stricken rural North Korea, a project that is being assisted by the U.S. government in the hope of lessening bilateral tensions over nuclear proliferation.[80]

The pace of micropower development

may in turn shape global developments. Bringing power back to the people could facilitate decentralizing tendencies within societies, empowering local communities and businesses. It may also help minimize the kind of rampant corruption in power contracts, such as those between western vendors and Indonesian power companies that were revealed in the wake of that country's financial collapse. And the move to micropower may accelerate the evolution to a carbon-free hydrogen economy, as proliferating fuel cells create a growing demand for hydrogen as an energy carrier. Governments, businesses, and citizens concerned about climate change share an interest in seeing this shift unfold.[81]

Micropower has many of the characteristics of what Joseph Bower and Clayton Christensen of the Harvard Business School have called "disruptive technologies." The term refers to new technologies such as small copiers or personal computers that carry important new attributes that customers do not value at first. Initially neglected by the companies that dominate a given market, and that are heavily invested in the existing infrastructure, the invading technologies are often spearheaded by hungry newcomers who gain early footholds and grow at steep trajectories, eventually pushing aside the market leaders.[82]

The key question regarding disruptive technologies is not whether they can outperform existing technologies according to conventional criteria, but whether they can meet future market needs more effectively. Market research may grossly underestimate the potential. Polaroid's president was informed by his marketers that only 100,000 of his instant cameras would be sold over their lifetime, as few foresaw the uses of instant photography. Likewise, most utility managers have difficulty envisioning a large market for small-scale power. But this is changing quickly, as environmental and economic forces reveal their inherent advantages—and the bigger firms scramble to keep up with their nimble competitors.[83]

As athletes "go for the gold" at the 2000 Games in Sydney, the shimmering solar cells of the Olympic Village symbolize another competition that will occupy our global village in the new millennium. The test is to see how quickly we can leave behind the large-scale, centralized electricity model as a distinctly twentieth-century relic, and "go for" a small-scale, decentralized system that is cheaper, cleaner, and more reliable, and that serves many more of the planet's people. Should this race continue to pick up speed, micropower might eventually become a household expression—in more ways than one.

Chapter 9

Creating Jobs, Preserving the Environment

Michael Renner

From November 1811 to January 1813, organized bands of English artisans in the wool and textile trades attacked the shearing frames, spinning jennies, gig mills, and other pieces of textile-manufacturing machinery that were beginning to replace them in the early stages of the Industrial Revolution. Acting in the name of "General Ned Ludd"—by all accounts an imaginary, yet immensely inspiring, figure—the "Luddites" conducted nighttime raids, ransacked factories, and even committed arson and burglary in their desperate struggle against the onset of an industrialism that they saw as a mortal threat to their livelihoods and the world they knew: craft, custom, and community.[1]

Kirkpatrick Sale, who has chronicled the Luddite struggle, observes that "by 1813 there were an estimated 2,400 textile looms operating by steam, but that burgeoned to 14,150 by 1820 and exploded to more than 100,000 just a decade later." Age-old crafts were obliterated, skills rendered obsolete, communities torn asunder. Some 100,000 handloom weavers in the cotton industry were impoverished. As the Industrial Revolution gathered force, hundreds of thousands of self-employed people all over Britain were driven out of work. Suddenly in need of money to secure food and other basic needs, they had no choice but to depend on wage labor. Factory towns emerged that became infamous for their reckless pollution of air and water and their inhumane treatment of workers who received no more than poverty wages. Despite deplorable conditions in factories, laws passed in 1799 and 1800 made it illegal for workers to organize.[2]

In support of the budding manufacturers, the British government unleashed the most severe episode of repression in the country's memory. Because recorded history is usually about the winners, the term Luddite entered the modern vocabulary with thoroughly negative connotations. Although Luddites are now often derided as "opposed to technological change," the machine breakers of early nineteenth-century England were not hostile to technology per se. Rather, they protested against "machinery hurtful to Commonality," as a contemporary letter put it: an onrush of technology beyond their control, introduced without regard to the social dislocation and environmental destruction it caused.[3]

The particular technologies that triggered the Luddite movement may seem quaint today, but the issues remain salient. The economic system that was first brought into existence by the Industrial Revolution embodies a turbulence—a constant changing and churning—that, even as it propels society forward, inserts a degree of uncertainty into people's jobs and lives. In today's Europe, as elsewhere in the industrial world, more than 10 percent of all jobs vanish each year, replaced by different jobs in new occupations and sometimes in new companies.[4]

This turbulence is amplified, and more socially disruptive, in times of fundamental structural change. New technologies and industries arise and old ones wither away; some regions prosper and others become rustbelts; new jobs and skills emerge and others fall victim to the march of technology. In a transformation perhaps as momentous as the one 200 years ago, the rise of an "information economy" and the trend toward economic globalization are today spawning considerable anxiety about job security, skills obsolescence, and wage trends.

But even as computers, the Internet, and associated technologies remake the economy, another challenge beckons: the growing urgency to move toward a sustainable economy. With the spreading industrial system, environmental concerns grew from localized (though severe) pollution in early-industrial Britain to planetary-scale degradation and alteration of natural systems. The implications for a more sustainable

development path are clear: we need to reduce the reliance on fossil fuels dramatically, curtail mining and logging of virgin areas, restructure the transportation and utility sectors, and alter industrial processes to minimize waste generation. The fear is that such measures will cause grave economic disruptions and massive job loss, and such worries are eagerly cultivated by industries resistant to change.

In today's Europe, more than 10 percent of all jobs vanish each year, replaced by different jobs.

Does sustainability need to be synonymous with economic insecurity? Primarily for reasons other than the pursuit of sustainability, logging, mining, and heavy industries like iron and steel are already far less important today than they once were for economic growth, and even less so for employment. For a long time, it was an article of faith among economists that energy and materials consumption move in lockstep with the gross national product, meaning that reduced resource use equaled lower growth and less employment. But this direct link has already been severed, and even greater resource efficiency will make it possible to produce more goods and services while reducing the burden on the natural world. The rapid rise of information and communications technologies could help forge a more sustainable economy. (See Chapter 8.)[5]

Energy and materials efficiency, renewable energy, recycling, waste avoidance, and "clean production" methods offer substantial employment opportunities—typically more jobs than in traditional industries. In place of the present resource-intensive, high-throughput economy, a sustainable economy will manufacture products so that they are energy- and materials-efficient, are durable, upgradable, and repairable, and can be remanufactured when their useful life comes to an end. All these characteristics promise new kinds of jobs. A sustainable economy will also emphasize the intelligent use of products rather than mere ownership; services built around the idea of extending the life of a product and maximizing its utility will offer additional job opportunities.

A new economy that provides sufficient employment without exacting massive environmental damage is possible. However, as with any fundamental economic transformation, there are transition costs. The lessons of the Luddite resistance will need to be heeded: people caught in the maelstrom of change will resist if they don't see a future for themselves. Affected workers, communities, and regions—particularly those dependent on resource extraction—will need help to master new skills, technologies, and industries. Creative policies are required to boost job creation, enhance workers' education and update their skills, and smooth the transition process. A new economy will not be viable unless it is both environmentally and socially sustainable.

The World of Work

In modern economies, wage employment is the primary source of income for most people. The world's labor force—those employed or available for work—has grown from 1.2 billion people in 1950 to an estimated 2.9 billion in 1998. And because of strong population growth, it will continue to swell: during the next half-century, the world will need to create nearly 30 million additional jobs each year. If past is pro-

logue, then there is reason for concern: Employment has expanded, but less so than the working-age population. Unemployment and uncertainty about future prospects allow those opposed to strong environmental policies to play on workers' fears about their jobs.[6]

Worldwide, at least 150 million people were unemployed at the end of 1998. Long-term structural unemployment—that is, joblessness that will not be easily reduced merely by cyclical upswings in the economy—accounts for a significant portion of the total. In western industrial countries, more than a quarter of the unemployed in 1997 had been jobless for a year or longer. In addition, as many as 900 million people are "underemployed"—involuntarily working substantially less than full-time, or working full-time but earning less than a living wage.[7]

Then there are the "discouraged" workers, who have given up hope of finding a job; since they no longer actively seek work, they are usually not even counted as unemployed. In the European Union (EU), for instance, some 18 million persons are officially unemployed, but at least another 9 million are "discouraged." And yet, even as millions are out of work, many workers end up putting in large amounts of overtime— the equivalent of at least 2 million full-time jobs in the EU.[8]

What economist Joseph Schumpeter dubbed the "creative destruction" of capitalist economies implies an ever-present changing and churning that, while creating greater wealth, at least as traditionally measured, is also bound to infuse labor markets with great uncertainty and all the attendant social and psychological impacts on individuals, families, and communities.[9]

At the turn of the millennium, the nature of work is changing dramatically, perhaps on a scale comparable only to the Industrial Revolution. Increasing international trade and investment and a new wave of automation are reshaping virtually every kind of human economic activity and speeding up the pace of change. (See also Chapter 10.) Conventional economics praises the process in which mature industries shed jobs and new industries emerge to provide employment. But it is unclear whether computers and microelectronics will render jobs more interesting or more stressful, whether they will entail jobs that are mostly routine tasks instead of requiring problem-solving skills that stimulate human creativity, and whether they will lead to a growing polarization of the work force between well-paid and poorly paid employees. Just like the earlier revolution, this transformation brings with it a de-skilling process, as existing abilities, expertise, and proficiencies lose in value and importance, and new skills and requirements rise in response.[10]

Technological development and increased capital mobility—the flow of money, technology, and machinery across borders—allow growing numbers of companies to embrace measures such as temporary or part-time hiring, parceling out components of the work process (subcontracting and "outsourcing") and tapping into a large pool of cheap labor in developing countries to either supplement or replace higher-paid workers in industrial countries. Products are now routinely made from components produced in far-flung places around the planet. Although these measures allow companies to be highly flexible and adapt rapidly to fast-changing market conditions, they also make job security more tenuous and weaken the bargaining power of labor unions.[11]

Disparities between skilled workers and those lacking skills or possessing outdated ones grow more noticeable. Manufacturing employment in western industrial countries stayed roughly even for skilled workers between 1970 and 1994, but declined 20 percent for unskilled workers. Likewise, the gap between those with full-time jobs and those working involuntarily in temporary or part-time jobs is becoming more prominent. Under the right circumstances, part-time work can be part of the solution to the employment and social challenges of our time. For the moment, however, it means mostly jobs with low pay and few benefits, limited career prospects, and no assurance that the position will still be available next week or next month. In Britain, part-time employment accounted for 15 percent of all jobs in 1971 but 25 percent in 1997. In Germany, 15 percent of employees were in "insecure" jobs (defined as part-time, temporary, or insufficient work) in 1970; by 1995, the figure had risen to 30 percent.[12]

The nature of work is changing dramatically, perhaps on a scale comparable only to the Industrial Revolution.

If current trends continue, the work force will become more polarized. A relatively small group of employees may emerge as "winners"—highly skilled, with secure, well-paid jobs, and more likely than not working substantial overtime in high-stress conditions—whereas many workers will probably confront episodes of unemployment or have to accept irregular, less secure work arrangements. The real losers may face more or less permanent exclusion from gainful employment because their skills, age, or other attributes are judged as inadequate or unneeded in a fast-paced, merciless labor market. In developing countries, too, growing disparities are likely. Employees in small high-tech enclaves such as one in Bangalore, India, are likely to benefit from world market integration. Free trade zones like Mexico's *maquiladoras* attract foreign investment and jobs, though wages and working conditions are often little better than in English factories during the early stages of the Industrial Revolution.[13]

Since the 1970s, unemployment in advanced industrial countries has been on the upswing. In Western Europe, it climbed from a little over 2 percent in 1970 to about 12 percent in the late 1990s; in Japan, from about 1 percent to above 4 percent. By contrast, after it rose from 5 percent in 1970 to above 10 percent in the early 1980s, the unemployment rate in the United States is now back to slightly below the 1970 level. (See Table 9–1.)[14]

But higher job creation has come at a

Table 9–1. Unemployment Rates by Region and Selected Countries, 1987 and 1997

Region or Country[1]	1987	1997
	(percent)	
Europe	10.4	10.5
Japan	2.8	3.4
United States	6.2	4.9
Latin America and Caribbean	5.7[2]	7.4
China	2.0	3.0[3]
India	3.4	2.3[4]
Other Asian countries	4.3[2]	4.2[3]
Central and Eastern Europe	7.2[4]	9.6[3]

[1]No comprehensive data for Africa are available.
[2]1990. [3]1996. [4]1993.
SOURCE: International Labour Organization, *World Employment Report 1998–99* (Geneva, 1998).

cost: Almost 28 percent of all U.S. workers now have jobs that pay wages at or below the official poverty level. From a peak in 1978, real hourly earnings for all production and nonsupervisory workers outside agriculture declined 9 percent by 1997. On average, U.S. wages are lower, and unemployment benefits and the social safety network far less generous, than in most other industrial nations. In 1997, manufacturing production workers on average received $18.24 an hour in the United States, $19.37 in Japan, $20.24 in the European Union, and $28.28 in Germany.[15]

Employment concerns are also high on the agenda in other parts of the world, as countries attempt to navigate the treacherous terrains of economic transition and developmental catching-up. Since the end of the cold war, most states in Eastern Europe and the former Soviet Union have seen a rapid rise in unemployment, from near zero to close to 10 percent. Joblessness has been accompanied by lower real wages and dramatic increases in income inequality. In Russia, where the economy has severely contracted, real wages plummeted by 58 percent between 1989 and 1996, and people often receive their wages months behind schedule.[16]

The East Asian economic crisis that broke out in 1997 has added at least 10 million people to the world's unemployment rolls and thrown substantial portions of the population there into renewed poverty, as unemployment benefits and other protective measures are sparse. Several other Asian countries—Cambodia, China, Laos, Mongolia, and Viet Nam— face serious labor market problems resulting from the excess labor in state and collective enterprises. In China, layoffs in 1998 alone affected 3.5 million workers, bringing the official unemployment rate to

5–6 percent. Perhaps as many as 30 million more workers are going to lose their jobs.[17]

In Latin America, the International Labour Organization (ILO) projected unemployment in the formal sector to rise from about 6 percent in the early 1990s to 9.5 percent in 1999, despite an upturn in the region's macroeconomic performance. Real wages have stagnated and minimum wages have, on average, fallen 27 percent since 1980.[18]

Since job creation in the formal sector is limited in many developing countries, much employment takes place in the informal sector. According to the ILO, this accounted for more than 60 percent of sub-Saharan Africa's urban work force in 1990 and 58 percent of Latin America's. The informal sector is an amalgam of economic activities, including family enterprises, that are not captured by traditional categories. It generates demand for semiskilled and unskilled labor, is more likely to adopt appropriate technologies and local resources, plays an important role in recycling and reusing waste materials, and provides a major source of income for women. But working conditions are frequently poor, social security is mostly nonexistent, and wages are often very low, typically below the official minimum wage.[19]

One of the most unsettling aspects of the jobs crisis is large-scale youth unemployment, which virtually everywhere is substantially higher than for the labor force as a whole. The ILO estimates that there are about 60 million people worldwide between the ages of 15 and 24 who are in search of work but cannot find it. In developing countries, high rates of population growth translate into massive pressure on job markets. In China, 26 percent of the population is 15 or younger; in the rest of Asia, the figure is 35 percent; in Latin America, 33 per-

cent; and in Africa, 43 percent.[20]

Although developing countries clearly face a growing jobs challenge, labor market data are relatively scarce; scarcer still are studies addressing the employment-environment linkage in the developing world. This chapter focuses on industrial countries. But developing countries must inevitably grapple with similar issues. If anything, they face a challenge of even greater magnitude. They need to find work for growing numbers of young people entering the job market, agriculture is still the most common occupation, and large-scale rural-urban migration is placing increasing burdens on urban job markets. Hence there is an urgent need for sustainable agricultural and rural industry jobs in order to lessen the pressure. The challenge for developing countries is to not follow the siren song of unsustainable development—pollute first, clean up later—but to exploit opportunities to leapfrog to sustainable technologies and to develop labor-intensive industries.

Boosting Resource Productivity

Ever since the beginnings of the Industrial Revolution, businesses have sought to economize on their use of labor, whereas land and natural resources were seen as boundless and cheap. While companies have emphasized raising labor productivity—using fewer workers for each car, refrigerator, or computer, they have largely neglected the issues of energy and materials productivity—using less oil, electricity, aluminum, and copper for each unit of output.

This may once have made perfect sense, when skilled labor was indeed scarce and when substituting machines for humans promised rapid economic progress. But today, given the environmental crisis and the growing abundance of human labor, particularly in developing countries, it is time to reevaluate these priorities. Not only is nature scarce today, there is no substitute for it once it is depleted: no matter what technologies human ingenuity dreams up, pure air and water, intact forests and fisheries, and a stable global climate are irreplaceable.

As labor productivity grows, output and consumption must grow at least as fast in order to maintain steady employment levels, and faster if the number of jobs is to expand. But as long as economic growth is predicated on burning large quantities of fossil fuels, using copious amounts of materials, and generating huge waste flows, this is a formula for growing environmental degradation. A sustainable economy must break the work-consumption-environmental degradation connection.

Capital, energy, and materials are steadily replacing labor. Although this section relies primarily on U.S. data for illustration purposes, the same holds true for other industrial countries and for newly industrializing nations as well. Total output by the U.S. manufacturing sector rose by about 440 percent between 1950 and 1996. Labor inputs (measured by the total number of hours worked) increased by about 40 percent between 1950 and 1969, then remained stagnant for a decade before beginning to decline slightly. By contrast, inputs of capital—that is, buildings and equipment—jumped by 525 percent during these 47 years, energy inputs rose by 369 percent, and materials inputs by 335 percent.[21]

As a result, the productivities of these individual inputs—the output of manufactured items for each unit

of input—diverged dramatically. Labor productivity more than tripled. In the auto industry, for instance, this means that where a worker was once able to produce one car in a given stretch of time, one individual can now produce three cars in the same period. By contrast, capital productivity has declined almost throughout the entire postwar period, so that it takes a growing investment in buildings and equipment to produce a dollar's worth of manufactured goods. Until rising oil prices in the 1970s forced the development of more-efficient motors, lighting, and production processes, energy productivity also declined: growing amounts of oil, gas, coal, and electricity were needed to produce a dollar's worth of output. By the mid-1990s energy productivity was only marginally higher than in 1950. Materials productivity rose until the early 1970s, but then lost some ground. (See Figure 9–1.)[22]

Energy and materials productivities could be boosted substantially, since more resource-saving technologies are already available, even better ones are on the drawing boards, and opportunities for redesigning whole systems as opposed to individual

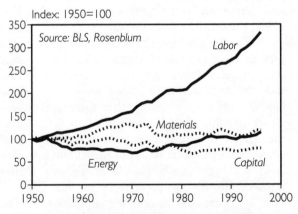

Index: 1950=100

Figure 9–1. Selected Factor Productivities in U.S. Manufacturing, 1950–96

products remain largely unexplored. For instance, eco-business pioneer Paul Hawken and Amory and Hunter Lovins of the Rocky Mountain Institute argue that just using existing technologies (including advanced polymer composites, better aerodynamic design, and fuel cells) can reduce new-car fuel consumption by as much as 85 percent, and slash materials use in car manufacturing up to 90 percent (by weight). What this means is that manufacturing industries now use far more energy and materials, and cause greater environmental damage, than need be.[23]

Just a handful of industries are responsible today for the bulk of environmentally damaging activities while at the same time providing only limited (though often well-paid) employment. Four U.S. manufacturing industries—primary metals, paper, oil refining, and chemicals—accounted for 21 percent of manufacturing value-added in the mid-1990s. They absorbed 78 percent of primary energy use in all U.S. manufacturing and were responsible for 64 percent of the amount of toxics released from manufacturing operations, but represented only 12 percent of all jobs and about 14 percent of total hours worked and payroll.

(See Table 9–2.) Outside the manufacturing sector, mining and utilities share these attributes: few jobs but substantial environmental impact. Thus, change in the environmentally most damaging sectors of the economy will affect only limited numbers of workers.[24]

Growing labor productivity explains why manufacturing employment in industrial countries has either stayed flat or even declined, although output almost doubled. In fact, relative to output, manufacturing employment has declined almost sevenfold since 1960 in Japan, 4.5-fold in France, and threefold in Germany and Britain.[25]

Employment is increasingly shifting into the "service" sector. (See Table 9–3.) Overall, services employment has roughly doubled in western industrial countries, and almost quadrupled in the United States since 1950. (See Figure 9–2.) For every manufacturing job, there are now almost five service jobs in the United States; three to four in Japan, France, and the United Kingdom; and more than two in Germany.[26]

Both in employment and environment terms, however, the shift toward services is an ambiguous development. The term "services" encompasses vastly disparate

Table 9–2. Value-Added, Employment, Energy Use, and Toxics Releases, Selected U.S. Manufacturing Industries, Mid-1990s

Industry	Value-Added	Number of Jobs	Hours Worked	Payroll	Energy Use	Toxics Released
	(percent within all manufacturing industries)					
Paper	4	3	4	4	12	11
Chemicals	11	4	4	6	25	36
Oil Refining	2	1	1	1	29	3
Primary Metals	4	4	5	4	11	15
Total	21	12	14	14	78	64

SOURCE: Worldwatch calculation, based on sources cited in endnote 24.

Table 9–3. Total Labor Force, Industrial and Developing Countries, by Economic Sector, 1960 and 1990

Sector	Agriculture		Industry		Services	
	1960	1990	1960	1990	1960	1990
	(percent)					
Industrial Countries	26	10	35	33	38	57
Developing Countries	76	61	9	16	15	23
World	61	49	17	20	22	31

Note: Categories may not add up to 100 percent due to rounding.
SOURCE: U.N. Development Programme, *Human Development Report 1996* (New York: Oxford University Press, 1996).

activities, including wholesaling and retailing, hotels and restaurants, health care, banking and finance, utilities, communications, and transportation. These sectors include some highly skilled and extremely well-paying jobs, but also many unskilled, low-paying ones. In the United States, a considerable share of job growth is taking place in the retail sector; that part of the economy, however, is increasingly characterized by low wages and insecure employment. In mid-1999, the average hourly wage in retail was a mere $9.02, compared with $12.61 in all services and $13.94 in manufacturing.[27]

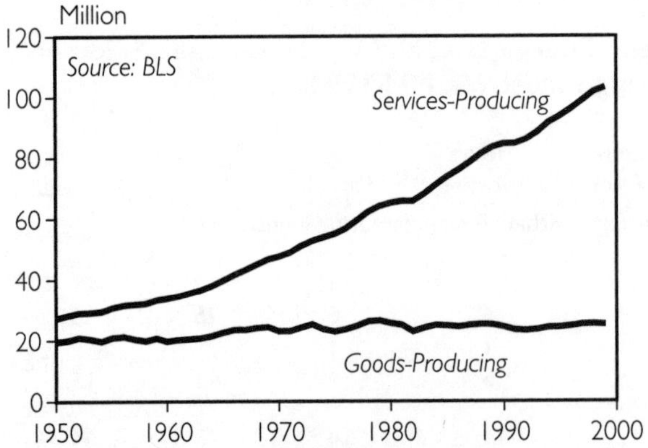

Service jobs are also by no means immune to the turbulence of change that has taken hold in mining and manufacturing. A study by the University of Würzburg concluded that computerization and information technologies in Germany may eventually do away with 61 percent of jobs in banking, 51 percent in wholesale and retail, and 74 percent in transportation and logistics.[28]

Most service establishments are directly responsible for very little pollution and environmental degradation. But many are inextricably linked to oil drilling, strip mining, forest clearcutting, paper pulping, and aluminum smelting—either by coordinating, facilitating, and financing resource extraction and processing, or by providing transport and distribution (that is, wholesale and retail) services. In essence, they are a part of the high-throughput economy. The challenge will be to generate service jobs that facilitate a shift away from our current resource-intensive forms of production and consumption—for instance, selling heating and cooling

Figure 9–2. U.S. Goods and Services-Related Jobs, 1950–99

services instead of fossil fuels, or transportation services instead of motor vehicles.[29]

Environment Policy: Job Killer or Creator?

Business leaders have long argued that environmental regulations would render them uncompetitive, forced to close plants, and compelled to delay or cancel new projects. The upshot: lost jobs. The "job killer" argument has lost some of its potency, however, for three reasons. First, many dire predictions have failed to come to pass: job loss due to environmental regulations has been extremely limited. Second, it has become clear that environmental regulations can have "technology-forcing" effects that actually give companies a competitive edge. And third, environmental regulations have spawned a sizable and rapidly growing industry (mostly focused on pollution control) that employs perhaps 11 million people worldwide.[30]

Still, as limited pollution control slowly gives way to farther-reaching pollution prevention measures and cutting-edge "clean production," and as the threat of climate change increasingly points to the need for a substantial restructuring of the energy economy, the belief in an economy-versus-environment tradeoff finds new adherents and is eagerly stoked by businesses opposed to change. But what are the impacts of environmental policy? Before looking at any specific cases, it is useful to undertake a brief conceptual assessment.

Like any other economic activity, investment in renewable energy sources, energy efficiency, railroads and public transit, less-polluting industrial production equipment, and other environment-friendly activities generates a certain number of jobs directly, plus indirect jobs in supplier industries. The crucial question is, Do these investments support more or fewer jobs for each dollar laid out than expenditures in more polluting and waste-generating industries? Countless studies suggest strongly that less damaging ways of producing, transporting, consuming, and disposing of goods tend to be more labor-intensive than the more damaging ways.[31]

Beyond specific comparisons of direct employment potential lies the larger issue of how well and efficiently an economy carries out its activities. For example, if energy services such as heating and cooling buildings, generating electricity, or powering motor vehicles can be provided more cheaply through boosted efficiency or other measures, then the money saved by businesses and households—the avoided costs—can be "re-spent" elsewhere in the economy. To the extent that this re-spending benefits segments of the economy that are more labor-intensive than the energy sector, it generates additional employment. (And because most countries import the bulk of their energy consumption, this re-spending would in effect substitute imported energy inputs with more local, decentralized labor—although oil-exporting countries would suffer accordingly.) Similar re-spending effects may also occur in other parts of the economy, as we restructure transportation, waste management, and other sectors.

When prices do not tell the truth, however, it is difficult in a market economy to realize opportunities for avoided costs and for redirecting investments and expenditures to sectors that will provide greater environmental and employment benefits. Phasing out subsidies that favor fossil fuel industries and other polluters and introducing environmental taxes will help to move toward full-cost accounting and to unveil re-spending opportunities.[32]

Energy again serves as a useful example. Presumably, the move toward greater efficiency will be brought about in part by higher energy taxes. Some of these tax revenues may go to financing the equipment and infrastructure for a more sustainable economy—creating jobs in energy efficiency technologies and public transit systems, for instance. Governments may decide to return the remainder to taxpayers, and that money would then be re-spent across the entire economy, replicating existing patterns of demand for goods and services—and creating more jobs than would have been supported in the fossil fuel industry. Alternatively, however, these funds could be used to reduce labor costs. Studies suggest that lowering employers' contributions to national health or social security funds can be a powerful stimulant for job creation, as discussed later in the chapter.

Although the losers are likely to be far outnumbered by the winners, some workers will be hurt in the economic restructuring toward sustainability—primarily those in mining, fossil fuels, and smokestack industries. At least some and perhaps many of the displaced individuals will not have the requisite skills for the new jobs without retraining, or the new jobs may primarily arise in other locations. Regions and countries that depend heavily on extractive and polluting industries will confront a substantial challenge to diversify their economies.

The process of industry restructuring is inherently painful. Because a job provides not only economic security but often also identity and meaning, job loss—even if temporary—can be a traumatic experience. For affected individuals and families, it is little consolation that environment-related job loss is likely to be insignificantly small in comparison with job loss due to "normal"

change in a market economy. Public policy must facilitate the transition to a sustainable economy by assisting individuals and communities; this may involve retraining and skill-enhancing programs and special regional development programs.

But most important, policy changes designed to make the economy more sustainable need to have a clear time horizon so that companies, communities, and individual employees know what they are up against. At the same time, however, the longer that necessary changes are postponed, the greater the urgency later on to move speedily—and the more damaging the social and economic impacts. Resistance to policies to avert climate change and to rein in other forms of environmental degradation will turn out to be a far greater job killer than embracing such policies in strategic fashion.

Restructuring Energy, Creating Jobs

Reducing fossil fuel use is one of the most central goals in moving toward a sustainable economy. The combustion of these fuels on a massive scale causes serious air pollution problems and is responsible for global climate change. Businesses opposed to serious efforts to avert this have sought to attract labor union support by arguing that an alternative energy policy would be a job killer. The AFL-CIO Executive Council, for example, issued a statement in February 1999 reaffirming its opposition to the Kyoto Protocol, arguing that it "could have a devastating impact on the U.S. economy and American workers." But even in the absence of an alternative energy policy, the number of jobs in many of these industries is already declining, often even as output continues to rise. Avoiding or postponing an environmentally responsible policy will do nothing to save these jobs; instead, it may even hasten their demise.[33]

Coal mining is a case in point, although similar stories could be told about oil refining, utilities, and energy-intensive industries like primary metals and steel. The coal industry is increasingly characterized by bigger and fewer companies, larger equipment, and less and less need for labor. Worldwide, it is estimated that only about 10 million jobs remain, accounting for just one third of 1 percent of the global work force. In the United States, coal production increased 35 percent between 1980 and 1998, but coal mining employment declined 63 percent, from 242,000 to 90,000 workers. (See Figure 9–3.)[34]

In Europe, jobs in this field have dropped even more, since production is falling substantially. In Germany, productivity gains and rising coal imports may cut employment from 265,000 in 1991 to less than 80,000 by 2020. British coal production has fallen to less than half its 1980 level, and employment fell from 224,000 to just 10,000 miners. China—the world's largest coal producer—has cut some

870,000 jobs in the past five years and will lay off another 400,000 workers in a bid to cut subsidies and to reduce output by about one fifth to bring it more in line with demand.[35]

While coal and other polluting industries are offering declining job opportunities, renewable energy and energy efficiency are beginning to make their mark. The European Wind Energy Association projects that up to 40 gigawatts of wind power capacity could be installed in Europe by 2010, creating between 190,000 and 320,000 jobs. Although no global job figure is available, some rough estimates can be made. The Danish wind turbine industry provided about 16,000 jobs (including 4,000 in installation) in 1995. Because Danish wind turbine manufacturers supply about half of the generating capacity in the world, the European Commission estimated worldwide employment in the wind power industry at 30,000–35,000 direct jobs in the mid-1990s.[36]

European wind energy companies accounted for about 90 percent of worldwide sales in 1997, and presumably will continue to garner the majority of jobs in the near future. But India, China, and other developing countries have considerable wind energy potential and could generate substantial employment by building a strong indigenous base. India already has 14 domestic turbine manufacturers.[37]

The European Commission notes that, as a rough rule of thumb, 1 megawatt of wind power generating capacity installed creates jobs for 15–19 people under present European market conditions and perhaps

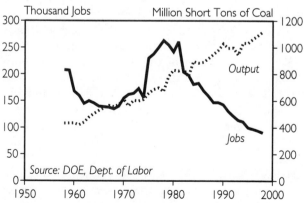

Figure 9–3. U.S. Coal Mining, Output and Jobs, 1958–98

double that in countries with higher labor intensity. Since this includes manufacturing, sales, installation, operations, and maintenance, it encompasses both permanent and temporary jobs. Applying this formula, there may have been 92,000–117,000 direct and indirect jobs worldwide in the mid-1990s; if installed capacity roughly doubles by 2001, as the European Commission projects, this could rise to 170,000– 216,000 jobs.[38]

A variety of studies confirm that wind power compares favorably in its job-creating capacity with coal- and nuclear-generated electricity. Wind power generation is mostly decentralized and small-scale, and the manufacturing of rotor blades and other components requires skilled labor input to ensure quality. Still, as the size of wind turbines and economies of scale increase, helping to make wind power a cheaper source of energy, the number of jobs per dollar invested will decrease somewhat in coming years.[39]

Like wind power, solar energy use, particularly in the form of photovoltaics (PV), is growing rapidly. U.S. solar industries directly employ nearly 20,000 people now and support more than 150,000 indirect jobs in diverse areas such as glass and steel manufacturing, electrical and plumbing contracting, architecture and system design, and battery and electrical equipment. The Solar Energy Industries Association (SEIA) claims that 3,800 jobs are created for every $100 million in PV cell sales, translating into 12,160 PV jobs in the United States in 1995. PV jobs in Europe are still very limited in number, but the European Photovoltaic Industry Association projects that the production, installation, and maintenance of PVs could directly employ up to 294,000 people there by 2010.[40]

Meanwhile, the European Solar Industry Federation, a group of about 300 solar thermal companies, employed more than 10,000 people in 1997 in designing, manufacturing, marketing, installing, and maintaining systems. Just under current market growth trends, the federation projects the creation of 70,000 additional jobs in the next 10 years, and a far larger number, perhaps up to 250,000, if strong governmental support for solar energy materializes.[41]

As a group, renewables have the potential to become a significant source of jobs. The U.S. industry association, SEIA, asserts that more than 350,000 net jobs will be added by 2010—a number equal to the employment provided by the largest U.S. car manufacturer. In a 1997 report, the European Commission lays out the objective of doubling the current share of renewable energy sources from 6 to 12 percent by 2010. Taking job losses in fossil fuel energy sectors into account, a half-million net additional jobs could be created in the renewable energy sector and in supplier industries, and another 350,000 jobs through exports of renewables.[42]

Like renewables, energy efficiency has considerable job potential awaiting mobilization. The American Council for an Energy-Efficient Economy (ACEEE) has assessed the impact of a "high-efficiency scenario," assuming cost-effective improvements throughout the U.S. economy. These run the gamut from better-insulated windows to more-efficient lighting to highly fuel-efficient cars. Average annual investments of $46 billion during 1992–2010 yield a 20-percent reduction in energy consumption below a business-as-usual scenario and a 24-percent reduction in carbon emissions. The study estimates that almost 1.1 million net jobs could be created by 2010. Just 10 percent of these are direct

Table 9–4. Job Impact Findings, Selected Studies on Climate Policy

Country	Policy Change	Years	Carbon Reduction	Employment Gain
			(million tons)	(net number of jobs)
Austria	Cogeneration, energy efficiency, renewables, alternative transportation	1997–2005	70	+ 12,200
Austria	Biomass, higher taxes on fossil fuels	1997–2005	20	+ 30,000
Denmark	Greater natural gas use, district heating, cogeneration, energy efficiency, renewables; total energy consumption stable	1996–2015	82	+ 16,000
Germany	Boosting efficiency, phasing out nuclear power, less oil and coal use, renewables to account for 10 percent of primary energy use, alternative transportation policies	1990–2020	518	+ 208,000
Netherlands	Efficiency gains in transport, industry, electric equipment, buildings; greater use of wind power	1995–2005	440	+ 71,000
United Kingdom	Accelerated uptake of cogeneration, efficiency, and renewables technologies	1990–2010	206	+ 537,000
European Union	Installation of high-performance double-pane windows in 60 percent of dwellings	10-year period	940	+ 126,000
United States	Improved efficiency in transportation, industry, power generation, buildings	1990–2010	188	+ 870,000

SOURCE: Worldwatch Institute, based on sources cited in endnote 44.

jobs in efficiency and in supplier industries; the rest are jobs created as consumers and businesses re-spend the money they save through avoided fuel costs on other goods and services that are more labor-intensive than the fossil fuel industry.[43]

Since the ACEEE study was published in 1992, other assessments have been undertaken in different industrial countries, spurred by the Kyoto Protocol on climate change and the growing urgency to deal with this issue. (See Table 9–4.) Although they rely on different methodologies, assumptions, and econometric models, making them difficult to compare directly with each other, these studies support the overall conclusion that pursuing energy alternatives will generate more jobs than the fossil fuel industries can.[44]

While this discussion has been focused on industrial countries, there are implications for developing countries as well. Given the substantial potential for wind and solar in developing countries, these energy sources could become important job creators. But there, too, a key employment benefit of moving away from energy-intensive, fossil-fuel-focused patterns of development lies in spending less of a society's financial resources on oil, coal, and natural gas (much of which must be imported) and more on labor-intensive sectors of the economy—the so-called re-spending effect. Seeking out investment and consumption choices that promise greater job creation than the traditional energy industries is of particular interest in countries that have surging numbers of job seekers and scarce economic resources.

Durability and Remanufacturing

Energy and materials productivity can be boosted by moving the economy away from the throwaway treadmill that churns out mass-produced items designed to fall apart easily or be rendered passé by fashion cycles. Durability, repairability, and "upgradability" are key to achieving sustainability.

In today's industrial economies, many products, even some that are nominally durable, have become "commodified": large quantities can be manufactured with such ease and at such relatively little monetary cost that there is considerable incentive to regard them as throwaways rather than to produce them for true durability. If durability is not a top consideration, then human dexterity, skill, and workmanship are also likely to be given low priority by management, and labor input will be regarded more as a cost factor than a way to ensure quality.[45]

Many of today's consumer products are made in such a way as to discourage repair

and replacement of parts, and sometimes even to render it impossible. And even when repair is possible, the cost is often too high relative to a new item. If repair and maintenance are not "worth the trouble," then jobs in such occupations are condemned to all but vanish. Although consumers have an obvious interest in cheap products, the price must be sufficiently high to justify ongoing maintenance, repair, or upgrading, and hence to make jobs in these occupations viable. In any event, a durable product with a higher up-front cost of purchase may well turn out, over time, to be economically more advantageous to consumers than individually cheaper, planned-obsolescence products that must be replaced frequently. If a $100 wristwatch lasts a lifetime, it represents a lower expenditure than a series of $10 or $20 watches that fall apart relatively quickly.

Products can no doubt be made to last longer, and their useful life can be stretched by making it easy to maintain and repair them. Interface, a carpet manufacturer, has developed a new material called solenium. This lasts four times as long as traditional carpets but uses 40 percent less material, reducing materials intensity by 86 percent. In addition, the material can be completely remanufactured—all the material in used floor covering can go into producing a new carpet. The company has boosted its employment by 73 percent between 1993 and 1998.[46]

For products subject to high wear and tear, such as carpets, remanufacturing is crucial; they should be easy to take apart so as much of the components can be reused as possible. For "non-consumable" goods like cars, refrigerators, washing machines, or computers, on the other hand, it is important that products be designed for easy refurbishing and upgrading, so that durability does not translate into technological obsolescence. This calls in particular for a "modular" approach that permits an easy replacement of parts and components. Computers serve as an example here: standardized slots commonly accept components such as modems, sound cards, or memory chips virtually irrespective of which company made them.

An economy that embraces durability will require a transportation system different in its structure and mix of modes. Instead of today's "making-using-disposing" system, with its one-way flow of raw materials, products, and waste, it would instead be a "making-unmaking-remaking" system—able to collect and take back products that need to be repaired or upgraded and then redistributed to consumers, as well as those disassembled for remanufacturing or for salvaging of parts and materials. Such a system would probably be focused less on long-distance supplies and deliveries and more on interchanges within local and regional economies.[47]

What are the job implications? When goods do not wear out rapidly, they need not be replaced as frequently. An obvious implication is that fewer goods will be produced. Would this mean that fewer employees are needed—compounding the unemployment challenge? Not necessarily. Producing longer-lasting, higher-quality products, using more robust materials, and processing and assembling them into final products with greater care than might suffice for "disposable" products implies a more craft-oriented, smaller-batch production process than the current mass-manufacturing practices; it takes more labor, and particularly more skilled labor. Plus, there will be greater opportunity and incentive to maintain, repair, upgrade, and remanufacture products, and thus associated job

potential throughout the life of a product. In the United States, remanufacturing is already a $53-billion-a-year business, employing some 480,000 people directly. Since products and production processes will need to undergo extensive redesign for durability and easy upgrading and disassembly, there is additional job creation stimulus in design, engineering, architecture, and other fields.[48]

In the United States, remanufacturing is already employing some 480,000 people directly.

Today's retail jobs depend on large-scale purchases of "stuff"—in principle, anything that sells, no matter what the quality and durability. Discount retailers in particular have led the trend toward a part-time, low-paid sales force; in such a quantity-focused environment, fewer consumer purchases translate into fewer retail jobs. In a sustainable economy, there is a need to move toward "quality retail," in which the salesperson knows how to sell intelligent use rather than mere ownership: advising consumers on the quality and upkeep of products; counseling them on how to extend usefulness with the least amount of energy and materials use; explaining the advantages of leasing versus ownership. Because such a system would not be geared to increasing throughput—focusing merely on getting products out of the showroom or off the store shelf—but instead to ensuring consumer utility and satisfaction, it entails higher-skill jobs.[49]

Truly durable products are likely to be more expensive than throwaways. For some items, the upfront cost could be steep, and this calls for the development of innovative financing plans. Where consumer credit is now geared to maintaining the hyper-throughput economy, allowing people to carry high personal debts and to rebound from insolvency in order to keep consuming, finance in a durable product economy will need to devise ways to make possible—and to reward—the purchase of long-life products. This may involve longer repayment periods, for instance.

A fuller evaluation of the employment implications of a shift toward durability would require detailed assessments of the specific changes, and how they translate into job losses and opportunities for new employment. Table 9–5 provides a rough conceptual exploration. Generally, however,

it seems clear that a durability strategy would accelerate the shift in employment from resource extraction and primary industries to the provision of services.

Shifting Taxes

Ecological tax reform is key to addressing both the challenge of adequate job creation and environmental protection. Depending on their design and scope, eco-taxes—such as landfill fees, taxes on nonrenewable energy, and emissions charges—promise several benefits. They can help reinforce the "polluter pays" principle, provide incentives for

Table 9–5. Likely Employment Effects of Durable, Repairable, and Upgradable Products

Product Life-Cycle Phase	Observation	Employment Effect
Design and Engineering	• Intense redesign of products (and production processes) required	Positive
Energy and Materials Inputs	• Fewer products; therefore fewer inputs needed, though more robust materials required	Negative
Manufacturing/Assembly	• Fewer products • Production more attentive to durability, and so on	Negative?
Distribution/Transport	• Fewer products • More (local) circulation from users to repair shops, remanufacturing, and so on, and back	Mixed
Maintenance	• Revitalizing neglected functions; labor-intensive	Positive
Re-Manufacturing	• Currently limited	Positive
Upgrading	• Currently limited; labor-intensive	Positive
Consulting	• Advice on product-life extension and substituting services for goods	Positive
Disposal at End of Life-Cycle	• Fewer products to be disposed of • Greater recycling, plus disassembly of parts and components for reuse; more labor-intensive than landfilling and incineration	Positive

SOURCE: Worldwatch Institute.

boosting energy and materials efficiency, help trigger technological innovation, and raise revenues to fund environmentally benign alternatives.

In the context of the environment-employment nexus, however, another aspect is key: using eco-tax revenues to reduce the payroll taxes that fund social security programs. In effect, some of the tax burden now falling on labor would thus be shifted, and levied instead on resource use and pollution. This shift is based on the recognition that current tax systems are severely out of balance; they make energy and natural resources far too cheap (inviting inefficiency and waste), but render labor too expensive (discouraging new hiring). The predictable result is an overuse of natural resources and underuse of human labor.[50]

In western industrial countries, payroll taxes and mandatory social security contributions accounted on average for 25 percent of all tax revenues in 1995, up from 18 percent in 1965. Energy taxes and nonenergy environmental taxes, in contrast, account for only about 7.5 percent, and taxes on capital have decreased in most countries. Given this situation, it is little wonder that companies have put far greater emphasis on boosting labor productivity than on enhancing capital, energy, and materials productivity—with the result that unemployment and environmental degradation are both higher than they would otherwise be. The potential impact of a tax shift is likely to be greatest in countries where labor taxation is particularly high, as it is in most of Europe. The United States, by contrast, has less leeway to decrease the tax burden on labor, but it has greater opportunities to effect a tax shift by raising its extremely low taxes on resource use.[51]

During the 1990s, a growing number of studies, principally in Europe, modeled the economic and employment impacts of ecological tax reform. Although the underlying assumptions about the nature and size of eco-taxes, as well as the precise ways in which the tax revenue would be used, vary widely, the key conclusion was that a tax shift is clearly good news for job creation. For instance, an array of U.K. energy and environment taxes to shift 6 percent of the tax burden from labor to environmentally damaging operations could generate some 717,000 additional jobs during 1997–2005. An influential German study undertaken in 1994 modeled the impact of a tax on all nonrenewable sources of energy and on electricity that would be imposed and increased by 7 percent annually over 15 years. Energy consumption and carbon emissions would decline by 21 percent, and some 600,000 new jobs be created.[52]

Discussed theoretically since the late 1970s, eco-tax shifting started to become a reality in the 1990s, as Denmark, Finland, Germany, the Netherlands, Norway, Sweden, and the United Kingdom linked a variety of such taxes to reductions in income taxes or social security contributions. The tax shifts have amounted to anywhere from 0.2 percent to 2.5 percent of these countries' total tax revenues.[53]

In the countries that have initiated a tax shift, eco-taxes are still quite modest, and energy-intensive industries are partially exempted from the eco-tax (either by paying a reduced rate or by receiving reimbursements). In the German case, all manufacturing firms are assessed at only 20 percent of the full tax rate, and coal and jet fuels are not taxed at all. This is because governments are reluctant to be seen as weakening energy-intensive industries' ability to compete internationally. But unless this preferential treatment is phased out over time, and national policies harmonized so that competitive fears are eased, the incentive to cut energy use and carbon emissions will be diminished considerably. Less progress toward energy efficiency also means that money continues to be bound up in the energy sector that could, if invested elsewhere, create more jobs.[54]

Rethinking Work

Ecological tax reform can help shift economic priorities from increasing labor productivity to boosting energy and materials productivity, and hence is an important component of any policy aimed at ensuring that economic progress is not synonymous with job destruction and heavy environmental damage.

But there is a danger that gains in energy and materials efficiency and in other environmentally beneficial measures may simply be offset and perhaps even overwhelmed by the rising tide of consumption. To illustrate, fuel savings through more efficient car engines and other improvements in automotive technology since the early 1970s have largely been erased by the trend toward larger vehicles and the continuing growth in the amount of driving.

In this context, a key question arises: should we channel future gains in productivity principally toward wage increases and hence growing consumption, or toward reduced work time? To date, the fruits of technological progress (in the form of boosted productivity) have primarily been translated into material rewards—an unprecedented quantity and variety of consumer goods affordable for greater numbers of people than ever before in human history. Only to a limited extent has greater productivity been translated into reduced working hours and more time spent pursu-

ing hobbies, caring for family members or friends, or doing volunteer work. Indeed, for all the mighty leaps in productivity, full-time employment still seems like an essential condition: For some, the ability to acquire a steady stream of material possessions seems to promise status, and perhaps even a fleeting happiness. But clearly others have to work full-time to make ends meet and pay the bills.[55]

The question then is not only, How can work and leisure be shaped without wrecking the planet in the process? But how can existing work be shared more equitably so that society is not condemned to polarization between the overworked and the underemployed, the haves and have-nots, the highly stressed and the alienated? The challenge is to develop a new understanding of work in modern societies, one that would help to break the work-consumption-environmental degradation dynamic.

Determining the "appropriate" length of the workday and week is an issue as old as the industrial system, and the answers have fluctuated with shifting economic structures and with the changing balance of power among employers, workers, and the state. At the last turn of the century, it was not unusual for a worker's total life-time work to tally some 100,000 hours, or well over 60 hours per week. By the 1950s, the figure had dropped by roughly one third; additional reductions have taken place since in most industrial countries, though the pace has slowed markedly during the last two decades. (See Table 9–6.)[56]

Persistent unemployment in Europe has reinvigorated the debate there over work time reductions. A variety of measures can be taken. These range from taking steps to chisel away at the hours spent in factories and offices, to providing for earlier retirement, lengthening vacations, and permitting sabbaticals or unpaid leave. In France, a mandatory 35-hour week is now being phased in. Germany has so far relied on collective bargaining agreements, such as one concluded at a Volkswagen auto plant in Wolfsburg. In 1993, the company and the labor union agreed to reduce weekly working hours to an average of 28.8, to introduce much greater work time flexibility, and to guarantee job security during the transition to shorter working hours. Denmark offers employees the option of vacating their job for up to one year to pursue adult education or to care for a child; during leave, the government pays them the equivalent of 100 percent or 60 percent (for child care) of the compensation an unemployed person would receive.[57]

The growth of part-time jobs in industrial nations can, in principle, be part of the answer, too. In Denmark, Sweden, and the United Kingdom, roughly one quarter of all employment is part-time; in Germany, one fifth. In the Netherlands, it has climbed to more than 38 percent. Growing numbers of people want to work part-time. Almost one third of German full-time employees have expressed a desire to work fewer hours. And in a 1996 U.S. poll, 45 percent of respondents said they would trade a day's pay for a day off.[58]

The problem today is that part-time work, and other flexible work arrangements, often involve low pay, few benefits, limited job security, and virtually no career opportunities. As part-time work becomes a major component of the labor market, governments may decide to adopt rules and regulations to make it a socially more acceptable work option.

In a 1996 U.S. poll, 45 percent of respondents said they would trade a day's pay for a day off.

Another approach is embodied in the idea of a guaranteed basic income. Fred Block, professor of sociology at the University of California, has put forward the idea that "all citizens, whether employed or not, would receive a monthly grant large enough to sustain a minimal standard of living, including housing, food, and other basic necessities. Such grants could substitute for the elaborate systems of welfare, unemployment insurance, and social security that exist in developed capitalist societies."[59]

Clearly, this is a controversial concept whose full implications need to be given considerable thought. But the idea entails benefits for employees as well as employers. A basic income scheme would allow individuals to pursue as much paid work as they desire to supplement their basic income. It would not force anyone to work less time than they wanted, but it would permit people to volunteer, rear children, care for family members, and pursue other, traditionally unpaid, forms of work with greater ease than today. But Block also argues that

Table 9–6. Average Workweek for Manufacturing Employees, Selected Industrial Countries, 1950–98

Country	1950	1980	1998
		(hours)	
Germany[1]	45	33	30
Sweden	38	28	32
France	39	34	32
Italy	38	34	35
United Kingdom	41	35	36
Japan	46[2]	41	37
United States	38	36	38

[1]Western Germany only.   [2]1955.
SOURCE: See endnote 56.

greater opportunities would emerge for individuals to pursue additional education and training—rendering workplaces more dynamic because new ideas and skills would percolate and spread more readily. In a modern, "knowledge-based" economy that more and more emphasizes the importance of continuing education and a frequent upgrading of skills, this would be of benefit to most businesses.[60]

Creating opportunities for affected workers to learn new skills will be key.

It is true that such a system would in some ways be a disincentive to work. However, employers would primarily find it difficult to fill the more poorly paid, unsatisfying jobs. That would be all for the better, argues Block, because "employers would have a strong incentive either to automate those jobs out of existence or to make them more attractive—by raising pay levels, improving working conditions, raising the problem-solving component of the work."[61]

Creating the right conditions for a sustainable economy to emerge presents one set of challenges for governments. But policymakers will also need to be attentive to the transition costs, and work to avoid the kind of severe social disruptions that confronted the Luddites and their contemporaries some 200 years ago. From a human well-being perspective, it is not enough to say that, in the end, economic transitions often leave societies better off. "Environmental measures that do not recognize a worker's right to a fair chance in the new economy," writes Alan Durning, executive director of Northwest Environment Watch in Seattle, "are equally menacing to our future" as jobs that depend on

despoiling nature.[62]

As in the days of the Luddites, it is Britain that offers an illustration of how not to proceed. In the mid-1980s, the British government restructured the coal industry, closing large numbers of mines and slashing coal subsidies—though motivated more by the intent to break the power of labor unions than the desire to avert climate change. While this policy did reduce carbon emissions, it also caused high unemployment and unleashed an array of associated social ills in coal mining regions, not least because the bitterly disputed policy was forced through in a short stretch of time.[63]

If individuals and communities have reasonable hope that the transition to a sustainable economy does not translate into social pain for them, they will be far less likely to oppose change. Creating opportunities for affected workers to learn new skills and providing assistance in their shift to new careers will be key. This may entail financial support to help pay tuition for vocational and other training programs, transition income support, and career counseling and placement services. The more that the economy moves from resource extraction and mass production to services and a "knowledge" economy, in which skill requirements change more frequently, the more do training and retraining become issues for the economy as a whole.

Important as they are, educational and skill-building programs by themselves are an inadequate response to the transition challenge. Measures to spur job creation and build a sustainable economic base are equally important. Because the transition challenge is especially pronounced in areas where logging, mining, and other heavily polluting industries play a disproportionate economic role, governments will need to design programs to assist regions with unsustainable and declining industries. This means helping to diversify and broaden the economic base and to build infrastructures that can support such a shift.

Governments can also adopt measures that reward job creation by companies, and particularly well-paying jobs. Favorable tax treatment for job creation would be part of a broader re-calibration of fiscal tools to shift the emphasis from labor productivity to resource productivity—from promoting resource extraction to supporting new employment.

But most important, policies must be pursued proactively instead of as an afterthought. The earlier that transition strategies are formulated, the greater the likelihood of success. As indicated earlier, employment is already declining in industries like coal mining, oil refining, utilities,

logging, and primary metals processing, even as output continues to grow. The time to act is now. Strengthening labor unions and building labor-environment coalitions would seem to be essential for policies to preserve jobs and the environment.

In the end, the jobs-and-environment nexus touches on basic questions of how society goes about generating wealth without destroying the environment, whether it translates economic prowess into more material rewards or more "leisure" time, and whether it can reduce the extremes of wealth and poverty, and of overemployment as well as underemployment. The implications of decoupling job creation from environmental destruction are in some ways no less revolutionary than the changes that confronted the Luddites nearly 200 years ago.

Chapter 10

Coping with Ecological Globalization

Hilary French

As the twenty-first century dawns, the planet seems to be steadily shrinking. Goods, money, people, ideas, and pollution are traveling around the world at unprecedented speed and scale, overwhelming financial managers, political leaders, and ecological systems. The "global commons," including the atmosphere and the oceans, is under environmental assault. The globalization of commerce is further internationalizing environmental issues, with trade in natural resources such as fish and timber soaring, and with private capital surges giving international investors a growing stake in distant corners of the globe.

Environmental problems are climbing ever higher on the international political agenda, at times preoccupying international diplomats almost as much as arms control negotiations did during the cold war. Industrial countries are increasingly arguing, with the European Union (EU) and the United

States now at odds on issues from global climate change to genetically modified organisms (GMOs). Environmental issues have also become acrimonious in North-South relations, with rich and poor countries divided over how to apportion responsibility for reversing the planet's ecological decline.

"Globalization" has become a common buzzword. But it means vastly different things to different people. To some, it is synonymous with the growth of global corporations whose far-flung operations transcend national borders and allegiances. To others, the term is closely linked with the information revolution, and the mobility of money, ideas, and labor that computers and other new technologies have been instrumental in bringing about. In this chapter, globalization is taken to mean a broad process of societal transformation in which numerous interwoven forces are making national borders more permeable than ever before, including growth in trade, investment, travel, and computer

This chapter is based on *Vanishing Borders* (New York: W.W. Norton & Company, in press).

networking. "Ecological globalization" is used here to refer to the collective impact that these diverse processes have on the health of the planet's natural systems.[1]

Ecological globalization in its many guises poses enormous challenges to traditional governance structures. National governments are ill suited for managing environmental problems that transcend borders, whether via air and water currents or through global commerce. Yet international environmental governance is still in its infancy, with the treaties and institutions that governments turn to for global management mostly too weak to put a meaningful dent in the problems. Nations are granting significant and growing powers to economic institutions such as the World Trade Organization (WTO) and the International Monetary Fund (IMF), but environmental issues remain mostly an afterthought in these bodies.

Although nation-states are losing ground in the face of globalization, other actors are moving to the fore, particularly international corporations and nongovernmental organizations (NGOs). New information and communications technologies are facilitating international networking, and innovative partnerships are being forged between NGOs, the business community, and international institutions.

Despite these hopeful developments, the world economy and the natural world are both in precarious states as we enter the new millennium, provoking fears that an era of global instability looms on the horizon. Over the course of the twentieth century, the global economy stretched the planet to its limits. The time is now ripe to build the international governance structures needed to ensure that the world economy of the twenty-first century meets

peoples' aspirations for a better future without destroying the natural fabric that underpins life itself.[2]

Trading on Nature

International movements of goods, money, and people play a major role in today's unprecedented biological losses. Yet the emerging rules of the global economy pay little heed to the importance of reversing the biological impoverishment of the planet. This mismatch between ecological imperatives and prevailing economic practice will need to be bridged if the world is to halt an unraveling of critical ecological systems in the early decades of this century.

The world's forests are a particularly important reservoir of biological wealth. They harbor more than half of all species on Earth and provide a range of other important natural services, including flood control and climate regulation. But the planet's forest cover is steadily shrinking. Nearly half of the forests that once covered Earth have already been lost, and almost 14 million hectares of tropical forest—an area almost three times the size of Costa Rica—are being sacrificed each year.[3]

The role of international trade in global deforestation has been a subject of controversy over the years. International timber trade is far from the only culprit in forest loss: the clearing of land for agriculture and grazing is also a major cause, as is fuelwood gathering in some regions and the felling of trees for commercial timber for domestic use. Yet the draw of international markets can be an inducement for countries to cut down trees far faster than would be required to meet domestic demand alone. Indonesia and Malaysia, for example, have both pushed plywood exports heavily in recent years, contributing in no small measure to rapid deforestation in both countries. Plywood exports from the two countries combined exploded from just 233,000 cubic meters in 1975 to 12 million cubic meters in 1998. These two countries now account for nearly 60 percent of world plywood exports, up from just 4 percent in 1975.[4]

The value of global trade in forest products has risen steadily over the last few decades, climbing from $47 billion in 1970 to $139 billion in 1998. (See Figure 10–1.) Recent years have seen particularly rapid growth in trade in more finished types of forest products such as plywood, pulp, and paper. (See Chapter 6.) Exports of industrial roundwood (raw logs), in contrast, have remained relatively constant. For all products other than logs, exports as a share of total world production increased significantly over this period—an important indication of the growing globalization of the industry.[5]

Mining and energy extraction also imperil the health of forests, as well as mountains, waters, and other sensitive ecosystems. Vast areas are often disturbed for the sake of a relatively small quantity of bounty. For every kilogram of gold pro-

duced in the United States, for example, some 3 million kilograms of waste rock are left behind. Prime extraction sites are often located in previously undisturbed forests or wilderness areas. According to the Washington-based World Resources Institute, mining, energy development, and associated activities represent the second biggest threat to frontier forests after logging, affecting nearly 40 percent of threatened forests. Besides disturbing valuable ecosystems, this activity also can be devastating for the indigenous peoples who inhabit them: current exploration targets suggest that an estimated 50 percent of the gold produced in the next 20 years will come from indigenous peoples' lands. Toxic byproducts of mining poison the rivers that local people drink from, and the mining operations themselves destroy the forests and fields that people rely on for sustenance.[6]

Industrial countries are large consumers of minerals, accounting for more than 90 percent of bauxite imports, nearly 100 percent of nickel imports, over 80 percent of zinc imports, and roughly 70 percent of copper, iron, lead, and manganese imports. But developing countries are home to much of the world's mineral production, along with the associated environmental damage. Collectively, developing countries account for 76 percent of all exports of bauxite and nickel ore, 67 percent of copper, 54 percent of tin, and 45 percent of iron ore.[7]

Even something as basic as our food supply is now deeply integrated into the global economy. The value of world agricultural trade skyrocketed in recent decades, more than doubling between 1972 and 1997 alone—from $224 billion to $457 billion. (See Figure 10–2.)

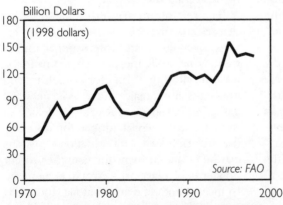

Figure 10–1. **World Trade in Forest Products, 1970–98**

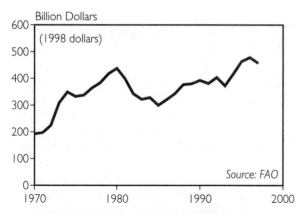

Figure 10–2. World Exports of Agricultural Products, 1970–97

Agriculture accounts for 11 percent of the value of all world exports. For some continents, this share is significantly higher—25 percent of Latin America's exports are agricultural, as are 18 percent of Africa's. Trade in basic food grains such as wheat, rice, and corn dominates international agricultural exports in volume terms. Nearly 240 million tons of grain were exported in 1998—some 13 percent of total world production. Global aggregates mask great variations in export and import dependence among countries and regions. Australia, for instance, exports 63 percent of its grain production, and Japan imports 75 percent of its consumption.[8]

The developing world is a net importer of basic foodstuffs such as grain and meat, but it is a major exporter of many cash crops, such as bananas, coffee, cotton, soybeans, sugarcane, and tobacco. As of 1997, developing countries accounted for 97 percent of cocoa exports, 92 percent of palm oil, 88 percent of coffee, and 86 percent of bananas. Although these crops are the mainstays of many national economies, heavy reliance on them can entail substantial social and environmental costs, including the displacement of subsistence farmers from their

land and the promotion of chemical-intensive agriculture. Recent decades have seen particularly rapid growth in so-called nontraditional exports—principally flowers, fruits, and vegetables. These crops tend to command far higher prices than those of traditional agricultural exports, which have been in decline in recent decades. But there are risks associated with these crops as well, one of the most serious of which is exposure to harmful levels of pesticides. A study of nearly 9,000 workers in Colombia's flower plantations indicated exposure to 127 different pesticides, some 20 percent of which are either banned or unregistered in the United Kingdom or the United States.[9]

The fishing industry is also increasingly linked into the global marketplace. Global fish exports have grown nearly fivefold in value since 1970, reaching $52 billion in 1997. (See Figure 10–3.) By volume, nearly half of the fish caught today are traded, up from only 32 percent in 1970. But the steady expansion of the catch, as well as habitat destruction and pollution, are taking a heavy toll on the world's fish stocks: the U.N. Food and Agriculture Organization estimates that 11 of the world's 15 major fishing grounds and 70 percent of major fish species are either fully or overexploited.[10]

Industrial countries dominate global fish consumption, accounting for more than 80 percent of all imports by value. Developing countries, on the other hand, contribute nearly half of all exports. Their share of the total has climbed steadily in recent decades as fleets have turned south in search of fish in response to the overfishing of northern waters. In 1970, developing countries accounted for 37 percent of all fish exports, measured by value; by 1997, their share had

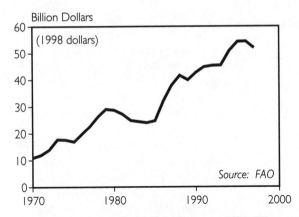

Figure 10–3. World Fish Exports, 1970–97

risen to 49 percent. Thailand, China, Chile, and Indonesia are all now among the world's top 10 fish exporters. Exports from these four countries have nearly quadrupled in value since 1980.[11]

With many Third World fisheries now becoming depleted as well, overfishing for export markets means depriving small-scale fishers of their bounty. It also drives up the price of domestically available fish to the point where it is beyond the means of local people. In Senegal, for instance, species once commonly eaten throughout the country are now either exported or available only to the elite. This trend has serious implications for food security, as nearly 1 billion people worldwide, most of them in Asia, rely on fish as their primary source of protein.[12]

Trade in more exotic forms of wildlife is also a booming business, placing a number of commercially valuable species at growing risk of overexploitation and even extinction. Each year, some 40,000 monkeys and other primates are shipped across international borders, along with some 2–5 million live birds, 3 million live farmed turtles, 2–3 million other live reptiles, 10–15 million raw reptile skins, 500–600 million ornamental fish, 1,000–2,000 raw tons of corals, 7–8

million cacti, and 9–10 million orchids. The wildlife trade is valued at some $10–20 billion annually, at least a quarter of which is thought to be illegal.[13]

The world community is just now beginning to awaken to a related though far more pervasive threat—the international spread of non-native "exotic" species, which is known as "bioinvasion." For most of history, natural boundaries such as mountains, deserts, and ocean currents have served to isolate ecosystems and many of the species they contain. But over the centuries these natural barricades have gradually broken down as people and organisms have spread around the globe. This process has accelerated exponentially in recent decades, as trade and travel have skyrocketed. Once exotics establish a beachhead in a given ecosystem, they often proliferate, suppressing native species. Taken as a whole, invasive species now pose the second largest threat to the diversity of life on Earth, after habitat destruction.[14]

Besides endangering the health of ecosystems, the spread of microbes around the globe also threatens the health of people. Airplanes carry people to the other side of the world in far less time than the incubation period for many ailments, facilitating the unwitting introduction of foreign microbes into vulnerable populations. More than 30 new infectious diseases have emerged over the past 20 years, including AIDS, Ebola, Hantavirus, and hepatitis C and E. According to the World Health Organization, "environmental changes have contributed in one way or another to the appearance of most if not all" of these diseases. Land use changes such as deforestation or the conversion of grasslands to agriculture that alter long-established equilibria between microbes and their hosts are sometimes to blame. In other cases, changes in human behavior are the culprit, such as careless disposal of food and beverage containers or car tires, which can create new breeding sites for disease-carrying organisms such as mosquitoes.[15]

As world trade continues its steady upward climb, it is placing unprecedented strains on the health of the planet's ecosystems. But today's emerging global governance structures for the most part give short shrift to the urgent need to halt global environmental decline. The World Trade Organization is a case in point.

The WTO Meets the Environment

The WTO came into being in 1994 as an outgrowth of the "Uruguay Round" of global trade talks under the General Agreement on Tariffs and Trade (GATT). Although many people in the business community, government, and academia hailed its creation as an enlightened step toward a new era of global prosperity, critics charged that the new organization elevated corporate rights to a new plane while devastating local communities and the environment. The intervening years have done little to cool the passions on either sides of the debate.

In late November 1999, trade ministers from around the world gathered in Seattle to launch a millennium round of global trade talks. Thousands of NGO activists were also in Seattle, many of them protesting what they see as the WTO's environmental blindness. Critics particularly decry the secrecy that shrouds WTO activities. Many important documents are unavailable to the public, and most WTO committees, as well as all dispute resolution proceedings,

are conducted in closed sessions dominated by trade rather than environmental experts.[16]

The text of the agreement that created the WTO ran to an astounding 26,000 pages and covered a bewildering array of issues, including agriculture, intellectual property rights, investment, and services. The organization was charged with overseeing the implementation of the new rules of world trade, including settling any disputes among nations related to their terms. Member-countries granted the WTO unprecedented powers for an international body, including a binding dispute resolution mechanism and provisions for stiff trade penalties to enforce its rulings.[17]

In a concession to the concerns of environmentalists, the preamble to the WTO agreement includes environmental protection and sustainable development among the organization's goals. The accord also included a commitment to create a Committee on Trade and Environment charged with analyzing the relationship between trade liberalization and environmental protection and with recommending any changes to WTO rules that might be needed to make the two goals "mutually supportive." But more than five years later, the committee has produced much talk but no concrete action.[18]

Widespread public concern about the environmental impact of GATT dates back to September 1991, when a dispute resolution panel shocked U.S. environmentalists by ruling that an embargo against Mexican tuna imposed under the U.S. Marine Mammal Protection Act violated GATT. The United States had imposed the embargo after determining that Mexicans were fishing for tuna with purse-seine nets that often have the unintended effect of ensnaring dolphins as well. In concluding that the tuna embargo contravened GATT rules, the panelists emphasized a key though controversial distinction between import restrictions aimed at the characteristic of products themselves and those keyed to production processes. The panelists decreed that the U.S. law was illegal under GATT because the United States was rejecting the process by which the tuna was harvested rather than the tuna itself.[19]

Although GATT, and later the WTO, contains a specific provision that ostensibly protects the right of countries to pursue environmental protection policies that might otherwise contradict trade rules, the panelists ruled that this exception pertains only to efforts by countries to protect the environment within their own borders. Because the Mexican tuna fishing took place outside of U.S. waters, the panelists viewed the embargo as tantamount to the United States foisting its environmental laws and values on the rest of the world. This point of view resonated with many people, particularly in the developing world, who looked to the rule-based GATT as a check on the U.S. tendency to wield its economic power unilaterally.[20]

But the decision exposed some glaring inconsistencies between the rules of the world trading system and emerging international environmental principles and practices. The trading system's aversion to process-related trade restrictions struck many people as particularly arbitrary, as environmental policy is moving increasingly toward focusing on the environmental impacts of products throughout their lifecycle—including production, distribution, use, and disposal. Products such as gold or timber may be harmless or beneficial as products, but enormously costly to human or environmental health in the ways they are processed. Reform of extraction and manufacturing processes is essential to making real environmental advances, yet trade rules put up a sizable hurdle to pursuing such efforts in a world economy that is becoming steadily more integrated.[21]

Also worrisome was the ruling's failure to acknowledge the right of countries to take action to protect the atmosphere, the oceans, and other parts of the global commons—a failure that raised questions about the legality under GATT of an array of other environmental policies besides the one aimed at protecting dolphins. What would become of policies aimed at reducing the use of harmful drift nets in fishing, protecting primary forests, or staving off ozone depletion or global warming? By the panel's reasoning, it seemed that even provisions of international environmental agreements designed to protect global resources could be ruled GATT-illegal. This clash between two different spheres of international law presented the world with a major legal challenge, as it is not always clear which agreement trumps the other in cases where two treaties are in conflict.[22]

Despite the furor over the tuna-dolphin decision, the WTO struck against another law in 1998, ruling against a U.S. measure aimed at reducing unintended sea turtle mortality as a byproduct of shrimp trawling. Sea turtles are both extremely endangered and highly mobile, making international action to protect them a high priority. The provisions of the U.S. law in question closed the lucrative U.S. shrimp market to countries that do not require their shrimpers to use turtle excluder devices (TEDs), simple but highly effective pieces of equipment that prevent turtles from getting ensnared in shrimp nets, or that do not have comparable policies in place. Spurred by the threat of U.S. trade restrictions, 16 nations, including 13 in Latin America plus

Indonesia, Nigeria, and Thailand, have by now moved to require the use of TEDs. India, Malaysia, and Pakistan chose a different tack, however, deciding to launch a WTO challenge rather than meeting the U.S. requirement. (Thailand joined them in this effort as a matter of principle, even though it had adopted TEDs.)[23]

Although the environmental effectiveness of the U.S. law was clear, both the initial WTO dispute resolution panel and a subsequent appeals panel concluded in 1998 that the measure violated WTO rules. The legal reasoning of the appeals panel was an improvement over earlier rulings, as it acknowledged that countries may in some circumstances be justified in using trade measures to protect global resources. But the panel nonetheless took issue with the way in which the U.S. law had been implemented, arguing that it was applied in an arbitrary manner that failed to treat countries evenhandedly. The bottom line was that the U.S. law would have to be changed in order to comply with WTO rules. This outcome was particularly alarming for environmentalists, as the Uruguay Round had strengthened the rules of dispute resolution proceedings to make rulings binding, and to provide for tougher trade retaliation in cases where countries are unwilling to change offending laws in order to adhere to panel findings.[24]

In response to the ruling, the U.S. government altered the way it was implementing the law without seeking any changes to the statute itself. The new guidelines provide for the import of specific shipments of shrimp that have been approved as turtle-safe even if the country as a whole has not met the certification requirements. It remains to be seen whether this response will satisfy the WTO, thus precluding the imposition of trade sanctions against the United States.[25]

In any case, many U.S. environmentalists are unhappy with the government's response. Their primary concern is that the shipment-by-shipment method will be less effective in safeguarding turtles than the earlier blanket restriction, as it will not compel countries to mandate the use of TEDs when fishing for shrimp not destined for the U.S. market. A turtle might thus survive an encounter with a TED-equipped boat only to later fall prey to a TED-free vessel. Environmentalists also worry that the new policy may facilitate the entry of "laundered" shrimp into the United States. Several environmental groups filed suit against the government at the U.S. Court of International Trade, charging that the revised guidelines were inconsistent with provisions of the Endangered Species Act that stipulate adequate protection for sea turtles. In a preliminary ruling in April 1999, the court sided with the environmental groups, placing national law and international trade rules on a possible collision course.[26]

The clash between two different spheres of international law presents the world with a major legal challenge.

Besides mandating strengthened enforcement, the package that created the WTO also included an Agreement on the Application of Sanitary and Phytosanitary (SPS) Measures that imposes new restrictions on laws designed to protect human, animal, and plant health. Trade specialists had argued that legislators were passing disingenuous laws that lacked a scientific rationale, with the primary goal of keeping foreign products off of their shelves. Although the ostensible reason for the SPS agreement was to prevent countries from using health and safety standards as disguised trade barriers, the worry is that legitimate laws will also run afoul of its rules.[27]

The European Union argues that this is what is happening in its ongoing dispute with the United States over an EU law that forbids the sale of meat produced using growth hormones. The European Community and the United States have been locking horns over the EU's beef-hormone ban for several years now. Since it went into effect in the late 1980s, the law has always applied equally to domestically raised and imported livestock, and has thus passed the WTO's bedrock test of nondiscrimination. The EU insists the ban is not an intentional trade barrier at all, but only a prudent response to public concern that eating hormone-treated beef might cause cancer and reproductive health problems. But the hormone-hooked U.S. livestock industry was threatened by the law, which blocks hundreds of millions of dollars worth of U.S. beef exports, and it prevailed upon the U.S. government to take up its cause at the WTO.[28]

This effort culminated in February 1998 when a WTO appeals panel ruling upheld an earlier dispute panel ruling that the European law violated WTO rules. In July 1999, the U.S. government imposed WTO-approved retaliatory sanctions on the EU for its refusal to accept U.S. hormone-treated beef, slapping 100-percent tariffs on $116.8 million worth of European imports, including fruit juices, mustard, pork, truffles, and Roquefort cheese. The U.S. sanctions were greeted with widespread consternation in Europe, particularly in France, where a number of McDonald's restaurants were targeted for protests. So far, the EU has refused to back down.[29]

The panelists' primary argument against the EU law was that it was based on inadequate risk assessment. They explicitly rejected the EU's defense that the import restriction was justified by the precautionary principle—a basic tenet of international environmental law that is steadily gaining ground. The Rio Declaration on Environment and Development, for example, which was agreed to at the June 1992 Earth Summit, declares that: "Where there are threats of serious or irreversible damage, lack of full scientific certainty shall not be used as a reason for postponing cost-effective measures to prevent environmental degradation."[30]

The WTO's provisions, on the other hand, require that health and safety laws be based on scientific principles and not be maintained with insufficient scientific evidence. Although on the face of it these requirements sound reasonable enough, in practice countries often disagree about how much evidence is "sufficient" to justify preventative measures. The WTO shifts the burden of proof to in effect require that chemicals and other food additives be proved harmful before their use can be restricted. The problem with this approach is that extensive testing, sometimes over a period of years, is required to know if a substance has long-term cumulative effects that might cause cancer, damage to the immune system, or other serious ailments.[31]

The beef hormone dispute is widely viewed as just a warm-up for a more serious trade controversy now brewing over genetically modified organisms. Once again, the European Union and the United States are the primary antagonists. Prompted by public concern over the health and ecological effects of GMOs, the EU passed legislation in 1998 requiring all food products containing genetically modified soybeans or corn to be labeled as such. Several other countries, including Australia, Brazil, Japan, and South Korea, are now following suit. A large share of food products made by U.S. companies—breads, salad oils, and ice cream, among them—now contain GMOs. Many European producers, in contrast, are steering clear of GMOs in the face of public concern. U.S. companies complain that the labeling requirements amount to trade barriers, and the U.S. and Canadian governments are now making this same point at the WTO and in other international forums.[32]

In February 1999, a proposed biosafety protocol to the U.N. Convention on Biological Diversity became the first major victim of the growing international trade war over GMOs. Negotiations under way for a few years had been aimed at putting in place a system of prior consent for the transport of genetically engineered seeds and products. The talks were scheduled to wrap up in Cartagena, Colombia, in February, but six major agricultural exporters—Argentina, Australia, Canada, Chile, the United States, and Uruguay—put a monkey wrench into these plans by blocking adoption of the accord. One of the main U.S. arguments against the protocol was a claim that its provisions ran counter to the rules of the WTO. As of November 1999, negotiators were still hoping to bridge the differences.[33]

The environmental impact of freeing trade in forest products is another controversial issue looming on the trade horizon. Under a proposed agreement now under consideration, most industrial countries would eliminate tariffs on pulp and paper by 2000, and on wood and other forest products such as furniture by 2002. Developing countries would be given an additional two years to meet these terms. The precise effects of these steps are difficult to predict, but studies suggest that the higher prices paid to producers as a result of tariff reductions will boost production significantly in some countries. A recent U.S. government report concluded that the agreement would likely increased production by nearly 3 percent in Malaysia and over 4 percent in Indonesia, although the report also forecasts production declines in some countries, including Mexico and Russia. With so little of today's timber industry based on sustainable practices, production increases often translate into increased forest destruction.[34]

The beef hormone dispute is just a warm-up for a serious trade controversy over genetically modified organisms.

Although the proposed accord would initially take aim at tariffs alone, its scope might well be expanded in the future to include nontariff barriers to trade. Over the longer term, these provisions might pose a greater threat to the health of the world's forests, and to the diversity of species that inhabit them. What looks to one country or company to be a "nontariff barrier to trade" is often viewed elsewhere as a legitimate environmental law. Even under existing agreements, concern is rising that measures designed to minimize the introduction of harmful exotic species will run afoul of WTO rules. Forest certification initiatives, aimed at creating a market for sustainably harvested timber, could also run head-on into WTO rules in the years ahead. (See also Chapter 6.)[35]

As opposition to the WTO continues to mount, many governments are beginning to acknowledge, rhetorically at least, that reforms are needed to make the world trading system environmentally sound. One idea gaining support is to enlist the WTO in an effort to reduce environmentally harmful subsidies. World trade rules have long discouraged subsidies, as they distort the economic playing field. The United States and six other nations have suggested building on this tradition by making the elimination of fishing subsidies an objective for the upcoming round of trade talks. These subsidies, which add up to some $14–20 billion annually, help propel overcapacity in the world's fishing fleet, which is itself a powerful driving force behind today's depleted fisheries. Other environmentally harmful payouts could also be tackled at the WTO, including multibillion-dollar agricultural, energy, and forestry subsidies.[36]

Taking on environmentally harmful subsidies would be an important step forward, but it does not let governments off the hook for amending existing WTO rules to buffer environmental laws from trade challenges. Among the priorities for reform are clearly incorporating the precautionary principle into WTO rules, protecting consumers' right to know about the health and environmental impact of products they purchase by safeguarding labeling programs, recognizing the legitimacy of distinguishing among products based on how they were produced, providing deference to multinational environmental agreements in cases where they conflict with WTO rules, ensuring the right of countries to use trade measures to protect the global commons, and opening the WTO to meaningful public participation. These changes are imperative if the WTO is to gain the public support it needs to stay in business.[37]

Greening the International Financial Architecture

During the 1990s, money became increasingly mobile in response to a range of factors, including the takeoff in computerized trading as well as the deregulation of international capital markets. International investment surged in response, particularly into the newly established stock markets of the developing world.[38]

Private capital inflows into developing countries and into the former Eastern bloc increased from only $53 billion at the beginning of the 1990s to an all-time high of $302 billion in 1997. (See Figure 10–4.) Large parts of Asia and Latin America were suddenly transformed in the minds of international investors from poor, "developing countries" into glistening "emerging markets." At the same time, new financial instruments such as hedge funds and derivatives created an explosion of foreign exchange trading, with an astounding $1.5 trillion now changing hands every day.[39]

But in 1997 the bubble burst. Thailand was the first economic domino to fall, when it was forced to devalue the country's currency sharply after it came under sustained speculative attack. The crisis soon spread to Indonesia, Malaysia, the Philippines, and South Korea. International investors lost their nerve and raced for the exits. Some $22 billion in 1997 and $30 billion in 1998 flowed out of the Asian countries in crisis. Banks failed and stock markets collapsed, sending the economies of the region into a tailspin. And the crisis did not stop at the continent's edge. Shaken by the Asian experience, investors began to pull money out of emerging markets everywhere. Russia's currency and stock market went into a free fall in late August 1998, forcing the country to default on $40 billion in international loans. Brazil appeared to be the next domino waiting to fall, prompting the International Monetary Fund to step in with a $42-billion bailout plan.[40]

The social and environmental fallout from the crisis was severe. In battered Asia,

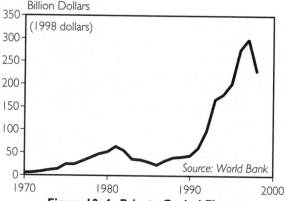

Figure 10–4. Private Capital Flows to Developing Countries, 1970–98

tens of millions of people fell into poverty as jobs were slashed, and as many as a million children were pulled out of school, with some of them pushed into prostitution by their desperate families. Growing poverty tied to the crisis also had environmental costs, such as a surge in Indonesia in the poaching of endangered monkeys, tigers, and other species as destitute people sought quick cash. And environmental spending was slashed to the bare bones by governments and businesses in crisis countries in order to stave off bankruptcy.[41]

The global economic crisis now appears to be in at least temporary remission. International capital has begun to return to most of the countries affected by the crisis, and economic growth rates are again headed upwards, although poverty rates have yet to respond. But many commentators warn against a false sense of complacency. They predict that the next jolt is not far off, as the globalization of international finance has outpaced the ability of governments and international institutions to manage the system effectively.[42]

If there was a silver lining to the crisis, it was the wake-up call it provided about the risks of rapid globalization, and the subse-

quent launching of a critical international dialogue about how to reform the international financial architecture to meet the demands of the twenty-first century. As Professor Dani Rodrik of Harvard University puts it: "markets are sustainable only insofar as they are embedded in social and political institutions. . . . It is trite but true to say that none of these institutions exists at the global level." As the process of devising international policies for a globalizing world proceeds, the need to protect the natural resource base that underpins the global economy merits a prominent place on the agenda.[43]

The logical place to begin any discussion of a new financial architecture is with existing structures, principally the International Monetary Fund and the World Bank. The IMF came under particularly close scrutiny in the wake of the economic crisis. The Fund's high-profile role as a conduit for multibillion-dollar bailout packages for the crisis-stricken countries was a clear demonstration of the organization's formidable powers. But it also stirred controversy, as prominent economists took issue with the wisdom of the institution's financial advice and the secrecy in which its operations are shrouded.[44]

Although the World Bank maintained a lower profile during the crisis, it has also been active in channeling funds into the crisis-ridden countries, often in close cooperation with the IMF. Over the last few years, the Bank has substantially boosted both its total lending and the share of its funds that are spent on cash infusions for "structural adjustment." Total Bank lending reached $29 billion in 1999, up from just $19 billion two years earlier. And more than half of the Bank's total lending in 1999 was for structural adjustment, compared with only 27 percent in 1997. Under conventional "structural adjustment" loans as well as the crisis-generated bailout packages, countries receiving funds agree to implement a long and specific list of policy changes intended to restore them to economic health and thus to creditworthiness. Privatization, price and exchange rate stability, and trade liberalization are among the policies commonly recommended.[45]

But the World Bank and the IMF pay insufficient heed to the profound effects of these policies on the ecological health and the social fabric of recipient countries. One important component of most adjustment loans is policies aimed at boosting exports in order to generate foreign exchange with which to pay back debts. Yet the pressure to export can lead countries to liquidate natural assets such as fisheries and forests, thereby undermining longer-term economic prospects. Intensive export-oriented agriculture is also often promoted, sometimes at the expense of small-scale farmers and indigenous peoples. At the same time that structural adjustment loans promote exports of environmentally sensitive commodities, they also often require countries to make Draconian cuts in government spending, causing the budgets of already overburdened environment and natural resource management ministries to plummet.[46]

All these effects are evident in the recent bailout packages. In Indonesia, the IMF encouraged more palm oil production as part of its broader strategy for pulling the country out of its economic crisis, mandating that the country remove restrictions on foreign investment in this sector. Yet rapid growth in palm oil exports has been a major contributor to the decimation of Indonesia's biologically rich tropical forests in recent years, raising profound questions

about the wisdom of pushing such exports further still. Exports of palm oil climbed from 1.4 million to nearly 3 million tons between 1991 and 1997. The devastating Indonesia wildfires of recent years were sparked in part by fires deliberately set to clear land for oil palm and pulpwood plantations.[47]

Environmental spending has declined markedly in the crisis-ridden countries, including Indonesia, the Philippines, South Korea, and Thailand. In Russia, the budget for protected areas was recently cut by 40 percent. And Brazil agreed as part of a recent pact with the IMF to cut its environmental spending by two thirds. A key international program aimed at protecting the Amazonian rainforest from destruction by ranchers, loggers, farmers, and miners is one of the programs to face the chopping block. The timing of these cuts was particularly poor in light of the unusually high rates of deforestation in the Brazilian Amazon over the last few years.[48]

Although structural adjustment programs often lead to environmental harm, they have also been used in a few cases to promote environmentally beneficial policy changes. In 1996 and 1997, the IMF suspended loans to Cambodia after government officials awarded logging concessions to foreign firms that threatened to open up the country's entire remaining forest area to exploitation—while funneling tens of millions of dollars into the bank accounts of the corrupt officials. And despite its worrisome provisions for stepped-up natural resource exports, the recent Indonesian bailout plan also included several provisions intended to benefit forest management in the country.[49]

As part of an assault on the country's tradition of "crony capitalism," the Indonesian bailout plan required a number of reforms to the country's corruption-laden forestry sector, including tighter control over a government reforestation fund, the revenues of which had more often been used to line the pockets of President Suharto's political allies than to plant trees. The bailout package also included several measures aimed specifically at protecting forests, such as reducing land conversion targets to environmentally sustainable levels, instituting an auctioning system for handing out concessions, and imposing new "resource rent" taxes on timber sales. Although these reforms were a step in the right direction, their ultimate effect on deforestation rates in the country remains to be seen.[50]

Besides using their influence to discourage unsustainable levels of natural resource exploitation, the IMF and the World Bank are well placed to promote environmentally beneficial fiscal reforms, such as cuts in environmentally harmful subsidies or the imposition of pollution taxes. They could both also help promote improvements in environmental accounting, such as incorporating the depletion of natural resources into national income figures. As things now stand, the destruction of natural assets such as forests, fisheries, and minerals is not typically included in national income figures, which means that policymakers are working from an incomplete set of books. And the IMF could include environmental issues in its mandate to conduct "surveillance" of the economic prospects of its member countries, in part by tracking environmental spending levels and structural adjustment–mandated legislative changes that affect the environment.[51]

Despite the clear links between economic and environmental health, the IMF has long resisted the idea that environmental issues have much to do with its mission.

When the organization was first created, its primary role was to help to tide countries over when they faced short-term liquidity problems rather than to help them meet longer-term development goals. But the abandonment of fixed exchange rates in the 1970s deprived the IMF of much of its original mandate. Since then it has become increasingly involved with issues of longer-term development, such as its prominent role in brokering debt restructuring deals in the 1980s. The Fund now accepts that issues such as fighting corruption and alleviating poverty intersect with its mission. It is difficult to see why environmental protection should be any different.[52]

The IMF has long resisted the idea that environmental issues have much to do with its mission.

On paper, the development-oriented World Bank has been far more open than the IMF to the idea that environmental concerns should be integrated into its structural adjustment lending. The Bank's policy that governs adjustment lending stipulates that the environmental impact of these loans should be fully considered as they are prepared, with a view toward promoting possible synergies and avoiding environmentally harmful results. But an internal review of more than 50 recent loans found few that paid much heed to environmental and social matters. Whereas a 1993 Bank report found that some 60 percent of adjustment loans included environmental goals, the recent study concluded that this share has now plummeted to less than 20 percent. An added problem is the fact that the Bank's policy on environmental impact assessment does not cover broad-based structural adjustment lending, although it is supposed to be applicable to adjustment loans aimed at specific sectors, such as agriculture or energy. IMF loans are also not subject to environmental impact assessment.[53]

Despite the World Bank's growing role in adjustment lending, project lending remains a mainstay of its activities. The Bank has traditionally made loans only to governments, but in the last few years it has increasingly emphasized supporting the private sector. It has done this both by using its own funds to guarantee private-sector projects and by stepping up the operations of two affiliated agencies, the International Finance Corporation (IFC) and the Multilateral Investment Guarantee Agency (MIGA). The IFC lends directly to private enterprises, while MIGA insures against political risks, such as expropriation, civil disturbance, and breach of contract. At last count, in 1995, the World Bank estimated that some 10 percent of all private-sector investment in the developing world was supported at least indirectly by its various private-sector programs.[54]

After more than a decade of pressure from NGOs and determined efforts by committed insiders, the World Bank now has an extensive set of environmental and social policies, which among other things cover environmental impact assessments of projects, forestry lending, involuntary resettlement, protection of wilderness areas, the rights of indigenous peoples, and pest management. The IFC and MIGA both recently issued their own parallel policies, and the World Bank published an updated Pollution Prevention and Abatement Handbook, which provides detailed pollution reduction guidelines for nearly 40 industries. In theory, Bank agencies are bound by their own policies, although the Bank admits that it has a tarnished history in following its own rules. The importance of the World Bank's standards is magnified by the fact that they are often looked to by private investors as the prevailing international norm.[55]

As World Bank environmental and social standards were strengthened over the last decade, private investors turned increasingly to bilateral export credit agencies to find support for projects that no longer pass muster at the Bank. Export credit support climbed from $24 billion in 1988 to $105 billion in 1996. All told, bilateral export promotion in the form of loans and investment insurance now underwrites more than 10 percent of all world trade. Bilateral export promotion often supports environmentally disruptive projects, including mines, pipelines, and hydroelectric dams.[56]

The U.S. government has had environmental policies in place for several years at its main export promotion agencies, the U.S. Overseas Private Investment Corporation and the U.S. Export-Import Bank. But in a global economy, tough national standards can easily be undermined by laggards abroad. The United States learned this lesson the hard way a few years back when its Export-Import Bank refused on environmental grounds to extend credits to companies such as the heavy equipment manufacturer Caterpillar that wanted to participate in China's controversial Three Gorges dam project. The Bank's counterparts in Canada, France, Germany, Japan, and Switzerland stepped into the breach. Stung by the experience, the United States is working to persuade other donor countries to develop environmental guidelines for their export finance agencies. Several countries are now in the process of developing such standards, including Canada, Japan, Norway, and the United Kingdom. Negotiations are also under way to create common environmental standards for the export finance agencies of the major industrial countries. Nongovernmental activists are pushing for these to be set at a high level.[57]

But even if this initiative succeeds, private capital markets can still be tapped for environmentally problematic projects. In the Three Gorges case, a number of prominent investment banks—including Lehman Brothers, Morgan Stanley, and Smith Barney—have sponsored bond offerings over the last few years to help the Chinese government raise funds for the dam. Although convincing private financiers to pay attention to the environment is substantially more difficult than lobbying public institutions such as the World Bank and export credit agencies, several efforts are afoot to encourage a heightened environmental consciousness on the Wall Streets of the world.[58]

A U.N. Environment Programme (UNEP) initiative launched in 1992 encourages commercial banks to incorporate environmental considerations into their lending programs. So far, 162 banks from 43 countries have signed onto the initiative's Statement by Banks on the Environment and Sustainable Development. The signatories underscore their expectation that borrowers must comply with "all applicable local, national, and international environmental regulations." They also pledge to update their accounting procedures to reflect environmental risks, such as the potential for chemical accidents or hidden hazardous waste dumps, and to develop banking products and services that promote environmental protection. Although laudable in its goals, the UNEP statement is short on specific commitments. In fact, several signatories were involved with the recent Chinese bond offerings that activists charge are helping to finance the Three Gorges project. In

order to minimize gaps between rhetoric and reality, the U.K.-based Green Alliance suggests strengthening the initiative by transforming the statement into a document whose expected standards of performance are clear enough to be subjected to the scrutiny of an audit.[59]

In a global economy, tough national standards can easily be undermined by laggards abroad.

Environmental liabilities could also be better incorporated into the way stock markets are regulated. Companies operating in the United States are required to disclose large environmental risks on the forms they file with the Securities and Exchange Commission. But the information varies widely in quality, with many companies submitting no data at all. Developing countries are particularly well placed to write environmental rules into the regulations governing newly established stock markets. Thailand, for one, requires companies listed on the Stock Exchange of Thailand to undergo an environmental audit that includes an environmental impact assessment as well as a site visit.[60]

If financial markets are to reflect environmental risks adequately, transparent information about corporate environmental performance is essential. The last several years have seen an explosion of interest in environmental reporting, but existing efforts have been poorly coordinated, leading to a proliferation of "non-standardized information reported in non-uniform formats," according to the Boston-based Coalition for Environmentally Responsible Economies (CERES). In an effort to address this deficiency, CERES has launched a Global Reporting Initiative in which corporations, NGOs, professional accounting firms, and UNEP are working together to produce a global set of guidelines for corporate sustainability reporting. The goal of the initiative is to elevate environmental reporting to the same plane as financial reporting, making it standard business practice worldwide.[61]

Innovations in Global Environmental Governance

The new rules of the global economy are for the most part being set by institutions such as the World Trade Organization and the International Monetary Fund, where the mindset of traditional economists prevails and where the "rules" are generally aimed at unshackling global commerce rather than harnessing it for the common good. But forging an environmentally sustainable society is about more than economics, and farsighted economics is about more than reducing restrictions on the movement of goods and money. Creating a global society fit for the twenty-first century will thus require not only reform of economic institutions, but also a strengthening of international environmental institutions so that they can act as an ecological counterweight to today's growing economic powerhouses.[62]

A good place to start is with the hundreds of agreements, declarations, action plans, and international treaties on the environment that now exist. Environmental treaties alone number more than 230; agreement on more than three fourths of them has been reached since the first U.N. conference on the environment was held in Stockholm in 1972. (See Figure 10–5.) These accords cover atmospheric pollution, ocean despoliation, endangered species protection, haz-

ardous waste trade, and the preservation of Antarctica, among other issues.[63]

Judging from the number of treaties, environmental diplomacy over the past few decades appears to have been a spectacular success. And many of these accords have in fact led to important results, such as the facts that chlorofluorocarbon emissions dropped 87 percent from their peak in 1988 as a result of the 1987 Montreal Protocol on ozone depletion, the killing of elephants plummeted in Africa following a 1990 ban on commercial trade in ivory, the annual whale take declined from more than 66,000 in 1961 to some 1,100 today as a result of agreements forged by the International Whaling Commission, the volume of oil spilled into the ocean has declined by 60 percent since 1981 even with a near doubling in oil shipments in response to International Maritime Organization regulations, and mining exploration and development have been forbidden in Antarctica for 50 years under a 1991 accord.[64]

Yet even as the number of treaties climbs, the condition of the biosphere continues to deteriorate. Carbon dioxide levels in the atmosphere have reached record highs, scientists are warning that we are in the midst of a period of mass extinction of species, the world's major fisheries are depleted, and water shortages loom worldwide. The notoriously slow pace of international diplomacy needs to be reconciled with the growing urgency of protecting the planet's life-support systems.

The main reason that environmental treaties have so far mostly failed to turn around today's alarming environmental trends is because the governments that created them have generally permitted only vague commitments and lax enforcement. Governments have also for the most part failed to provide sufficient funds to imple-

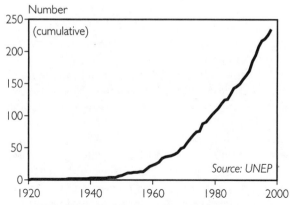

Figure 10–5. International Environmental Treaties, 1921–98

ment treaties, particularly in the developing world. Ironically, environmentalists need to take a page from the WTO and push for international environmental commitments that are as specific and as enforceable as trade accords have become.

One idea now gaining political currency is to upgrade the U.N. Environment Programme into a World Environment Organization (WEO) on a par with the WTO. Although UNEP has had some important successes since it was founded in 1972, it has suffered from meager resources and a limited mandate. Upgrading the status of environmental issues within the U.N. system is long overdue. Still, it is important that debates over form not distract from the ultimately far more important questions of function. A world environment organization could usefully serve as an umbrella organization for the current scattered collection of treaty bodies, just as domestic environment agencies oversee the implementation of national environmental laws. But in order to do so effectively, the treaties themselves would need to stipulate strong enforcement capacities, and the WEO would need to be endowed with sufficient financial resources to catalyze innovative

programs.[65]

Tomorrow's international environmental institutions may turn out to be vastly different in character than the bureaucratic bodies that predominate in many quarters today. A nascent system of international environmental governance is now emerging from diverse quarters, proving that governance is no longer just for governments.[66]

In recent years, the private sector has itself become increasingly and often controversially enmeshed in the standard-setting business. One prominent example is the voluntary environmental management guidelines forged by the Geneva-based International Organization for Standardization, a worldwide federation of national standards-setting bodies. Although these guidelines should not to be confused with actual environmental performance standards, they are nonetheless a useful tool. But the credibility of the process has suffered from the fact that it is widely perceived to be industry-dominated.[67]

Another type of international standard setting is embodied in the numerous eco-labeling initiatives now beginning to take hold. One strength of these efforts is the diverse range of stakeholders they bring to the table. The organic agriculture community was an early pioneer. As far back as the early 1970s, it came together through the International Federation of Organic Agriculture Movements to stipulate conditions that farmers must meet in order to claim organic credentials. More recently, the Forest Stewardship Council was founded in 1993 to set standards for sustainable forest production through a cooperative process involving timber traders and retailers as well as environmental organizations and forest dwellers. And a Marine Stewardship Council has now been formed to devise criteria for sustainable fish harvesting.[68]

A particularly encouraging development of recent years has been the steady growth of the international nongovernmental movement. Environmental activists are flourishing at the national and grassroots level in most corners of the globe. Growth has been particularly rapid in the developing world and in Eastern Europe, where democratization over the last decade has opened up political space for NGOs. The number of NGOs working across international borders has also soared over this century, climbing from just 176 in 1909 to more than 23,000 in 1998. Environmental groups have risen steadily as a share of the total, climbing by one count from just 2 percent of transnational social change NGOs in 1953 to 14 percent in 1993.[69]

Empowered by e-mail and the Internet, environmental activists have gradually organized themselves into a range of powerful international networks. To name but a few, the Climate Action Network links more than 250 international groups and national organizations active on climate change; the Pesticide Action Network includes at least 500 consumer, environment, health, labor, agriculture, and public interest groups worldwide; the World Forum of Fish Workers & Fish Harvesters brings together people from small-scale fishing communities on six continents; the International POPs Elimination Network coordinates hundreds of NGOs worldwide in their push for an effective treaty to limit persistent organic pollutants; and the Women's Environment and Development Organization links activists from around the world who are committed to increasing female participation in decisionmaking at the United Nations and in other international forums where issues of concern such as population stabilization are discussed.[70]

Among their many accomplishments, NGOs have educated millions of people about environmental issues, and then effectively harnessed the power of a knowledgeable citizenry to pierce the veil of secrecy that all too often surrounds both international negotiations and corporate decisionmaking. Recent NGO successes include exposing the environmental deficiencies of a proposed Multilateral Agreement on Investment and thereby bringing it to a halt, and slowing the introduction of genetically modified organisms worldwide until their effects on both human and ecological health are better understood.[71]

Thirty years ago, photographs of Earth taken from space by the Apollo expeditions indelibly impressed on all who saw them that our planet, while divided by political boundaries, is united by ecological systems. These photos helped inspire the first Earth Day, which in turn motivated numerous countries to pass environmental laws and create environmental ministries. This year, the world will celebrate Earth Day 2000. The time has come for a comparable groundswell in support of the international governance reforms that are needed to safeguard the health of the planet in the new millennium.

Notes

Chapter 1. Challenges of the New Century

1. Wright brothers flight from *Encyclopedia Britannica* on-line, <www.britannica.com>, viewed 25 October 1999.

2. Martin Campbell-Kelly and William Aspray, *Computer: A History of the Information Machine* (New York: Basic Books, 1996).

3. Nicholas Denton, "Microsoft Capitalisation Exceeds $200bn," *Financial Times,* 26 February 1998.

4. Dow Jones Industrial Average from Dow Jones & Company, <averages.dowjones.com>, viewed 25 October 1999.

5. Cropland in Kazakhstan from U.S. Department of Agriculture (USDA), *Production, Supply, and Distribution,* electronic database, Washington, DC, updated November 1999; Atlantic swordfish from Lisa Speer et al., *Hook, Line and Sinking: The Crisis in Marine Fisheries* (New York: Natural Resources Defense Council, February 1997); Aral Sea from Sandra Postel, *Pillar of Sand* (New York: W.W. Norton & Company, 1999); forest products industries from Janet N. Abramovitz, *Taking a Stand: Cultivating a New Relationship with the World's Forests,* Worldwatch Paper 140 (Washington, DC: Worldwatch Institute, April 1998); Chesapeake Bay from John Jacobs, Maryland Department of Natural Resources, unpublished printout sent to author, 3 August 1994.

6. Population of sub-Saharan Africa from Population Reference Bureau (PRB), "1999 World Population Data Sheet," wall chart (Washington, DC: June 1999); prevalence and number infected in sub-Saharan Africa from Joint United Nations Programme on HIV/AIDS (UNAIDS), *AIDS Epidemic Update: December 1998* (Geneva: December 1998).

7. Water pumping in India from David Seckler et al., "Water Scarcity in the Twenty-First Century" (Colombo, Sri Lanka: International Water Management Institute, 27 July 1998).

8. United Nations, *World Population Prospects: The 1998 Revision* (New York: December 1998).

9. Carbon Dioxide Information Analysis Center, *Trends: A Compendium of Data on Global Change* (Oak Ridge, TN: Oak Ridge National Laboratory, 1998), <cdiac.esd.ornl.gov>, viewed 25 October 1999; C.D. Keeling and T.P. Whorf, "Atmospheric CO_2 Concentrations (ppmv) Derived From In Situ Air Samples Collected at Mauna Loa Observatory, Hawaii," Scripps Institution of Oceanography, La Jolla, CA, July 1999.

10. Figure 1–1 from James Hansen et al., Goddard Institute for Space Studies, Surface Air Temperature Analyses, "Global Land-Ocean Temperature Index," <www.giss.nasa.gov/

update/gistemp>, viewed 4 November 1999.

11. Tom M.L. Wigley, *The Science of Climate Change: Global and U.S. Perspective* (Arlington, VA: Pew Center on Global Climate Change, June 1999).

12. Paul Epstein et al., *Marine Ecosystems: Emerging Diseases as Indicators of Change*, Health, Ecological and Economic Dimensions of the Global Change Program (Boston: Center for Health and Global Environment, Harvard Medical School, December 1998).

13. Melting ice caps and glaciers from Greenpeace International, "Climate Change and the Earth's Mountain Glaciers: Observations and Implications," May 1998, <www.greenpeace.org/~climate/arctic99/reports/glaciers2.htm>, viewed 25 October 1999; "Melting Himalayan Glaciers Pose Flooding Dangers," *Reuters*, 3 June 1999; Liu Jun, "Cold Comfort for Yangtze Glaciers," *China Daily*, 20 January 1999; Richard Monastersky, "Sea Change in the Arctic," *Science News*, 13 February 1999; Joby Warrick, "As Glaciers Melt, Talks on Warming Face Chill," *Washington Post*, 2 November 1998; William K. Stevens, "Dead Trees and Shriveling Glaciers as Alaska Melts," *New York Times*, 18 August 1998; U.S. and British report from "Melting of Antarctic Ice Shelves Accelerates," *Environment New Service*, 9 April 1999.

14. Alps discovery from John Noble Wilford, "Move Over, Iceman! New Star From the Andes," *New York Times*, 25 October 1995; Yukon discovery from James Brooke, "Remains of Ancient Man Discovered in Melting Canadian Glacier," *New York Times*, 25 August 1999.

15. Postel, op. cit. note 5; 480 million being fed with food produced with the unsustainable use of water based on an average annual diet of 300 kilograms, or 0.3 tons, of grain.

16. Seckler et al., op. cit. note 7; annual addition to India's population from United Nations, op. cit. note 8; malnourished children in India from U.N. Food and Agriculture Organization (FAO), *The State of Food Insecurity in the World*

(Rome: 1999); 6 million children from World Health Organization, *The World Health Report 1998* (Geneva: 1998).

17. Liu Yonggong and John B. Penson, "China's Sustainable Agriculture and Regional Implications," paper presented to the symposium on Agriculture, Trade and Sustainable Development in Pacific Asia: China and Its Trading Partners, Texas A&M University, College Station, TX, 12–14 February 1998; for a more detailed analysis, see Lester R. Brown and Brian Halweil, "China's Water Shortage Could Shake World Food Security," *World Watch*, July/August 1998.

18. World grainland from USDA, op. cit. note 5; population from United Nations, op. cit. note 8.

19. Grainland from USDA, op. cit. note 5; population from United Nations., op. cit. note 8.

20. USDA, op. cit. note 5.

21. FAO, *Yearbook of Fishery Statistics: Catches and Landings* (Rome: various years), with 1990–97 data from Maurizio Perotti, fishery statistician, Fishery Information, Data and Statistics Unit, Fisheries Department, FAO, Rome, letter to Anne Platt McGinn, Worldwatch Institute, 10 November 1998; marine biologists from FAO, *The State of World Fisheries and Aquaculture, 1996* (Rome: 1997).

22. United Nations, op. cit. note 8.

23. Forested area from Dirk Bryant, Daniel Nielsen, and Laura Tangley, *The Last Frontier Forests, Ecosystems and Economies on the Edge* (Washington, DC: World Resources Institute, 1997); population from United Nations, op. cit. note 8.

24. Jonathan Baillie and Brian Groombridge, eds., *1996 IUCN Red List of Threatened Animals* (Gland, Switzerland: World Conservation Union–IUCN, 1996).

25. Thomas Kuhn, *The Structure of Scientific Revolutions*, 3rd ed. (Chicago, IL: University of

Chicago Press, 1996).

26. Overpumping from Postel, op. cit. note 5; 480 million being fed with food produced with the unsustainable use of water based on an average annual diet of 300 kilograms, or 0.3 tons, of grain.

27. Share of world's food from irrigated land from Postel, op. cit. note 5.

28. Annual growth in global economy from Worldwatch update of Angus Maddison, *Monitoring the World Economy 1820–1992* (Paris: Organisation for Economic Co-operation and Development, 1995), and from International Monetary Fund, *World Economic Outlook* (Washington, DC: December 1998).

29. Centers for Disease Control and Prevention (CDC), "History of the 1964 Surgeon General's Report," CDC's Tobacco Information and Prevention Sourcepage, <www.cdc.gov/nccdphp/osh>, viewed 27 October 1999.

30. Milo Geyelin, "Forty-Six States Agree to Accept $206 Billion Tobacco Settlement," *Wall Street Journal*, 23 November 1998; David Stout, "Justice Dept. Plans Tobacco Suit Seeking Billions in Health Costs," *New York Times*, 22 September 1999.

31. Amy Ridenour, "We Should All Be Worried by Foreign Tobacco Lawsuits," *Houston Chronicle*, 9 June 1999; Saundra Terry, "Cigarette Firms Suited by Foreign Governments," *Washington Post*, 17 January 1999.

32. Mike R. Bowlin, ARCO, "Clean Energy: Preparing Today for Tomorrow's Challenges," presented at Cambridge Energy Research Associates 18th Annual Executive Conference: Globality & Energy: Strategies for the New Millennium, Houston, TX, 9 February 1999.

33. Corporate plans and Shell hydro station from Fred Pearce, "Iceland's Power Game," *New Scientist*, 1 May 1999.

34. Road moratorium from USDA, Forest Service, "Forest Service Limits New Road Construction in Most National Forests," press release (Washington, DC: 11 February 1999).

35. Tom Kenworthy, "Major Change Sought in Forest Regulations," *Washington Post*, 1 October 1999.

36. Wang Chuandong, "Logging Ban to Transform Timber Industry," *China Daily*, 7 September 1998; "Forestry Cuts Down on Logging," *China Daily*, 26 May 1998.

37. Addition of nearly 80 million from PRB, op. cit. note 6; Zhang Yan, "Reliance on Foreign Lumber to Increase," *China Daily*, 7 September 1998.

38. Ration of 1,000 tons of water for 1 ton of grain from FAO, *Yield Response to Water* (Rome: 1979).

39. Water required to produce the grain and other farm commodities imported into the region from USDA, op. cit. note 5, with conversion as described in note 38; Nile river flow from Postel, op. cit. note 5.

40. Lester R. Brown, *Who Will Feed China?* (New York: W.W. Norton & Company, 1995).

41. For a discussion of this positive reinforcing cycle, see Rodolfo A. Bulatao, *The Value of Family Planning Programs in Developing Countries* (Santa Monica, CA: RAND, 1998); imminent population stability in South Korea, Taiwan, and Thailand from United Nations, op. cit. note 8.

42. Continuing rapid population growth in sub-Saharan Africa and South Asia from United Nations, op. cit. note 8.

43. UNAIDS, op. cit. note 6.

44. Life expectancies from United Nations, op. cit. note 8.

45. Share of beds in South Africa from Robert C.-H. Shell, Population Research Unit, Rhodes University, East London, South Africa, discussion with author, 15 March 1999; Stella Mapenzauswa, "AIDS Weighs on Zimbabwe's Health System," *Reuters*, 21 July 1999.

46. Ssewankambo quoted in Mercedes Sayagues, "How AIDS is Starving Zimbabwe," *Mail & Guardian*, 16 August 1999.

47. Ibid.

48. International Labour Organization (ILO), "Social and Labour Implications of HIV/AIDS," background document prepared for regional tripartite workshop on Strategies to Tackle Social and Labour Implications of HIV/AIDS, organized by ILO in collaboration with UNAIDS, Windhoek, Namibia, 11–13 October 1999.

49. Ibid.; Durban-Westville from Shell, op. cit. note 45.

50. Denmark from Birger Madsen, BTM Consult, Ringkobing, Denmark, letter to Christopher Flavin, Worldwatch Institute, 29 February 1999; Schleswig-Holstein from Andreas Wagner, Fordergesellschaft Windenergie, Hamburg, Germany, e-mail to Christopher Flavin, Worldwatch Institute, 20 January 1999; Navarra from Energia Hidroelectrica de Navarra, S.A., "Projects and Scope of Action of Energia Hidroelectrica de Navarra" (Pamplona, Spain: August 1998); U.S. wind farms from Kent Robertson, American Wind Energy Association, Washington, DC, e-mail to Christopher Flavin, Worldwatch Institute, 26 February 1999.

51. India generating capacity from Madsen, op. cit. note 50; Dutch project in China from Niall Martin, "First Commercial Wind Farm in China," *Windpower Monthly*, September 1998.

52. Share of world electricity from hydroelectric is a Worldwatch estimate based on United Nations, *Energy Statistics Yearbook 1995* (New York: 1997), and on British Petroleum, *BP Statistics Review of World Energy 1997* (London: Group Media & Publications, 1997); wind potential in China is Worldwatch estimate based on World Bank, Asia Alternative Energy Unit,

"China: Renewable Energy for Electric Power" (Washington, DC: 11 September 1996), and on U.S. Department of Energy, Energy Information Administration, *International Energy Outlook 1999* (Washington, DC: March 1999); D.L. Elliott, L.L. Windell, and G.L. Gower, *An Assessment of the Available Windy Land Area and Wind Energy Potential in the Contiguous United States* (Richland, WA: Pacific Northwest Laboratory, 1991).

53. Electric Power Research Institute, *Renewable Energy Technology Characterizations* (Palo Alto, CA: 1997).

54. Paul Maycock, "1998 World Cell/Module Shipments," *PV News*, February 1999.

55. Paul Maycock, "Japan Expands '70,000 Roofs' Program," *PV News*, July 1998; Paul Maycock, "German '100,000 Roofs' Program Details," *PV News*, March 1999; Shell/Pilkington facility from Paul Maycock, Photovoltaic Energy Systems, Inc., Warrenton, VA, e-mail to Brian Halweil, Worldwatch Institute, 1 November 1999; Paul Maycock, "Italian 10,000 Roofs Program Takes Shape," *PV News*, July 1998.

56. Growth in oil and natural gas use from BP Amoco, *BP Amoco Statistical Review of World Energy* (London: Group Media & Publications, 1999); oil peak from Colin J. Campbell and Jean H. Laherrere, "The End of Cheap Oil," *Scientific American*, March 1998.

57. Internet growth from Network Wizards, "Internet Domain Surveys, 1981–1999," <www.nw.com>, updated January 1999.

58. Countries that are restructuring their tax system from David Malin Roodman, "Building a Sustainable Society," in Lester R. Brown et al., *State of the World 1999* (New York: W.W. Norton & Company, 1999); German tax shift of 1998 from Peter Norman, "SPD and Greens Agree German Energy Tax Rises," *Financial Times*, 19 October 1998; Michael Kohlhaas, "Ecological Tax Reform in Germany: Impact and Implications," presentation at American Institute for Contemporary German Studies,

Washington, DC, 21 April 1999.

59. N. Gregory Mankiw, "Gas Tax Now!" *Fortune*, 24 May 1999.

60. U.S. Bureau of the Census, *International Data Base*, electronic database, Suitland, MD, updated 30 November 1998; PRB, op. cit. note 6.

61. PRB, op. cit. note 6.

62. U.N. Population Fund (UNFPA), *The State of World Population* (New York: 1999); UNFPA, "Executive Director's Statement on Withdrawal of U.S. Funding from UNFPA," New York, 20 October 1998.

63. Madsen, op. cit. note 50.

64. UNAIDS, op. cit. note 6; Uganda from World Bank, *Confronting AIDS: Public Priorities in a Global Epidemic* (New York: Oxford University Press, Inc., 1997); funding for HIV/AIDS in Thailand from Phil Guest, Population Council, Horizons Program, Bangkok, Thailand, e-mail to Mary Caron, Worldwatch Institute, 5 March 1999.

Chapter 2. Anticipating Environmental "Surprise"

1. J. Almendares et al., "Critical Regions: A Profile of Honduras," *Lancet*, 4 December 1993; David Waltner-Toews, *Mad Cows and Bad Berries*, IIED Gatekeeper Series 84 (London: International Institute for Environment and Development, 1999).

2. Almendares et al., op. cit. note 1; Waltner-Toews, op. cit. note 1; Lori Ann Thrupp, *Bittersweet Harvests for Global Supermarkets* (Washington, DC: World Resources Institute, 1995).

3. Almendares et al., op. cit. note 1; Waltner-Toews, op. cit. note 1.

4. Almendares et al., op. cit. note 1.

5. Janet N. Abramovitz and Seth Dunn, "Record Year for Weather-Related Disasters," Vital Signs Brief 98-5 (Washington, DC: Worldwatch Institute, 27 November 1998); James C. McKinley, Jr., with William K. Stevens, "The Life of a Hurricane, The Death That It Caused," *New York Times*, 9 November 1998; Paul R. Epstein, "Climate and Health," *Science*, 16 July 1999.

6. For "super-problems," see Norman Myers, "Environmental Unknowns," *Science*, 21 July 1995; for the general lack of research, see Norman Myers, "Two Key Challenges for Biodiversity: Discontinuities and Synergisms," *Biodiversity and Conservation*, September 1996.

7. Forest extent from Emil Salim and Ola Ullsten, *Our Forests, Our Future: Report of the World Commission on Forests and Sustainable Development* (Cambridge, U.K.: Cambridge University Press, 1999); current net forest loss from U.N. Food and Agriculture Organization (FAO), *State of the World's Forests 1999* (Rome: 1999); extinctions from Edward O. Wilson, *The Diversity of Life* (New York: W.W. Norton & Company, 1992).

8. Annual deforestation in the tropics from FAO, op. cit. note 7; fires from Abramovitz and Dunn, op. cit. note 5.

9. William F. Laurance, "A Crisis in the Making: Responses of Amazonian Forests to Land Use and Climate Change," *TREE* (Trends in Ecology and Evolution), October 1998; Daniel C. Nepstad et al., "Large-Scale Impoverishment of Amazonian Forests by Logging and Fire," *Nature*, 8 April 1999; P.M. Fearnside, "Causes of Deforestation in the Brazilian Amazon," in Robert E. Dickinson, ed., *The Geophysiology of Amazonia: Vegetation and Climate Interactions* (New York: John Wiley and Sons, 1987).

10. B.J. Meggers, "Archaeological Evidence for the Impact of Mega-Niño Events on Amazonia During the Past Two Millennia," *Climate Change*, vol. 28 (1994), pp. 321–28; Stephen J. Pyne, *World Fire: The Culture of Fire on Earth* (Seattle: University of Washington Press, 1997).

11. Mark A. Cochrane and Mark D. Schulze, "Forest Fires in the Brazilian Amazon," *Conservation Biology*, October 1998; S. Milius, "Amazon Forests Caught in Fiery Feedback," *Science News*, 3 October 1998; Mark Cochrane et al., "Positive Feedbacks in the Fire Dynamic of Closed Canopy Tropical Forests," *Science*, 11 June 1999.

12. Cochrane and Schulze, op. cit. note 11; Cochrane et al., op. cit. note 11.

13. Nepstad et al., op. cit. note 9.

14. Ibid.; Cochrane et al., op. cit. note 11.

15. Nepstad et al., op. cit. note 9.

16. Laurance, op. cit. note 9.

17. Ibid.

18. Norman Myers, "Our Forestry Prospect: The Past Recycled or a Surprise-Rich Future?" *The Environmentalist*, vol. 17 (1997), pp. 233–44; Laurance, op. cit. note 9.

19. El Niño events and Amazon fires from Laurance, op. cit. note 9, and from Cochrane et al., op. cit. note 11; El Niño events and climate change from Richard A. Kerr, "Big El Niños Ride the Back of Slower Climate Change," *Science*, 19 February 1999, and from Michael J. McPhaden, "Genesis and Evolution of the 1997–98 El Niño," *Science*, 12 February 1999; major carbon-emitting countries from Christopher Flavin and Seth Dunn, *Rising Sun, Gathering Winds: Policies to Stabilize the Climate and Strengthen Economies*, Worldwatch Paper 138 (Washington, DC: Worldwatch Institute, November 1997).

20. Myers, op. cit. note 18.

21. John G. Robinson, Kent H. Redford, and Elizabeth L. Bennett, "Wildlife Harvest in Logged Tropical Forests," *Science*, 23 April 1999.

22. Feng Gao et al., "Origin of HIV-1 in the Chimpanzee *Pan troglodytes troglodytes*," and Robin A. Weiss and Richard W. Wrangham, "From *Pan* to Pandemic," *Nature*, 4 February 1999.

23. S.V. Smith, "Coral-Reef Area and the Contributions of Reefs to Processes and Resources of the World's Oceans," *Nature*, 18 May 1978; Norman Myers, "Synergisms: Joint Effects of Climate Change and Other Forms of Habitat Destruction," in Robert L. Peters and Thomas E. Lovejoy, eds., *Global Warming and Biological Diversity* (New Haven, CT: Yale University Press, 1992).

24. Thomas J. Goreau et al., "Coral Reefs and Global Climate Change: Impacts of Temperature, Bleaching, and Disease," *Sea Wind*, December 1998; Thomas J. Goreau and Ray L. Hayes, "Coral Bleaching and Ocean 'Hot Spots'," *Ambio*, May 1994; Christopher Flavin, "Global Temperature Goes Off the Chart," in Lester R. Brown, Michael Renner, and Brian Halweil, *Vital Signs 1999* (New York: W.W. Norton & Company, 1999); J.R. Petit et al., "Climate and Atmospheric History of the Past 420,000 Years from the Vostok Ice Core, Antarctica," *Nature*, 3 June 1999.

25. Earliest mention of bleaching from P.W. Glynn, "Coral Reef Bleaching: Ecological Perspectives," *Coral Reefs*, December 1993; bleaching and sea surface temperatures (SSTs) from Goreau and Hayes, op. cit. note 24, and from Ove Hoegh-Guldberg, "Climate Change, Coral Bleaching, and the Future of the World's Coral Reefs" (Sydney, Australia: University of Sydney, Coral Reef Research Institute, August 1999); Figure 2–1 from ibid., and from Goreau and Hayes, op. cit. note 24.

26. Recent trends in SSTs are described in Hoegh-Guldberg, op. cit. note 25; for descriptions of coral death in the Indian Ocean, see Joby Warrick, "Hot Year Was Killer for Coral," *Washington Post*, 5 March 1999, and Clive Wilkinson et al., "Ecological and Socioeconomic Impacts of 1998 Coral Mortality in the Indian Ocean: An ENSO Impact and a Warning of Future Change?" *Ambio*, March 1999; coral bleaching as a potentially chronic phenomenon

from Paul Epstein et al., *Marine Ecosystems: Emerging Diseases as Indicators of Change*, Health, Ecological and Economic Dimensions of the Global Change Program (Boston: Center for Health and Global Environment, Harvard Medical School, December 1998); for a description of bleaching in other marine invertebrates, see Osha Gray Davidson, *The Enchanted Braid: Coming to Terms with Nature on the Coral Reef* (New York: John Wiley and Sons, 1998).

27. Nutrient pollution from David Malakoff, "Death by Suffocation in the Gulf of Mexico," and T.D. Jickells, "Nutrient Biogeochemistry of the Coastal Zone," *Science*, 10 July 1998; effect of such pollution on the reefs from Clive Wilkinson and Robert Buddemeier, *Global Climate Change and Coral Reefs* (Gland, Switzerland: World Conservation Union–IUCN (IUCN), 1994), and from K.P. Sebens, "Biodiversity of Coral Reefs: What Are We Losing and Why?" *American Zoologist*, vol. 34 (1994), pp. 115–33.

28. Epstein et al., op. cit. note 26; Davidson, op. cit. note 26.

29. Descriptions of Australasian reef diversity from David Doubilet, "Coral Eden," *National Geographic*, January 1999, and from Dirk Bryant et al., *Reefs at Risk: A Map-Based Indicator of Threats to the World's Coral Reefs* (Washington, DC: World Resources Institute, 1998); mangrove loss from Clive Wilkinson et al., eds., *Proceedings of the Third ASEAN-Australian Symposium on Living Coastal Resources, Vol. 1: Status Reviews* (Townsville, Australia: Australian Institute of Marine Science, 1994).

30. Davidson, op. cit. note 26; P.B. Tomlinson, *The Botany of Mangroves* (Cambridge, U.K.: Cambridge University Press, 1995).

31. Andrew W. Bruckner and Robin J. Bruckner, "Emerging Infections on the Reefs," *Natural History*, December 1997/January 1998; S. Milius, "Bacteria Cause Plague in Coral Reef," *Science News*, 11 April 1998.

32. Thomas J. Goreau et al., "Rapid Spread of Diseases in Caribbean Coral Reefs," *Revista*

Biología Tropical, vol. 46.5 (1998), pp. 157–71; Epstein et al., op. cit. note 26; rapid wasting syndrome from Thomas Goreau, Global Coral Reef Alliance, Chappaqua, NY, discussion with author, 9 September 1999.

33. Goreau et al., op. cit. note 32; R.L. Hayes and N.I. Goreau, "The Significance of Emerging Diseases in the Tropical Coral Reef Ecosystem," *Revista Biología Tropical*, vol. 46.5 (1998), pp. 173–85; "Devastating Losses Affect Florida's Coral Reefs," *Environmental News Network*, <www.enn.com>, viewed 20 November 1997.

34. Laurie L. Richardson, "Coral Diseases: What Is Really Known?" *TREE* (Trends in Ecology and Evolution), November 1998; possible link between black band and pollution from Goreau, op. cit. 32.

35. G. Smith et al., "Caribbean Sea Fan Mortalities," *Nature*, 10 October 1996; G. Smith, "Response of Sea Fans to Infection with Aspergillus spp.," *Revista Biología Tropical*, vol. 46.5 (1998), pp. 205–08; interview with James Porter and Drew Harvell, "Living on Earth," National Public Radio, 29 January 1999; "Sea-Fan-Killing Fungus Caught in the Act," *Environmental News Network*, <www.enn.com>, viewed 15 July 1998.

36. Thomas Goreau, Global Coral Reef Alliance, Chappaqua, NY, e-mail to author, 18 October 1999; seawater concentrations of calcium from Joan A. Kleypas et al., "Geochemical Consequences of Increased Atmospheric Carbon Dioxide on Coral Reefs," *Science*, 2 April 1999.

37. John Pernetta et al., eds., *Impacts of Climate Change on Ecosystems and Species: Marine and Coastal Ecosystems* (Gland, Switzerland: IUCN, 1994).

38. Douglas Chadwick, "Coral in Peril," *National Geographic*, January 1999; Bryant et al., op. cit. note 29.

39. Thomas E. Graedel and Paul J. Crutzen, *Atmosphere, Climate, and Change* (New York: Scientific American Library, 1995).

40. Molly O'Meara, "Sulfur and Nitrogen Emissions Unchanged," in Lester R. Brown, Michael Renner, and Christopher Flavin, *Vital Signs 1997* (New York: W.W. Norton & Company, 1997); idem, "Acid Rain Threats Vary," in Lester R. Brown, Michael Renner, and Christopher Flavin, *Vital Signs 1998* (New York: W.W. Norton & Company, 1998); Odil Tunali, "A Billion Cars: The Road Ahead," *World Watch*, January/February 1996; Figure 2–2 from Dr. Jane Dignon, Lawrence Livermore National Laboratory, Livermore, CA, unpublished data series, letter to Molly O'Meara, 22 January 1997, and from Sultan Hameed and Jane Dignon, "Global Emissions of Nitrogen and Sulfur Oxides in Fossil Fuel Combustion 1970–86," *Journal of the Air and Waste Management Association*, February 1992.

41. G.E. Likens, C.T. Driscoll, and D.C. Buso, "Long-Term Effects of Acid Rain: Response and Recovery of a Forest Ecosystem," and Jocelyn Kaiser, "Acid Rain's Dirty Business: Stealing Minerals from the Soil," *Science*, 12 April 1996.

42. Increasing acidity from Likens, Driscoll, and Buso, op. cit. note 41, and from Kaiser, op. cit. note 41; aluminum from D.D. Richter, D.W. Johnson, and K.H. Dai, "Cation Exchange and Al Mobilization in Soils," in D.W. Johnson and S.E. Lindberg, eds., *Atmospheric Deposition and Forest Nutrient Cycling* (New York: Springer Verlag, 1992), and from Alan Wellburn, *Air Pollution and Acid Rain: The Biological Impact* (Harlow, U.K.: Longman Scientific and Technical, 1988).

43. Orie L. Loucks, "The Epidemiology of Forest Decline in Eastern Deciduous Forests," *Northeastern Naturalist*, May 1998.

44. James E. Gawel et al., "Role for Heavy Metals in Forest Decline Indicated by Phytochelatin Measurements," *Nature*, 2 May 1996; Jon R. Luoma, "Damage in Trees Tied to Heavy Metals in Air," *New York Times*, 7 May 1996.

45. Oak-hickory decline from Loucks, op. cit. note 43; Hubbard Brook from Likens, Driscoll,

and Buso, op. cit. note 41; William E. Sharpe, Bryan R. Swistock, and Troy L. Sunderland, "Soil Acidification and Sugar Maple Decline in Northern Pennsylvania," in William E. Sharpe and Joy R. Drohan, eds., *The Effects of Acidic Deposition on Pennsylvania's Forests* (University Park, PA: Pennsylvania State University Environmental Resources Research Institute, 1999); Jerry Jenkins, "The Vanishing Sugar Maple," *Wildlife Conservation*, January/February 1999; National Acid Precipitation Assessment Program, *NAPAP Biennial Report to Congress: An Integrated Assessment* (Silver Spring, MD: National Science and Technology Council, 1998).

46. Seth Dunn, "King Coal's Weakening Grip on Power," *World Watch*, September/October 1999; Zhu Bao, "Acid Rain Causes Heavy Losses," *China Daily*, 4 May 1999; deposition exported to Japan from O'Meara, "Acid Rain Threats Vary," op. cit. note 40; computer modeling study from David G. Streets et al., "Energy Consumption and Acid Deposition in Northeast Asia," *Ambio*, March 1999.

47. Streets et al., op. cit. note 46.

48. Peter M. Vitousek et al., "Human Alteration of the Global Nitrogen Cycle: Causes and Consequences," *Issues in Ecology 1* (Washington, DC: Ecological Society of America, spring 1997).

49. Ibid.

50. Nitrogen as a limiting nutrient from ibid.

51. "Scientific Consensus on Air Pollution Impacts on Appalachian Forests," summary of consensus findings from the Duke University Conference, "Acid Rain, Ozone and the Great Eastern Forests," 26–27 March 1999; John Aber, "Nitrogen Cycling and Nitrogen Saturation in Temperate Forest Ecosystems," *TREE* (Trends in Ecology and Evolution), July 1992; John Aber et al., "Forest Biogeochemistry and Primary Production Altered by Nitrogen Saturation," *Water, Air, and Soil Pollution*, vol. 85 (1995), pp. 1,665–70.

52. "Scientific Consensus on Air Pollution Impacts on Appalachian Forests," op. cit. note 51; Aber, op. cit. note 51; study showing nitrogen injury to conifers from John Aber et al., "Nitrogen Saturation in Temperate Forest Ecosystems," *BioScience*, November 1998.

53. Homogenizing effect from Vitousek et al., op. cit. note 48, and from David A. Wedin and David Tilman, "Influence of Nitrogen Loading and Species Composition on the Carbon Balance of Grasslands," *Science*, 6 December 1996; nitrogen loading and bioinvasion from Jeffrey S. Dukes and Harold A. Mooney, "Does Global Change Increase the Success of Biological Invaders?" *TREE* (Trends in Ecology and Evolution), April 1999.

54. For overviews of the potential effects of climate change on vegetation, see F. Ian Woodward, "A Review of the Effects of Climate on Vegetation: Ranges, Competition, and Composition," and Margaret B. Davis and Catherine Zabinski, "Changes in Geographical Range Resulting from Greenhouse Warming: Effects on Biodiversity in Forests," in Peters and Lovejoy, op. cit. note 23; for die-back from warming, see Allan N.D. Auclair et al., "Climatic Perturbation as a General Mechanism of Forest Dieback," in Paul D. Manion and Denis Lachance, eds., *Forest Decline Concepts* (St. Paul, MN: APS Press, 1992); for Siberia, see Masatoshi Yoshino et al., "Temperate Asia," in Robert T. Watson et al., eds., *The Regional Impacts of Climate Change: An Assessment of Vulnerability*, a special report of the Intergovernmental Panel on Climate Change (Cambridge, U.K.: Cambridge University Press, 1998), and Salim and Ullsten, op. cit. note 7.

55. Werner A. Kurz et al., "Global Climate Change: Disturbance Regimes and Biospheric Feedbacks of Temperate and Boreal Forests," in George M. Woodwell and Fred T. Mackenzie, eds., *Biotic Feedbacks in the Global Climate System* (New York: Oxford University Press, 1995); for a report of the spruce budworm infestation in Alaska, see Adam Markham, "Early Warning Signs Abound" (op ed), *Washington Post*, 2 November 1998.

56. Drew T. Shindell, David Rind, and Patrick Lonergan, "Increased Polar Stratospheric Ozone Losses and Delayed Eventual Recovery Owing to Increasing Greenhouse-Gas Concentrations," *Nature*, 9 April 1998; William J. Manning and Andreas V. Tiedemann, "Climate Change: Potential Effects of Increased Atmospheric Carbon Dioxide (CO_2), Ozone (O_3), and Ultraviolet-B (UV-B) Radiation on Plant Diseases," *Environmental Pollution*, vol. 88 (1995), pp. 219–45; foliar damage and decline from David R. Houston, "A Host-Stress-Saprogen Model for Forest Dieback-Decline Diseases," in Manion and Lachance, op. cit. note 54.

57. For an exploration of complexity in the natural world, see Robert E. Ulanowicz, *Ecology, the Ascendent Perspective* (New York: Columbia University Press, 1997); for an account of unpredictability in complex systems in general, see Charles Perrow, *Normal Accidents: Living with High-Risk Technologies* (New York: Basic Books, 1984).

58. For a critique of sustainable resource management in tropical forestry, see Richard E. Rice, Raymond E. Gullison, and John W. Reid, "Can Sustainable Management Save Tropical Forests?" *Scientific American*, April 1997.

59. On the virtual impossibility of reversing change in complex systems, see David J. Rapport and Water G. Whitford, "How Ecosystems Respond to Stress," *BioScience*, March 1999.

60. Robert Jervis, *System Effects: Complexity in Political and Social Life* (Princeton, NJ: Princeton University Press, 1997); Perrow, op. cit. note 57.

61. Obliqueness as a management principle from Edward Tenner, *Why Things Bite Back* (New York: Alfred A. Knopf, 1996); floodplain management from Janet N. Abramovitz, *Imperiled Waters, Impoverished Future: The Decline of Freshwater Ecosystems*, Worldwatch Paper 128 (Washington, DC: Worldwatch Institute, March 1996); polyculture and pest reduction from Stephen R. Gliessman, *Agroecology: Ecological Processes in Sustainable Agriculture* (Ann Arbor,

MI: Sleeping Bear Press, 1998).

62. Inevitability of multiple effects from Jervis, op. cit. note 60; organic agriculture as a means of reducing nutrient leaching from L.E. Drinkwater, P. Wagoner, and M. Sarrantonio, "Legume-Based Cropping Systems Have Reduced Carbon and Nitrogen Losses," *Nature*, 19 November 1998.

Chapter 3. Redesigning Irrigated Agriculture

1. For a description of irrigation's role in the rise and fall of early civilizations, see Chapter 2, "History Speaks," in Sandra Postel, *Pillar of Sand* (New York: W.W. Norton & Company, 1999).

2. Estimate of 40 percent is approximate, and is based on a 36-percent estimate in W. Robert Rangeley, "Irrigation and Drainage in the World," in Wayne R. Jordan, ed., *Water and Water Policy in World Food Supplies* (College Station, TX: Texas A&M University Press, 1987), and on a 47-percent estimate (just for grain) in Montague Yudelman, "The Future Role of Irrigation in Meeting the World's Food Supply," in Soil Science Society of America, *Soil and Water Science: Key to Understanding Our Global Environment* (Madison, WI: 1994); 17-percent figure from U.N. Food and Agriculture Organization (FAO), *1996 Production Yearbook* (Rome: 1997).

3. Growth rates in recent decades from Mark Rosegrant, *Water Resources in the Twenty-First Century: Challenges and Implications for Action* (Washington DC: International Food Policy Research Institute (IFPRI), 1997).

4. Drop of 17 percent by 2020 assumes global irrigated area expands by some 40 million hectares, as projected in Per Pinstrup-Anderson, Rajul Pandya-Lorch, and Mark W. Rosegrant, *The World Food Situation: Recent Developments, Emerging Issues, and Long-Term Prospects* (Washington, DC: IFPRI, 1997); 28-percent drop by 2020 assumes no net growth in irrigated area by

2020—that is, gains are offset by losses; irrigation trends from FAO, *Production Yearbook* (Rome: various years); population from United Nations, United Nations, *World Population Prospects 1950–2050—The 1998 Revision* (New York: December 1998).

5. One in five hectares from F. Ghassemi, A.J. Jakeman, and H.A. Nix, *Salinisation of Land and Water Resources: Human Causes, Extent, Management and Case Studies* (Sydney: University of New South Wales Press Ltd., 1995), and from Postel, op. cit. note 1.

6. National Environmental Engineering Research Institute (NEERI), "Water Resources Management in India: Present Status and Solution Paradigm" (Nagpur, India: undated, circa 1997).

7. "Alarming Ground Water Depletion in Haryana and Punjab," *IARI News* (Indian Agricultural Research Institute), October-December 1993; Mehsana district depletion from Tata Energy Research Institute, *Looking Back to Think Ahead: Executive Summary* (New Delhi: undated); David Seckler, David Molden, and Randolph Barker, *Water Scarcity in the Twenty-First Century,* Water Brief No. 1 (Colombo, Sri Lanka: International Water Management Institute (IWMI), 1998).

8. NEERI, op. cit. note 6.

9. "Alarming Ground Water Depletion in Haryana and Punjab," op. cit. note 7; Surendar Singh, "Some Aspects of Groundwater Balance in Punjab," *Economic and Political Weekly*, 28 December 1991.

10. Ruth Meinzen-Dick, *Groundwater Markets in Pakistan: Participation and Productivity,* Research Report 105 (Washington DC: IFPRI, 1996); L.R. Khan, "Environmental Impacts of Groundwater Development in Bangladesh," paper presented at the IX World Water Congress of the International Water Resources Association, Montreal, Canada, 1–6 September 1997; Rural Development Sector Unit, South Asia Region, *Water Resource Management in*

Bangladesh: Steps Toward a New National Water Plan (Dhaka, Bangladesh: World Bank, 1998); excess suffering of poor farmers from author's participation in seminar of Bangladeshi water experts, organized by International Development Enterprises, Dhaka, Bangladesh, 16 January 1998.

11. Zhang Qishun and Zhang Xiao, "Water Issues and Sustainable Social Development in China," *Water International*, vol. 20, no. 3 (1995); Liu Yonggong and John B. Penson, "China's Sustainable Agriculture and Regional Implications," paper presented to the Symposium on Agriculture, Trade and Sustainable Development in Pacific Asia: China and its Trading Partners (no place or date given); Sandia National Laboratories, China Infrastructure Initiative: Decision Support Systems, <www.igaia.sandia.gov/igaia/china/china model.html>.

12. River basin demand and supply data and projections from Dennis Engi, Sandia National Laboratories, Albuquerque, NM, discussion with Brian Halweil, Worldwatch Institute, April 1998; U.S. Department of Agriculture (USDA), *Production, Supply, and Distribution*, electronic database, Washington, DC, updated February 1999.

13. California Department of Water Resources, *California Water Plan Update*, vol. 1 (Sacramento, CA: 1994).

14. Figure of one fifth from National Research Council (NRC), *A New Era for Irrigation* (Washington, DC: National Academy Press, 1996); total and annual net depletion based on data in Edwin D. Gutentag et al., *Geohydrology of the High Plains Aquifer in Parts of Colorado, Kansas, Nebraska, New Mexico, Oklahoma, South Dakota, Texas, and Wyoming* (Washington, DC: U.S. Government Printing Office, 1984), and in Dork L. Sahagian, Frank W. Schwartz, and David K. Jacobs, "Direct Anthropogenic Contributions to Sea Level Rise in the Twentieth Century," *Nature*, 6 January 1994.

15. NRC, op. cit. note 14.

16. Aquifer volume from Jamil Al Alawi and Mohammed Abdulrazzak, "Water in the Arabian Peninsula: Problems and Perspectives," in Peter Rogers and Peter Lydon, eds., *Water in the Arab World* (Cambridge, MA: Harvard University Press, 1994); grain production from USDA, op. cit. note 12; peak water deficit from A.S. Al-Turbak, "Meeting Future Water Shortages in Saudi Arabia," paper presented at the IX World Water Congress of the International Water Resources Association, Montreal, Canada, 1–6 September 1997; current water deficit is author's estimate.

17. African Sahara depletion from Sahagian, Schwartz, and Jacobs, op. cit. note 14; Libya's depletion calculated from Rajab M. El Asswad, "Agricultural Prospects and Water Resources in Libya," *Ambio*, September 1995; "GMR Wins Prestigious Pipeline Award," *World Water and Environmental Engineering*, July 1997; "Libyan GMR Military Claims Ridiculed," *World Water and Environmental Engineering*, February 1998; Sandra Postel, *Last Oasis*, rev. ed. (New York: W.W. Norton & Company, 1997); 80-percent figure from Roula Khalaf, "Gadaffi Taps Desert Waters in Bid to Make a Big Splash," *Financial Times*, 11 September 1996; Fred Pearce, "Will Gaddafi's Great River Run Dry?" *New Scientist*, September 1991.

18. Twain quote from Barbara K. Rodes and Rice Odell, *A Dictionary of Environmental Quotations* (New York: Simon & Schuster, 1992).

19. Mark W. Rosegrant and Claudia Ringler, "Impact on Food Security and Rural Development of Reallocating Water from Agriculture for Other Uses," paper prepared for Harare Expert Group Meeting on Strategic Approaches to Freshwater Management, Harare, Zimbabwe, 28–31 January 1998.

20. Ibid.

21. Figure for 1949 from Zhang Zezhen et al., "Challenges to and Opportunities for Development of China's Water Resources in the 21st Century," *Water International*, vol. 17, no. 1 (1992); sources differ on both the present num-

ber of China's cities (generally between 600 and 640) and the number that are water-short (generally between 300 and 480); planning study cited in James E. Nickum, "Issue Paper on Water and Irrigation," prepared for the project on Strategy and Action for Chinese and Global Food Security, Millennium Institute, USDA, World Bank, and Worldwatch Institute, Arlington, VA, September 1997; United Nations, Population Division, *World Urbanization Prospects: The 1996 Revision* (New York: 1998).

22. Figure of 70 times from Lester R. Brown and Brian Halweil, "China's Water Shortage Could Shake World Food Security," *World Watch*, July-August 1998; Patrick E. Tyler, "China Lacks Water to Meet its Mighty Thirst," *New York Times*, 7 November 1993.

23. Urban increase calculated from Population Reference Bureau (PRB), *1998 Population Data Sheet*, wallchart (Washington, DC: 1998), and from United Nations, op. cit. note 21; Tirupur from Jan Lundqvist, "Food Production and Food Security in an Urbanising World," prepared for the IXth World Water Congress of the International Water Resources Association, Montreal, Canada, 1–6 September 1997.

24. Ganjar Kurnia, Teten Avianto, and Bryan Bruns, "Farmers, Factories and the Dynamics of Water Allocation in West Java," in Bryan Bruns and Ruth S. Meinzen-Dick, eds., *Negotiating Water Rights* (New Delhi: Sage Publishers, forthcoming).

25. Santos Gomez and Anna Steding, "California Water Transfers: An Evaluation of the Economic Framework and a Spatial Analysis of the Potential Impacts," Pacific Institute, Oakland, CA, April 1998.

26. "Water in California: Flowing Gold," *The Economist*, 10 October 1998; initial price of transferred water from Sue McClurg, "Cutting Colorado River Use: The California Plan," *Western Water*, November-December 1998.

27. "Water Capping under Discussion in Australia," *World Water and Environmental Engi-*

neering, June 1997; Sandra Postel, *Dividing the Waters: Food Security, Ecosystem Health, and the New Politics of Scarcity*, Worldwatch Paper 132 (Washington, DC: Worldwatch Institute, September 1996).

28. Sandra L. Postel, "Water for Food Production: Will There by Enough in 2025?" *BioScience*, August 1998.

29. Ibid.; annual net grain imports for these countries, averaged over 1994–96, totaled 48 million tons; World Food Programme, *Tackling Hunger in a World Full of Food: Tasks Ahead for Food Aid* (Rome: 1996); 1996–97 food aid from International Commission on Irrigation and Drainage, ICID News Update, New Delhi, India, November 1997.

30. Postel, op. cit. note 17.

31. Ibid.

32. S.K. Suryawanshi, "Success of Drip in India: An Example to the Third World," in Freddie R. Lamm, ed., *Microirrigation for a Changing World*, Proceedings of the Fifth International Microirrigation Congress (St. Joseph, MI: American Society of Agricultural Engineers, 1995); World Bank, *Gains That Might be Made from Water Conservation in the Middle East* (Washington, DC: 1993).

33. Sandra Postel, *Conserving Water: The Untapped Alternative*, Worldwatch Paper 67 (Washington, DC: Worldwatch Institute, September 1985); Postel, op. cit. note 17; Dale A. Bucks, "Historical Developments in Microirrigation," in Lamm, op. cit. note 32; current irrigated area is an estimate by Dale A. Bucks, USDA, Agricultural Research Service, Beltsville, MD, discussion with author, 1 July 1998.

34. Dan Rogers, Research and Extension Agricultural Engineer, Kansas State University, Manhatten, KS, discussion with author, 9 September 1999; Indian National Committee on *Irrigation and Drainage, Drip Irrigation in India* (New Delhi: 1994); R.K. Sivanappan, chair of the Centre of Agricultural Rural Development and

Environmental Studies in Coimbatore, discussions with author during field visits in Madhya Pradesh, India, January 1998.

35. California Department of Water Resources, "CIMIS: Fifteen Years of Growth and a Promising Future" (Sacramento, CA: December 1997).

36. Ibid.; the source does not distinguish the portion of water savings from reduced evapotranspiration.

37. Amy L. Vickers, *Handbook of Water Use and Conservation,* prepublication manuscript (Amherst, MA: Amy Vickers & Associates, 1999); Consultative Group on International Agricultural Research, "The World Water and Climate Atlas for Agriculture: A New Technology," press release (Washington, DC: 16 March 1997); the atlas is available at <www.cgiar.org/iwmi>.

38. Ramesh Bhatia, Rita Cestti, and James Winpenny, *Water Conservation and Reallocation: Best Practice Cases in Improving Economic Efficiency and Environmental Quality* (Washington, DC: World Bank, 1995).

39. Tamara Daniel, "TAES Research Focuses on Center Pivot Irrigation Management Decisions," *The Cross Section* (High Plains Underground Water Conservation District), June 1995; corn and cotton yield increases and sprinkler coverage from Ken Carver, Assistant Manager, High Plains Underground Water Conservation District No. 1, Lubbock, TX, discussion with author, 21 July 1998.

40. Carver, op. cit. note 39; author's visit to the High Plains Water District, Lubbock, TX, 26 October 1999.

41. Some 1.2 billion people do not get enough calories every day for an adequate diet; see the discussion of hunger in Chapter 4.

42. Alastair Orr, A.S.M. Nazrul Islam, and Gunnar Barnes, *The Treadle Pump: Manual Irrigation for Small Farmers in Bangladesh* (Dhaka, Bangladesh: Rangpur Dinajpur Rural Service, 1991); author's field visits in Bangladesh, January 1998.

43. Cost and returns from Paul Polak, "Tripling the Small Farmer Harvest: A Solution for Rural Hunger" (Denver, CO: International Development Enterprises (IDE), January 1998).

44. Figure of 1.2 million from Len Jornlin, IDE Country Director, Dhaka, Bangladesh, e-mail to author, 23 August 1998; economic impact based on estimated net returns of $100 per pump and a per-pump multiplier of 2.5; number of manufacturers, dealers, and installers as of October 1998 from Jeff Saussier, IDE, Denver, CO, letter to author, December 1998; potential market from Paul Polak, President, IDE, discussion with author during trip to Bangladesh, January 1998.

45. A. Kandiah, "Summary of Findings of Missions in Selected Countries in East and Southern Africa," in FAO, *Irrigation Technology Transfer in Support of Food Security,* Proceedings of a Subregional Workshop, Harare, Zimbabwe, 14–17 April 1997, Water Reports No. 14 (Rome: 1997).

46. For an excellent treatment of indigenous water management in Africa, see W.M. Adams, *Wasting the Rain: Rivers, People, and Planning in Africa* (London: Earthscan, 1992); FAO, *Irrigation in Africa in Figures,* Water Reports No. 7 (Rome: 1995).

47. Marc Andreini, "Bani Irrigation: An Alternative Water Use," in M. Rukuni et al., eds., *Irrigation Performance in Zimbabwe: Proceedings of Workshops in Harare and Juliasdale, Zimbabwe* (University of Zimbabwe: 1994); Ruth Meinzen-Dick, Research Fellow, IFPRI, Washington, DC, e-mail to author, December 1998; Zimbabwe *dambo* figures from K. Palanisami, "Economics of Irrigation Technology Transfer and Adoption," in FAO, op. cit. note 45.

48. Randall Purcell, "Potential for Small-Scale Irrigation in Sub-Saharan Africa: The Kenyan Example," in FAO, op. cit. note 45.

49. Area cultivated from Thayer Scudder, "The

Need and Justification for Maintaining Transboundary Flood Regimes: The Africa Case," *Natural Resources Journal*, winter 1991.

50. Michael M. Horowitz and Muneera Salem-Murdock, "Development-Induced Food Insecurity in the Middle Senegal Valley," *GeoJournal*, vol. 30, no. 2 (1993); quote from Michael M. Horowitz, "The Management of an African River Basin: Alternative Scenarios for Environmentally Sustainable Economic Development and Poverty Alleviation," in *Proceedings of the International UNESCO Symposium, Water Resources Planning in a Changing World* (Karlsruhe, Germany: Bundesanstalt für Gewässerkunde, 1994).

51. Number of tanks from R.K. Sivanappan, "Technologies for Water Harvesting and Soil Moisture Conservation in Small Watersheds for Small-Scale Irrigation," in FAO, op. cit. note 45; Anil Agarwal and Sunita Narain, eds., *Dying Wisdom* (New Delhi: Centre for Science and Environment, 1997).

52. Sivanappan, op. cit. note 51.

53. Figure of $33 billion from Norman Myers, "Perverse Subsidies: Their Nature, Scale and Impacts," a report to the MacArthur Foundation, Chicago, IL, October 1997.

54. Number of countries from Douglas L. Vermillion, *Impacts of Irrigation Management Transfer* (Colombo, Sri Lanka: IWMI, 1997).

55. Sam H. Johnson III, "Irrigation Management Transfer: Decentralizing Public Irrigation in Mexico," *Water International*, vol. 22 (1997), pp. 159–67.

56. San Joaquin Valley Drainage Implementation Program, *Drainage Management in the San Joaquin Valley: A Status Report* (Sacramento, CA: California Department of Water Resources, 1998).

57. Lawrence J. MacDonnell et al., "Using Water Banks to Promote More Flexible Water Use," Report to the U.S. Geological Survey

(Boulder, CO: Natural Resources Law Center, August 1994); California Department of Water Resources, *The California Water Plan Update*, Public Review Draft, Bulletin 160–98, vol. 1 (Sacramento, CA: January 1998).

58. J. Eduardo Mestre R., "Integrated Approach to River Basin Management: Lerma-Chapala Case Study—Attributions and Experiences in Water Management in Mexico," *Water International*, September 1997.

59. Water requirements of U.S. diet from FAO, *World Food Summit—Technical Background Documents*, Vol. 2 (Rome: 1996).

Chapter 4. Nourishing the Underfed and Overfed

1. Half the world from World Health Organization (WHO), *World Health Report 1998* (Geneva: 1998). Table 1 is based on the following sources: micronutrient deficiency from United Nations, Administrative Committee on Coordination, Sub-Committee on Nutrition (UN ACC/SCN), "Fourth Report on The World Nutrition Situation" (draft) (New York: July 1999); hunger and overconsumption, which are Worldwatch Institute estimates, from ibid., from WHO, *Obesity: Preventing and Managing the Global Epidemic*, Report of a WHO Consultation on Obesity (Geneva: 1997), from K.M. Flegal et al., "Overweight and Obesity in the United States, Prevalence and Trends, 1960–1994," *International Journal of Obesity*, August 1998, and from Rafael Flores, Research Fellow, International Food Policy Research Institute (IFPRI), Washington, DC, e-mail to Brian Halweil, 5 November 1999.

2. Corinne Cataldo et al., *Nutrition and Diet Therapy: Principles and Practices* (Belmont, CA: Wadsworth Publishing Company, 1999).

3. Ibid.; traditional diets from a multiyear, international conference series, "Public Health Implications of Traditional Diets," organized by Oldways Preservation & Exchange Trust, Cambridge, MA, <www.oldwayspt.com>.

4. U.N. Food and Agriculture Organization (FAO), *The State of Food Insecurity in the World* (Rome: 1999); Lisa Smith, *Can FAO's Measure of Chronic Undernourishment Be Strengthened?* Food Consumption and Nutrition Division (FCND) Discussion Paper No. 44 (Washington, DC: IFPRI, May 1998).

5. UN ACC/SCN, op. cit. note 1.

6. U.S. Department of Agriculture (USDA), "Household Food Security in the United States, 1995–1998" (Advance Report) (Washington, DC: July 1999); Graham Riches, ed., *First World Hunger* (New York: St. Martin's Press, 1997).

7. UN ACC/SCN, op. cit. note 1; WHO, *Obesity,* op. cit. note 1. BMI is calculated as a person's weight in kilos divided by the square of height in meters. Thus, a person standing 1.78 meters (5 feet, 10 inches) tall and weighing 80 kilos (175 pounds) has a BMI of 25, and is considered overweight; at 95 kilos (210 pounds), the same person would be considered obese.

8. Flegal et al., op. cit. note 1; childhood overweight and obesity from Rajen Anand, introductory presentation to "Childhood Obesity: Causes and Prevention," symposium sponsored by USDA, Center for Nutrition Policy and Promotion, Washington, DC, 27 October 1998; WHO, *Obesity,* op. cit. note 1.

9. Survey for United Nations from Flores, op. cit. note 1; China data from Catherine Geissler, "China: The Soya Pork Dilemma," *Proceedings of the Nutrition Society,* May 1999, and from Barry M. Popkin et al., "Body Weight Patterns among the Chinese: Results from the 1989 and 1991 China Health and Nutrition Surveys," *American Journal of Public Health,* May 1995; Brazil and Colombia from WHO, *Obesity,* op. cit. note 1; overweight exceeding underweight from Flores, op. cit. note 1.

10. WHO, "World Health Organization Sets Out to Eliminate Iodine Deficiency Disorder," press release (Geneva: 25 May 1999); Vitamin A from UN ACC/SCN, op. cit. note 1.

11. UN ACC/SCN, op. cit. note 1.

12. Exclusive breast-feeding in Latin America from ibid; UNICEF recommendation from Ted Greiner, Section for International Maternal and Child Health, Uppsala University, Sweden, e-mail to authors, 21 October 1999; breast-feeding in North America from Alan Ryan, Ross Products Division, Abbott Laboratories, Inc., Cleveland, OH, fax to authors, 26 October 1999.

13. Industrial-nation calorie breakdown from World Cancer Research Fund and the American Institute for Cancer Research (WCRF/AICR), *Food, Nutrition, and the Prevention of Cancer: A Global Perspective* (Washington, DC: AICR, 1997); refined grains from Judy Putnam and Shirley Gerrior, "Trends in the U.S. Food Supply, 1970–97," in Elizabeth Frazao, ed., *America's Eating Habits: Changes and Consequences* (Washington, DC: USDA, Economic Research Service (ERS), April 1999); french fries and potato chips from Linda Scott Kantor, "A Comparison of the U.S. Food Supply with the Food Guide Pyramid Recommendations," in ibid.

14. Amartya Sen, *Poverty and Famines: An Essay on Entitlement and Deprivation* (New York: Oxford University Press, 1981); 80 percent of all malnourished children from Smith, op. cit. note 4; Lisa C. Smith and Lawrence Haddad, *Explaining Child Malnutrition in Developing Countries: A Cross-Country Analysis,* FCND Discussion Paper No. 60 (Washington, DC: IFPRI, April 1999).

15. FAO, "Women Feed the World," prepared for World Food Day, 16 October 1998 (Rome: 1998).

16. Ibid.

17. Relationship between land distribution, rural poverty, and hunger from Klaus Deininger and Lin Squire, "New Ways of Looking at Old Issues: Inequality and Growth," *Journal of Development Economics,* December 1998, and from Radha Sinha, *Landlessness: A Growing Problem* (Rome: FAO, 1984); response to debt from Peter Uvin, *The International Organization of Hunger* (London: Kegan Paul International, 1994).

18. Benefits of food exports from Uvin, op. cit. note 17, from Lori Ann Thrupp, *Bittersweet Harvest for Global Supermarkets* (Washington, DC: World Resources Institute, 1995), from Walden Bello, "The WTO's Big Losers," *Far Eastern Economic Review*, 24 June 1999, and from Frances Moore Lappé et al., *World Hunger: Twelve Myths* (San Francisco: Grove Press, 1998).

19. Kevin Watkins, "Agricultural Trade and Food Security" (London: Oxfam, May 1996); Per Pinstrup-Andersen and Marc J. Cohen, "An Overview of the Future Global Food Situation," prepared for the Millennial Symposium on "Feeding a Planet," sponsored by and held at the Cosmos Club, Washington, DC, 12 February 1999; Sophia Murphy, "Trade and Food Security: An Assessment of the Uruguay Round Agreement on Agriculture" (London: Catholic Institute for International Relations, 1999).

20. Karen Lehman, "The Great Train Robbery of 1996" (Minneapolis, MN: Institute for Agriculture and Trade Policy, 25 June 1996); Watkins, op. cit. note 19.

21. Marc J. Cohen and Torsten Feldbruegge, "Acute Nutrition Crises and Violent Conflict" (Washington, DC: IFPRI, forthcoming).

22. Marion Nestle et al., "Behavioral and Social Influences on Food Choice," *Nutrition Reviews*, May 1998; B.J. Rolls and E.A. Bell, "Intake of Fat and Carbohydrate: Role of Energy Density," *European Journal of Clinical Nutrition*, April 1999; J.E. Blundell and J.I. MacDiarmid, "Fat as a Risk Factor for Overconsumption: Satiation, Satiety, and Patterns of Eating," *Journal of American Dietetic Association*, July 1997.

23. Barry M. Popkin, "The Nutrition Transition and Its Health Implications in Lower-Income Countries," *Public Health Nutrition*, January 1998; Adam Drewnowski and Barry M. Popkin, "The Nutrition Transition: New Trends in the Global Diet," *Nutrition Reviews*, February 1997; Geissler, op. cit. note 9; Figure 4–1 from USDA, *Production, Supply, and Distribution*, electronic database, Washington, DC, updated November 1999.

24. Urban population from United Nations, *World Urbanization Prospects: The 1996 Revision* (New York: 1997); Marie T. Ruel et al., *Urban Challenges to Food and Nutrition Security: A Review of Food Security, Health, and Caregiving in the Cities*, FCND Discussion Paper No. 51 (Washington, DC: IFPRI, October 1998); sweeteners and fat from Barry M. Popkin, Department of Nutrition, University of North Carolina at Chapel Hill, "Urbanization, Lifestyle Changes, and the Nutrition Transition," unpublished manuscript.

25. Popkin, op. cit. note 23; agricultural work force from FAO, *FAOSTAT Statistics Database*, <apps.fao.org>, viewed 5 October 1999.

26. Nestle et al., op. cit. note 22; Anthony E. Gallo, "Food Advertising in the United States," in Frazao, op. cit. note 13; Austria, Belgium, and France from NTC Publications Ltd., *European Marketing Pocket Book, 1999 Edition* (Oxfordshire: 1998); Goeff Tansey and Tony Worsley, *The Food System: A Guide* (London: Earthscan, 1995).

27. Consumers International, *A Spoonful of Sugar—Television Food Advertising Aimed at Children: An International Comparative Survey* (London: 1996); targeting of children from Victor C. Strasburger, *Adolescents and the Media* (Thousand Oaks, CA: Sage Publications Inc., 1995); U.S. fast-food ads from Gallo, op. cit. note 26; McDonald's and Coca-Cola from *Advertising Age*, <www.adage.com/dataplace>, viewed 5 October 1999; life-long craving from Nestle et al., op. cit. note 22, and from L.L. Birch, "Development of Food Acceptance Patterns in the First Years of Life," *Proceedings of Nutrition Society*, November 1998.

28. Michael Jacobson, Center for Science in the Public Interest, Washington, DC, discussion with Brian Halweil, 8 September 1999; Judy Putnam and Shirley Gerrior, "Trends in the U.S. Food Supply, 1970–97," in Frazao, op. cit. note 13.

29. Anna White, "The Cola-ized Classroom," *Multinational Monitor,* January/February 1999; franchises in American schools from Claire Hope Cummings, "Entertainment Food," *The Ecologist,* January/February 1999.

30. Michael F. Jacobson, "Liquid Candy: How Soft Drinks are Harming Americans' Health" (Washington, DC: Center for Science in the Public Interest, October 1998); Brian Wansink, "Can Package Size Accelerate Usage Volume?" *Journal of Marketing,* July 1996; Lisa R. Young and Marion Nestle, "Portion Size in Dietary Assessment: Issues and Policy Implications," *Nutrition Reviews,* June 1995.

31. Fast-food ads and nutrition education from Gallo, op. cit. note 26; Kellogg's from Michael Jacobson, Director, Center for Science in the Public Interest, Washington, DC, discussion with Brian Halweil, 26 October 1999; Kelly Brownell quote from Bonnie Liebman, "The Pressure to Eat," *Nutrition Action Healthletter* (Center for Science in the Public Interest), July/August 1998.

32. Coca-Cola Company, *1998 Annual Report* (Atlanta: 1999); four of five McDonald's from Brian Breuhaus, "Risky Business?" *Restaurant Business,* 1 November 1998; Steven A. Neff et al., *Globalization of the Processed Foods Market* (Washington, DC: USDA, ERS, October 1996).

33. Carlos A. Monteiro et al., "Shifting Obesity Trends in Brazil," *European Journal of Clinical Nutrition* (in press); Popkin, op. cit. note 23; Soowon Kim, Soojae Moon, and Barry M. Popkin "The Nutrition Transition in South Korea," *American Journal of Clinical Nutrition* (in press).

34. Table 4–4 based on the following: cancer from WCRF/AICR, op. cit. note 13; coronary heart disease from A.M. Wolf and G.A. Colditz, "Current Estimates of the Economic Costs of Obesity Research in the United States," *Obesity Research,* March 1998; adult-onset (Type II)

diabetes from WHO, *Obesity,* op. cit. note 1; childhood blindness from Dr. Alfred Sommer, Dean, School of Public Health, Johns Hopkins University, Baltimore, MD, discussion with Brian Halweil, 2 November 1999; mental retardation from Ministry of Foreign Affairs, *Nutrition: Interaction of Food, Health, and Care,* Sectoral and Theme Policy Document of Development Cooperation No. 10 (The Hague: Government of the Netherlands, September 1998), and from WHO, *World Health Report,* op. cit. note 1.

35. John Mason, Tulane University, letter to Brian Halweil, 16 September 1999; John Mason notes that the 22 percent for hunger and 18 percent for micronutrient deficiencies cannot simply be added, due to overlap in the population affected, as well as possible synergies between the two conditions; original study is Christopher J.L. Murray and Alan D. Lopez, eds., *The Global Burden of Disease* (Cambridge, MA: Harvard University Press, 1996); original study uses the term Disability-Adjusted Life Year (DALY), and one DALY as "one lost year of healthy life"; Dr. Lopez's estimate based on discussion with Brian Halweil, 15 August 1999.

36. UN ACC/SCN, op. cit. note 1.

37. D.J.P. Barker, *Mothers, Babies, and Disease in Later Life* (London: British Medical Journal Publishing, 1994).

38. UN ACC/SCN, op. cit. note 1; Barker quoted in Alan Berg, "Furthermore. . . ," *New and Noteworthy in Nutrition* (World Bank), 12 June 1999; 54 percent from Sonya Rabeneck, UN ACC/SCN, e-mail to authors, 5 October 1999.

39. Ministry of Foreign Affairs, op. cit. note 34.

40. UN ACC/SCN, op. cit. note 1; mental impairment *in utero* from Rafael Flores, IFPRI, Washington, DC, e-mail to authors, 14 October 1999.

41. "Recommendations to Prevent and Control Iron Deficiency in the United States, *Morbidity* *and Mortality Weekly Report*, 3 April 1998; Mary Murray, "Ads Raise Questions About Milk and Bones," *New York Times*, 14 September 1999; T. Remer and F. Manz, "Estimation of the Renal Net Acid Excretion by Adults Consuming Diets Containing Variable Amounts of Protein," *American Journal of Clinical Nutrition*, June 1994; D. Feskanich et al., "Milk, Dietary Calcium, and Bone Fractures in Women: A 12-Year Prospective Study," *American Journal of Public Health*, June 1997.

42. Popkin, op. cit. note 23.

43. China Health Survey from J. Chen et al., *Diet, Lifestyle, and Mortality in China: a Study of the Characterisitics of 65 Chinese Counties* (Oxford, U.K.: Oxford University Press, 1990), and from T. Colin Campbell, "Associations of Diet and Disease: A Comprehensive Study of Health Characteristics in China," presented at conference on "Social Consequences of Chinese Economic Reforms," Harvard University, Fairbank Center on East Asian Studies, Cambridge, MA, 23–24 May 1997; Framingham Study from Susan Brink, "Unlocking the Heart's Secrets," *U.S. News & World Report*, 7 September 1998.

44. WHO, *Obesity*, op. cit. note 1; industrial-nation mortality from WHO, op. cit. note 1, and from U.S. Centers for Disease Control and Prevention, National Center for Health Statistics, <www.cdc.gov/nchswww/fastats/deaths.htm>; Wolf and Colditz, op. cit. note 35; WCRF/AICR, op. cit. note 13.

45. WHO, *Obesity*, op. cit. note 1; D.C. Neiman et al., "Influence of Obesity on Immune Function," *Journal of American Dietetic Association*, March 1999.

46. International Diabetes Federation (IDF), "Diabetes Around the World" (Brussels: 1998), and <www.idf.org>, viewed 14 July 1999; Ginger Thompson, "With Obesity in Children Rising, More Get Adult Type of Diabetes," *New York Times*, 14 December 1998; Elizabeth Frazao, "The High Costs of Poor Eating Habits in the United States," in Frazao, op. cit. note 14.

47. IDF, op. cit. note 46; WCRF/AICR, op. cit. note 13; Popkin, op. cit. note 23.

48. World Bank study from Ministry of Foreign Affairs, op. cit. note 34; Spain and Indonesia from Beryl Levinger, *Nutrition, Health, and Education for All* (New York: U.N. Development Programme, 1994); Philippines study from M.E. Mendez and L.S. Adair, "Severity and Timing of Stunting in Infancy and Performance on IQ and School Achievement Tests in Late Childhood," *Journal of Nutrition*, August 1999. Equally interesting, this Filipino study found that the effects of stunting were reduced if an improved diet in later childhood allowed the infants' growth to catch up with their peers; however, the previously stunted children still performed more poorly than their peers who were never stunted.

49. Susan Horton, "The Economics of Nutrition Interventions" (draft, April 1999) in R.D. Semba, ed., *Nutrition and Health in Developing Countries* (Totowa, NJ: Humana Press, forthcoming).

50. Productivity losses from ibid; cost for entire developing world based on data in World Bank, *World Development Report 1998/99* (Washington, DC: June 1999); analysis of Horton's work from Jay Ross and Susan Horton, *Economic Consequences of Iron Deficiency* (Ottawa, ON, Canada: The Micronutrient Initiative, 1998).

51. U.S. statistics from Wolf and Colditz, op. cit. note 34, and from Anne Wolf, University of Virginia, Charlottesville, VA, discussion with Brian Halweil, 1 November 1999; Sweden study from WHO, *Obesity*, op. cit. note 1.

52. WHO, *Obesity*, op. cit. note 1.

53. Neal D.Barnard, Andrew Nicholson, and Jo Lil Howard, "The Medical Costs Attributable to Meat Consumption," *Preventive Medicine*, November 1995.

54. Graham Colditz, Harvard School of Public Health, "The Economic Costs of Obesity and Inactivity," unpublished manuscript; $33 billion

from Wolf and Colditz, op. cit. note 34.

55. Annual health expenditures per person from Arun Chockalingam and Ignasi Balaguer Vintró, eds., *Impending Global Pandemic of Cardiovascular Diseases* (Barcelona: Prous Science, 1999).

56. WCRF/AICR, op. cit. note 13.

57. Rodolfo A. Bulatao, *The Value of Family Planning Programs in Developing Countries* (Santa Monica, CA: RAND, 1998); WHO, *Obesity*, op. cit. note 1.

58. WHO, op. cit. note 10.

59. F. James Levinson, "Addressing Malnutrition in Africa," Social Dimensions of Adjustment in Sub-Saharan Africa Working Paper No. 13 (Washington, DC: World Bank, 1991); UN ACC/SCN, op. cit. note 1; Sonya Rabeneck, UN ACC/SCN, e-mail to authors, 10 October 1999; Hanny Friesen, et al., "Protection of Breastfeeding in Papua New Guinea," *Bulletin of the World Health Organization*, vol. 77, no. 3 (1999); Baby-Friendly Hospital Initiative (BFHI) from *BFHI News*, March/April 1999.

60. Marc J. Cohen and Don Reeves, "Causes of Hunger," *2020 Brief 19* (Washington, DC: IFPRI, May 1995); Barbara MkNelly and Christopher Dunford, "Are Credit and Savings Services Effective Against Hunger and Malnutrition?—A Literature Review," Research Paper Number 1 (Davis, CA: Freedom From Hunger, February 1996); IFPRI study from Lisa C. Smith and Lawrence Haddad, "Explaining Child Malnutrition in Developing Countries: A Cross-Country Analysis," FCND Discussion Paper 60 (Washington, DC: IFPRI, April 1999).

61. Cohen and Reeves, op. cit. note 60; Nancy Birdsall, "Macroeconomic Reform: Its Impact on Poverty and Hunger," in Ismail Serageldin and Pierre Landell-Mills, eds., *Overcoming Global Hunger* (Washington, DC: World Bank, 1993); Per Pinstrup-Andersen and Marc J. Cohen, "Aid to Developing-Country Agriculture: Investing in Poverty Reduction and New Export Opportunities," *2020 Brief 56* (Washing-

ton, DC: IFPRI, October 1998).

62. Kerala from C. R. Soman, Professor of Nutrition, College of Medicine, University of Kerala, e-mail to Brian Halweil, 29 October 1999; Cohen and Reeves, op. cit. note 60.

63. Uvin, op. cit. note 17; Aileen Kwa, "Will Food Security Get Trampled as the Elephants Fight over Agriculture?" Focus-on-Trade No. 36, electronic bulletin from Focus on the Global South, Bangkok, July 1999.

64. Denise Grady, "Doctors' Review of 5 Deaths Raises Concern about the Safety of Liposuction," *New York Times*, 13 May 1999.

65. WHO, *Obesity*, op. cit. note 1; additional Trim and Fit information from Barry Popkin, University of North Carolina, Chapel Hill, NC, e-mail to authors, 21 September 1999; U.S. program from P.R. Nader et al., "Three-Year Maintenance of Improved Diet and Physical Activity: The Child and Adolescent Trial for Cardiovascular Health (CATCH) Cohort," *Archives of Pediatrics and Adolescent Medicine*, July 1999.

66. Meredith May, "Lunch Going Organic, Berkeley Schools to OK Plan for Pesticide-Free Food," *San Francisco Chronicle*, 17 August 1999.

67. C. Howard Davis, "The Report to Congress on the Appropriate Federal Role in Assuring Access by Medical Students, Residents, and Practicing Physicians to Adequate Training in Nutrition," *Public Health Reports*, November-December 1994.

68. Pekka Puska, "Development of Public Policy on the Prevention and Control of Elevated Blood Cholesterol," *Cardiovascular Risk Factors*, August 1996.

69. Consumer survey from *The Shopper Report* (Consumer Research Network, Inc., Philadelphia, PA), September 1997.

70. P. Puska, "The North Karelia Project: From Community Intervention to National Activity in

Lowering Cholesterol Levels and CHD Risk," *European Heart Journal Supplements* (1999).

71. Tim Lang, "Food & Nutrition in the EU: Its Implications for Public Health," prepared for the European Union Main Public Health Issues Project, September 1998.

72. Kelly D. Brownell, "Get Slim with Higher Taxes," *New York Times* , 15 December 1994.

73. Ibid.

74. Slow Food Movement from <www.slow food. com>, viewed 23 October 1999; Oldways Preservation & Exchange Trust from <www.old wayspt.org>, viewed 23 October 1999.

Chapter 5. Phasing Out Persistent Organic Pollutants

1. A. Bernard et al., "Food Contamination by PCBs and Dioxins," *Nature*, 16 September 1999; Debora Mackenzie, "Indecent Exposure," *New Scientist*, 26 June 1999; toxicity of dioxin from Hugh Warwick, "Agent Orange: The Poisoning of Vietnam," *The Ecologist*, September/October 1998.

2. Bernard et al., op. cit. note 1; Charles Truehart, "EU Gives Order To Destroy Belgian 'Chicken a la Dioxin'," *Washington Post Foreign Service*, 3 June 1999; Debora MacKenzie, "Recipe for Disaster," *New Scientist*, 12 June 1999.

3. Truehart, op. cit. note 2; Raf Casert, "Belgium Food Scandal Spreads," *Associated Press,* 7 June 1999; "Dioxin Crisis Topples Belgian Government," *Environmental News Service*, 15 June 1999.

4. J.B. Opschoor and D.W. Pearce, "Persistent Pollutants: A Challenge for the Nineties," in J.B. Opschoor and David Pearce, eds., *Persistent Pollutants: Economics and Policy* (Boston, MA: Kluwer Academic Publishers, 1991).

5. Frank Wania and Don MacKay, "Global

Distillation," *Our Planet*, March 1997.

6. Amount of chemicals in humans from Theo Colborn, "Restoring Children's Birthrights," *Our Planet*, March 1997; Matthew P. Longnecker, Walter J. Rogan, and George Lucier, "The Human Health Effects of DDT (Dichlorodiphenyltrichloroethane) and PCBs (Polychlorinated Biphenyls) and an Overview of Organochlorines in Public Health," *Annual Review of Public Health,* vol. 18 (1997).

7. William H. van der Schalie et al., "Animals as Sentinels of Human Health Hazards of Environmental Chemicals," *Environmental Health Perspectives*, April 1999; Carson quote in Theo Colborn, Dianne Dumanoski, and John Peterson Myers, *Our Stolen Future* (New York: Penguin Group, 1996).

8. Donald MacKay, "Environmental Fate and Modeling of Long-Distance Atmospheric Transport of POPs," presentation at Harvard University School of Public Health, Center for Continuing Professional Education Symposium, "Persistent Organic Pollutants (POPs): A Public Health Perspective," Cambridge, MA, 16–18 June 1999; Donald MacKay, Environmental Modelling Center, Trent University, Peterborough, ON, Canada, e-mail to author, 24 September 1999; number used in commerce from Robert Ayres, "The Life-Cycle of Chlorine, Part I: Chlorine Production and the Chlorine-Mercury Connection," *Journal of Industrial Ecology*, vol. 1, no. 1 (1997).

9. Global data from René P. Schwarzenbach, Philip M. Gschwend, and Dieter M. Imboden, *Environmental Organic Chemistry* (New York: John Wiley & Sons, Inc., 1993), and from Jennifer Myers, "Nations Plan Phase-Out of Deadly Chemicals," *World Watch*, September/ October 1998; U.S. data from Dennis J. Paustenbach, "'A Survey of Health Risk Assessment," in Dennis J. Paustenbach, ed., *The Risk Assessment of Environmental Hazards* (New York: John Wiley & Sons, Inc., 1989); Figure 5–1 from U.S. International Trade Commission (ITC), *Synthetic Organic Chemicals: United States Production and Sales* (Washington, DC:

various years).

10. Schwarzenbach, Gschwend, and Imboden, op. cit. note 9; Environmental Research Foundation, "Chemical Industry Strategies, Part 1," *Rachel's Environment & Health Weekly*, 23 May 1996.

11. Ivan Amato, "The Crusade To Ban Chlorine," *Garbage*, Summer 1994, quoting W. Joseph Stearns, director of chlorine issues for Dow Chemical Company; Ayres, op. cit. note 8.

12. Estimates from Michael Walls, Senior Counsel, Chemical Manufacturers Association, Arlington, VA, discussion with author, 17 September 1999, and from Pat Costner, Senior Scientist, Greenpeace International, e-mail to author, 20 September 1999.

13. L. Ritter et al., *An Assessment Report on: DDT-Aldrin-Dieldrin-Endrin-Chlordane-Heptachlor-Hexachlorobenzene-Mirex-Toxaphene, Polychlorinated Biphenyls, Dioxins and Furans* prepared for the Inter-Organization Programme for the Sound Management of Chemicals (Guelph, ON: Canadian Network of Toxicology Centres, December 1995). Table 5–1 based on the following: production dates and cumulative totals from Paul Johnston, David Santillo, and Ruth Stringer, "Marine Environmental Protection, Sustainability and the Precautionary Principle," *Natural Resources Forum*, May 1999; introduction dates for aldrin, dieldrin, endrin, chlordane, and heptachlor, and 1977 production data for endrin and heptachlor, from Terry Gips, *Breaking the Pesticide Habit: Alternatives to 12 Hazardous Pesticides* (Penang, Malaysia: International Alliance for Sustainable Agriculture and International Organization of Consumers Unions, 1990); dioxins and furans from U.N. Environment Programme (UNEP), Chemicals Division, *Dioxin and Furan Inventories: National and Regional Emissions of PCDD/PCDF* (Geneva: Inter-Organization Programme for the Sound Management of Chemicals, May 1999); uses from Physicians for Social Responsibility, "International Effort Would Phase Out 12 Toxins," *PSR Monitor*, February 1998, and from Ritter et al., op. cit. this note.

14. "New Risks," *Audubon*, November/December 1998; World Wildlife Fund (WWF), *Resolving the DDT Dilemma: Protecting Human Health and Biodiversity* (Washington, DC: June 1998).

15. Shortage considered health threat from Amato, op. cit. note 11; producers and child deaths from malaria from WWF, op. cit. note 14; resistance from Amir Attaran, The Malaria Project, Washington, DC, e-mail to author, 20 September 1999; Graham Brown, "Fighting Malaria," *British Medical Journal*, 14 June 1997.

16. D. Pimentel and D. Kahn, "Environmental Aspects of 'Cosmetic Standards' of Foods and Pesticides," in David Pimentel, ed., *Techniques for Reducing Pesticide Use: Economic and Environmental Benefits* (New York: John Wiley & Sons, 1997); Bommanna G. Loganathan and Kurunthachalam Kannan, "Global Organochlorine Contamination Trends: An Overview," *Ambio*, May 1994; 26-fold from David Pimentel, Cornell University, Ithaca, NY, e-mail to author, 23 September 1999.

17. Brain Halweil, "Pesticide-Resistant Species Flourish," in Lester R. Brown, Michael Renner, and Brian Halweil, *Vital Signs 1999* (New York: W.W. Norton & Company, 1999); toxicity increase 10- to 100-fold from Pimentel and Kahn, op. cit. note 16.

18. Molecular structure from Frank Wania and Donald MacKay, "Tracking the Distribution of Persistent Organic Pollutants," *Environmental Science & Technology*, vol. 30, no. 9 (1996); Russia from Thomas W. Lippman, "Russian PCBs Complicate Toxics Treaty," *Washington Post*, 26 May 1998; uses of PCBs from Amato, op. cit. note 11, and from Yung-Cheng Joseph Chen et al., "Cognitive Development of Yu-Cheng ('Oil Disease') in Children Prenatally Exposed to Heat-Degraded PCBs," *Journal of the American Medical Association*, 9 December 1992.

19. Decline in environment from U.S. National Research Council (NRC), *Hormonally Active Agents in the Environment*, Committee on Hormonally Active Agents in the Environment, Board on Environmental Studies and Toxicology, Commission on Life Sciences (Washington, DC: National Academy Press, prepublication copy, July 1999); fat, blood, and milk and 70 percent from Loganathan and Kannan, op. cit. note 16; 70 percent also from R. Lloyd, "Some Ecotoxicological Problems Associated With the Regulation of PMPs," in Opschoor and Pearce, op. cit. note 4; stable compounds from Mitchell D. Erickson, *Analytical Chemistry of PCBs*, 2nd ed. (New York: CRC Press, Inc., 1997).

20. UNEP, op. cit. note 13.

21. Mari Yamaguchi, "Japan Facing Dioxin Pollution Woes," *Associated Press*, 3 July 1999; Andrew Pollack, "In Japan's Burnt Trash, Dioxin Threat," *New York Times*, 27 April 1997; differences among countries on sources of dioxin from Heidi Fiedler, UNEP Chemicals, Geneva, e-mail to author, 7 October 1999.

22. UNEP, op. cit. note 13.

23. Multinational Resource Center (MRC) and Health Care Without Harm, *The World Bank's Dangerous Medicine: Promoting Medical Waste Incineration in Third World Countries* (Washington, DC: MRC, June 1999); Indians from Ann Leonard, "Dangerous Medicine," *Multinational Monitor*, December 1996.

24. Quote from Linda Birnbaum, "POPs Toxicology," presentation at Harvard University School of Public Health, Center for Continuing Professional Education Symposium, Persistent Organic Pollutants (POPs): A Public Health Perspective, Cambridge, MA, 16–18 June 1999; Jane Ellen Simmons, "Chemical Mixtures: Challenge for Toxicology and Risk Assessment," *Toxicology*, December 1995.

25. Cod and turbot from WWF, op. cit. note 14; gulls from WWF, "Persistent Organic Pollutants: Hand-Me-Down Poisons that Threaten Wildlife and People," *Issue Brief* (Washington, DC: January 1999).

26. Theo Colborn, Frederick S. vom Saal, and

Ana M. Soto, "Developmental Effects of Endocrine-Disrupting Chemicals in Wildlife and Humans," *Environmental Health Perspectives*, October 1993.

27. Figure of 90 percent from Joint Canada-Philippines Planning Committee, *International Experts Meeting on Persistent Organic Pollutants: Towards Global Action*, meeting background report, Vancouver, Canada, 4–8 June 1995 (revised October 1995); India, Thailand, and Viet Nam from Loganathan and Kannan, op. cit. note 16; source of 90 percent of exposure from Arnold Schecter, "Exposure Assessment: Measurement of Dioxins and Related Chemicals in Human Tissues," in Arnold Schecter, ed., *Dioxins and Health* (New York: Plenum Press, 1994); fish advisories from Indigenous Environmental Network, "U.S. Department of State Bows to Chemical Industry; Will Propose Plan But No Action in Global POPs Treaty," news alert (Bemidji, MN: 31 August 1999).

28. Greenpeace International, *Unseen Poisons: Levels of Organochlorine Chemicals in Human Tissues* (Amsterdam: June 1998); 12 percent from David Moberg, "Unclean Bill of Health," *In These Times*, 13 June 1994; cows' milk from Pat Costner, *Dioxin Elimination: A Global Imperative* (Amsterdam: Greenpeace International, 1999); recommendations from Walter J. Rogan, "Pollutants in Breast Milk," *Archives of Pediatric and Adolescent Medicine,* September 1996; Walter J. Rogan et al., "Should the Presence of Carcinogens in Breast Milk Discourage Breast Feeding?" *Regulatory and Toxicological Pharmacology*, June 1991.

29. Process from Frank Wania and Donald MacKay, "Global Fractionation and Cold Condensation of Low Volatility Organochlorine Compounds in Polar Regions," *Ambio*, February 1993; Les Line, "Old Nemesis, DDT, Reaches Remote Midway Albatrosses," *New York Times*, 12 March 1996; Jennifer D. Mitchell, "Nowhere to Hide: The Global Spread of High-Risk Synthetic Chemicals," *World Watch*, March/April 1997, citing Rebecca Renner, "Researchers Find Unexpectedly High Levels of Contaminants in Remote Sea Birds,"

Environmental Science & Technology, January 1996; tree bark from Staci L. Simonich and Ronald A. Hites, "Global Distribution of Persistent Organochlorine Compounds," *Science*, 29 September 1995.

30. Robie Macdonald, "The Arctic Ocean and Contaminants: Pathways that Lead to Us," presentation at "Oceans Limited" conference, Simon Fraser University, Vancouver, Canada, available at <www.sfu.ca/oceans/macdonald.htm>; Wania and MacKay, op. cit. note 18; Wania and MacKay, op. cit. note 29.

31. PCBs from David VanderZwaag, "International Law and Arctic Marine Conservation and Protection: A Slushy, Shifting Seascape," *Georgetown International Environmental Law Review*, winter 1997; most exposed and chlordane amounts from "Poisoned Land," *Down To Earth*, 15 November 1998, and from Fred Pearce, "Northern Exposure," *New Scientist*, 31 May 1997; muktuk from Anthony DePalma, "An Arctic Meal: Seal Meat, Corn Chips, and PCB's," *New York Times*, 5 February 1999.

32. Sales in 1996 from U.N. Food and Agriculture Organization (FAO), "Cleaning Up Pesticides Nobody Wants," news release, <www.fao.org/news/1999/990504-e.htm>, viewed 4 June 1999; Figure 5–2 from FAO, *FAOSTAT Statistics Database*, <apps.fao.org>, viewed 15 July 1999, with figures deflated using the U.S. GNP Implicit Figure Deflator, provided in U.S. Department of Commerce, *Survey of Current Business*, August 1998.

33. Achila Imchen, "A Pest of a Problem," *Down To Earth*, 31 October 1998; Foundation for Advancements in Science and Education (FASE), *Exporting Risk: Pesticide Export from US Ports, 1995–1996* (Los Angeles: spring 1998),.

34. FASE, op. cit. note 33; lack of labeling from Carl Smith, "Exporting Risk—US Hazardous Trade, 1995–1996," *Pesticide News*, June 1998.

35. Ted Schettler et al., *Generations at Risk: Reproductive Health and the Environment*,

(Cambridge, MA: Massachusetts Institute of Technology Press, 1999); Colborn, Dumanoski, and Myers, op. cit. note 7; Colborn, vom Saal, and Soto, op. cit. note 26.

36. DDT biocidal properties from Loganathan and Kannan, op. cit. note 16, citing D.E. Howell, "A Case of DDT Storage in Human Fat," *Proceedings of the Oklahoma Academy of Sciences*, vol. 29 (1948); Philip G. Jenkins, "Ceaseless Vigilance for Chemical Safety," *World Health Forum*, vol. 19, no. 3 (1998); roosters from Colborn, Dumanoski, and Myers, op. cit. note 7, and from WWF, "A Warning from Wildlife," *Issues: Linking Theory and Practice in the Worldwide Conservation of Nature*, April 1996; 20 percent from "The Unintentional Global Experiment with Hormone-Disrupting Chemicals," *ENDS Report*, April 1996.

37. Colborn, Dumanoski, and Myers, op. cit. note 7.

38. Immunity from ibid., and from Peter S. Ross, Rik L. De Swart, and Albert D.M.E. Osterhaus, "Contaminant-related Suppression of Delayed-type Hypersensitivity and Antibody Responses in Harbor Seals Fed Herring from the Baltic Sea," *Environmental Health Perspectives*, February 1995; Mitchell, op. cit. note 29; Paul Epstein et al., *Marine Ecosystems: Emerging Diseases as Indicators of Change*, Health, Ecological and Economic Dimensions of the Global Change Program (Boston: Center for Health and Global Environment, Harvard Medical School, December 1998).

39. Schettler et al., op. cit. note 35; Colborn, Dumanoski, and Myers, op. cit. note 7; Colborn, vom Saal, and Soto, op. cit. note 26.

40. Figure of 45 from Colborn, vom Saal, and Soto, op. cit. note 26; total of 60 from Frederick S. vom Saal and Daniel M. Sheehan, "Challenging Risk Assessment," *Forum for Applied Research and Public Policy*, fall 1998; T. Colborn, "Endocrine Disruption from Environmental Toxicants," in W.N. Rom, ed., *Environmental and Occupational Medicine*, 3rd ed. (Philadelphia: Lippincott-Raven Publishers,

1998); F. Brucker-Davis, "Effects of Environmental Synthetic Chemicals on Thyroid Function," *Thyroid*, vol. 8, no. 9 (1998); P. Short and T. Colborn, "Pesticides Use in the U.S. and Policy Implications: A Focus on Herbicides," *Toxicology and Industrial Health*, vol. 15, no. 1–2 (1999); 250 and testing from "Industry Glimpses New Challenges as Endocrine Science Advances," *ENDS Report*, March 1999.

41. R.M. Giusti, K. Iwamoto, and E.E. Hatch, "Diethylstilbestrol Revisited: A Review of the Long-term Health Effects," *Annuals of Internal Medicine*, 15 May 1995; Colborn, vom Saal, and Soto, op. cit. note 26; Schettler et al., op. cit. note 35; Brian Halweil, "Sperm Counts Dropping," in Brown, Renner, and Halweil, op. cit. note 17.

42. Details of Taiwan incident and normal furan exposure level from Mei-Lin Yu et al., "Increased Mortality from Chronic Liver Disease and Cirrhosis 13 Years After the Taiwan 'Yucheng' ('Oil Disease') Incident," *American Journal of Industrial Medicine*, vol. 31 (1997); average U.S. PCB exposure from Longnecker, Rogan, and Lucier, op. cit. note 6.

43. Human health effects from Longnecker, Rogan, and Lucier, op. cit. note 6; increased liver disease from Yu et al., op. cit. note 42; 2.7-fold from Chen et al., op. cit. note 18; Yueliang L. Guo, George H. Lambert, and Chen-Chin Hsu, "Growth Abnormalities in the Population Exposed *in Utero* and Early Postnatally to Polychlorinated Biphenyls and Dibenzofurans," *Environmental Health Perspectives*, September 1995.

44. Joseph L. Jacobson and Sandra W. Jacobson, "Intellectual Impairment in Children Exposed to Polychlorinated Biphenyls in Utero," *New England Journal of Medicine*, 12 September 1996; S. Patandin et al., "Effects of Environmental Exposure to Polychlorinated Biphenyls and Dioxins on Cognitive Abilities in Dutch Children at 42 Months of Age," *Journal of Pediatrics*, January 1999.

45. Drop of 50 percent from Elisabeth Carlsen et al., "Evidence for Decreasing Quality of Semen During Past 50 Years," *British Medical Journal*, 12 September 1992; 1.5 and 3.1 percent from Shanna H. Swan et al., "Have Sperm Densities Declined? A Reanalysis of Global Trend Data," *Environmental Health Perspectives*, November 1997; male reproductive health problems from Leonard J. Paulozzi, "International Trends in Rates of Hypospadias and Cryptorchidism," *Environmental Health Perspectives*, April 1999, from J. Toppari et al., "Male Reproductive Health and Environmental Xenoestrogens," *Environmental Health Perspectives*, August 1996, from R. Bergstrom et al., "Increase in Testicular Cancer Incidence in Six European Countries: a Birth Cohort Phenomenon," *Journal of the National Cancer Institute*, June 1996, and from Halweil, op. cit. note 41.

46. NRC, op. cit. note 19; Paulozzi, op. cit. note 45; C. Alvin Paulsen, "Male Reproductive Health: Is It in Danger?" *Health & Environment Digest*, September 1998; Marla Cone, "Hormone Study Finds No Firm Answers," *Los Angeles Times*, 4 August, 1999; John Heinze, "Interregional Differences Undermine Sperm Trend Conclusions," *Environmental Health Perspectives*, August 1998; Gina Kolata, "Study Inconclusive on Chemicals' Effects," *New York Times*, 4 August 1999; Swan et al., op. cit. note 45; ongoing tests from Paul Smaglik, "Ambitious Plan to Screen for Endocrine Disruptors Unveiled," *The Scientist*, 14 September 1998, from C. Wu, "Huge Testing Planned for Hormone Mimics," *Science News*, 5 September 1998, from Davis Balz, "Implementing FQPA: U.S. EPA and Endocrine Disruptors," *Global Pesticide Campaigner*, April 1999, and from "Commission Cites Need to Address Environment Risks from Endocrine Disruptors," *International Environment Reporter*, 28 October 1998.

47. Robert Repetto and Sanjay S. Baliga, *Pesticides and the Immune System: The Public Health Risks* (Washington, DC: World Resources Institute, March 1996); E. Tielemans et al., "Pesticide Exposure and Decreased Fertilisation Rates in Vitro," *The Lancet*, 7 August 1999; Lennart Hardell and Mikael Eriksson, "A Case-Control

Study of Non-Hodgkin Lymphona and Exposure to Pesticides," *Cancer,* 15 March 1999; birth defects from Vincent F. Garry et al., "Pesticide Appliers, Biocides, and Birth Defects in Rural Minnesota," *Environmental Health Perspectives,* April 1996, and from Ida S. Weidner et al., "Cryptorchidism and Hypospadias in Sons of Gardeners and Farmers," *Environmental Health Perspectives,* December 1998; heart disease from J. Vena et al., "Exposure to Dioxin and Nonneoplastic Mortality in the Expanded IARC International Cohort Study of Phenoxy Herbicide and Chlorophenol Production Workers and Sprayers," *Environmental Health Perspectives,* April 1998.

48. Marian Burros, "High Pesticide Levels Seen in U.S. Food," *New York Times,* 19 February 1999; "Pesticides Report: How Safe Is Our Produce?" *Consumer Reports,* March 1999; Matthew L. Wald, "Citing Children, E.P.A. Is Banning Common Pesticide," *New York Times,* 3 August 1999.

49. Steven F. Arnold and John A. McLachlan, "Synergistic Signals in the Environment," *Environmental Health Perspectives,* October 1996; Leslie Lang, "Strange Brew: Assessing Risk of Chemical Mixtures," *Environmental Health Perspectives,* February 1995; Simmons, op. cit. note 24; Barry L. Johnson and Christopher T. DeRosa, "Chemical Mixtures Released from Hazardous Waste Sites: Implications for Health Risk Assessment," *Toxicology,* December 1995.

50. Sheldon Krimsky, "The Precautionary Approach," *Forum for Applied Research and Public Policy,* fall 1998; Johnston, Santillo, and Stringer, op. cit. note 13.

51. Jenkins, op. cit. note 36.

52. Growth in number of pesticide bans from Toni Nelson, "Efforts to Control Pesticides Expand," in Lester R. Brown, Christopher Flavin, and Hal Kane, *Vital Signs 1996* (New York: W.W. Norton & Company, 1996).

53. Bertil Hägerhäll, "Sluggish Conventions Gather Momentum," *Enviro,* April 1991; Ellen Hey, "The Precautionary Approach: Implications of the Revision of the Oslo and Paris Conventions," *Marine Policy,* July 1991; Independent World Commission on Oceans, *The Ocean, Our Future* (New York: Cambridge University Press, 1998); Costner, op. cit. note 28.

54. U.N. Sustainable Development Programme (UNSD), "Agenda 21 – Chapter 17: Protection of the Oceans, All Kinds of Seas, Including Enclosed and Semi-enclosed Seas, Coastal Areas and the Protection, Rational Use and Development of Their Living Resources," <www.un. org/esa/sustdeve/agenda21chapter17.htm>, viewed 11 October 1999; International Forum on Chemical Safety (IFCS) from Jenkins, op. cit. note 36, and from John Whitelaw, "Implementing the Plan," *Our Planet,* March 1997; UNSD, "Agenda 21 – Chapter 19: Environmentally Sound Management of Toxic Chemicals, Including Prevention of Illegal International Traffic in Toxic and Dangerous Products," <www.un. org/esa/sustdeve/agenda21chapter19.htm>, viewed 11 August 1999; ground-breaking from Clif Curtis, Director, Toxics Program, WWF, Washington, DC, discussion with author, 24 September 1999.

55. UNEP, "Decision 18/32 of the UNEP Governing Council: Persistent Organic Pollutants," May 1995, <irptc.unep.ch/pops/>, viewed 21 September 1999; UNEP, "Decisions Adopted by the Governing Council at Its Nineteenth Session," February 1997, <irptc.unep. ch/pops/gcpops_e.html>, viewed 17 September 1999; WWF, op. cit. note 25; International Institute for Sustainable Development (IISD), "The Second Session of the International Negotiating Committee for an International Legally Binding Instrument for Implementing International Action on Certain Persistent Organic Pollutants (POPS): 25–29 January 1999, A Brief History of the POPs Negotiations," *Earth Negotiations Bulletin,* 1 February 1999.

56. Töpfer quote from UNEP, "Report of the Intergovernmental Negotiating Committee for an International Legally Binding Instrument for Implementing International Action on Certain

Persistent Organic Pollutants on the Work of Its First Session, Montreal, 29 June–3 July 1998" (Geneva: 3 July 1998); UNEP, "Elimination of 10 Intentionally Produced Persistent Organic Pollutants Favoured by Treaty Negotiators: Health Need for DDT Exemption Seen," press release (Geneva: 13 September 1999); World Wide Fund for Nature, "Persistent Organic Pollutants Treaty: Eight Down, Four To Go," press release (Gland, Switzerland: 11 September 1999).

57. Walls, op. cit. note 12; Curtis, op. cit. note 54; "UN Progress Slow on 'Dirty Dozen' Pollutants Ban," *Reuters*, 15 September 1999; UNEP, "Report of the Intergovernmental Negotiating Committee for an International Legally Binding Instrument for Implementing International Action on Certain Persistent Organic Pollutants on the Work of Its Third Session, Geneva, 6–11 September 1999" (Geneva: 17 September 1999).

58. UNEP, "Experts Agree to Criteria, Procedure for Adding Persistent Organic Pollutants to Global POPs Treaty," press release (Geneva: 18 June 1999); Aarhus from UNEP, "Analysis of Selected Conventions Covering the Ten Intentionally Produced Persistent Organic Pollutants" (Geneva: 2 June 1999), and from Daniel Pruzin, "UN/ECE Draft Protocol on Heavy Metals, Persistent Organic Pollutants Concluded," *International Environment Reporter*, 18 February 1998; Sara Thurin Rollin, "Several Numerical Criteria Set for Adding Chemicals to U.N. Pops Treaty, Officials Say," *International Environment Reporter*, 9 December 1998; IISD, "Report of the First Session of the Criteria Expert Group for Persistent Organic Pollutants: 26–30 October 1998," *Earth Negotiations Bulletin*, 2 November 1998.

59. Lara Santoro, "Banning vs. Managing 'Dirty Dozen' Pollutants," *Christian Science Monitor*, 16 February 1999.

60. UNEP, "Possible Capacity-Building Activities and Their Associated Costs Under the International Legally Binding Instrument for Implementing International Action on Certain

Persistent Organic Pollutants: a Preliminary Review, First Revision" (Geneva: 23 July 1999).

61. Basel Action Network, "The Basel Ban—A Triumph for Global Environmental Justice," Briefing Paper No. 1 (Basel, Switzerland: May 1999); UNEP and FAO, *Report on Study on International Trade in Widely Prohibited Chemicals* (Rome: 16 February 1996); FAO, "Rotterdam Convention on Harmful Chemicals and Pesticides Adopted and Signed," press release (Rome: 11 September 1998).

62. Alemayehu Wodageneh, "Trouble in Store," *Our Planet*, March 1997; FAO, "FAO Warns of the Dangerous Legacy of Obsolete Pesticides," press release (Rome: 24 May 1999).

63. Charlie Clay, "The POPS Treaty," *Rachel's Environment & Health Weekly*, 4 June 1998.

64. U.S. Environmental Protection Agency (EPA), "Chapter 1: Toxics Release Inventory Reporting and the 1997 Public Data Release," *Toxics Release Inventory* (Washington, DC: 1997); EPA, "Toxics Release Inventory," <www.epa.gov/tri/general.htm>, viewed 6 August 1999; U.S. General Accounting Office, *Toxic Chemicals: EPA's Toxic Release Inventory Is Useful but Can Be Improved* (Washington, DC: June 1991).

65. Countries cited in EPA, "Chapter 1," op. cit. note 64, and in UNEP Chemicals, *National Inventories of Persistent Organic Pollutants: Selected Examples as Possible Models*, Preliminary Report (Geneva: July 1999); Czech Republic from Jindrich Petrlik, Deti Zeme (Children of the Earth), Prague, message posted on IPEN e-mail discussion group <pops-network@igc.org>, 20 June 1999.

66. UNEP, "National PRTR Activities: Challenges and Experiences," 1 March 1996, <irptc.unep.ch/prtr/nat01.html>, viewed 11 August 1999; "The Citizens' Right to Know," *New York Times*, 1 June 1999.

67. Radha Basu, "Landmark Law to Throw Out Waste Incineration," *Inter Press Service*, 7

May 1999, and Von Hernandez, Toxics Campaign Southeast Asia, Greenpeace International, message posted on IPEN e-mail discussion group <pops-network@igc.org>, 18 May 1999.

68. "Background Information on Ecology's Initiative," undated information sheet, Washington Toxics Coalition, Seattle, WA, received 17 August 1999; Carol Dansereau, "Statement of Carol Dansereau, Director, Washington Toxics Coalition, to Seattle Public Meeting on State PBT Strategy, 25 February 1999," Seattle, WA; Paul Engler, "Policing Corporate Polluters," *The Progressive*, March 1999.

69. Philippines from Basu, op. cit. note 67; Nike, Lego, and IKEA from Meagan Boltwood, "PVC: Going, Going, Gone?" *E Magazine*, May/June 1999; Greenpeace International, "General Motors Announces Plan to Go PVC-Free, Breaking Industry 'Code of Silence' on PVC Problems," press release (Amsterdam: 23 September 1999).

70. Transition fund from Joe Thornton and Jack Weinberg, "Dioxin Prevention: Focusing on Chlorine Chemistry," *New Solutions*, spring 1995; Valerie Frances et al., *F.A.C.T. Fair Agricultural Chemical Taxes: Tax Reform for Sustainable Agriculture* (Washington, DC: Friends of the Earth, May 1999).

71. Anders Emmerman, "Sweden's Reduced Risk Pesticide Policy," *Pesticides News*, December 1996; Olle Pettersson, "Pesticide Use in Swedish Agriculture: The Case of a 75% Reduction," in Pimentel, op. cit. note 16; U.K. Department of the Environment, Transport, and the Regions, "Executive Summary," in *Design of a Tax or Charge Scheme for Pesticides* (London: March 1999).

72. Karen L. Clark, *A Montreal Protocol for POPs?: An Evaluative Review of the Suitability of the Montreal Protocol as a Model for International Legally Binding Instruments Regarding the Control and Phase-Out of Persistent Organic Pollutants* (Gland, Switzerland: WWF, May 1996); IFCS Ad Hoc Working Group on POPs, "Some Relevant Aspects of Montreal Protocol" (Mani-

la: 21–22 June 1996); Hilary French and Molly O'Meara, "Learning from the Ozone Experience," in Lester R. Brown et al., *State of the World 1997* (New York: W.W. Norton & Company, 1997).

73. Table 5–5 from the following sources: dioxins from Barry Commoner et al., "Zeroing Out Dioxins: Conclusions of a CBNS Report," *New Solutions*, winter 1997; pesticides from Gips, op. cit. note 13; PCBs from Joint Canada-Philippines Planning Committee, op. cit. note 27, from Richard F. Addison, "PCB Replacements in Dielectric Fluids," *Environmental Science & Technology*, vol. 17, no. 10 (1983), and from WWF, "Successful, Safe, and Sustainable Alternatives to Persistent Organic Pollutants," *Issue Brief* (Washington, DC: September 1999); upgrade of incinerators from Swedish National Chemicals Inspectorate, *Alternatives to Persistent Organic Pollutants: The Swedish Input to the IFCS Expert Meeting on Persistent Organic Pollutants in Manila, the Philippines, 17–19 June 1996* (Solna, Sweden: July 1996).

74. Autoclaves and estimate of 1,500 facilities from MRC and Health Care Without Harm, op. cit. note 23.

75. Thornton and Weinberg, op. cit. note 70.

76. Commoner et al., op. cit. note 73; Center for the Biology of Natural Systems, Queens College, "New Reports Show Cost-Effective Ways to a Dioxin-Free Great Lakes," press release (Flushing, NY: 26 June 1996).

77. Geoffrey A.T. Target and Brian M. Greenwood, "Impregnated Bednets," *World Health*, May-June 1998; costs from World Health Organization (WHO), *World Health Organization Report on Infectious Diseases: Removing Obstacles to Healthy Development* (Geneva: 1999); Brown, op. cit. note 15.

78. WWF, op. cit. note 14; Sheryl Gay Stolberg, "DDT, Target of Global Ban, Has Defenders in Malaria Experts," *New York Times*, 28 August 1999; Sarah Boseley, "Malaria Fears Over Planned DDT Ban," *The Guardian*, 30

August 1999; 1.1 and 275 million from WHO, op. cit. note 77; Brown, op. cit. note 15.

79. Halweil, op. cit. note 17; Gary Gardner, "IPM and the War on Pests," *World Watch*, March/April 1996.

80. Gary Gardner, "Organic Farming Up Sharply," in Brown, Flavin, and Kane, op. cit. note 52; growth rate of 20 percent from Joel Bourne, "The Organic Revolution," *Audubon*, March/April 1999; U.S. sales figures from Organic Trade Association, industry statistics, <www.ota.com/business%20facts.htm>, viewed 11 October 1999.

Chapter 6. Recovering the Paper Landscape

1. Advertisement for UPM-Kymmene, *Financial Times*, 7 December 1988.

2. Number of grades from International Institute for Environment and Development (IIED), *Towards a Sustainable Paper Cycle* (London: 1996).

3. Figure for 1997 from Miller Freeman, Inc., *International Fact and Price Book 1999* (San Francisco, CA: 1998); 1950 from IIED, op. cit. note 2; projected demand from Shushuai Zhu, David Tomberlin, and Joseph Buongiorno, *Global Forest Products Consumption, Production, Trade and Prices: Global Forest Products Model Projections to 2010* (Rome: Forest Policy and Planning Division, U.N. Food and Agriculture Organization (FAO), December 1998).

4. FAO, *FAOSTAT Statistics Database*, <apps.fao.org>, viewed 28 October 1999; projection in Figure 6–1 from Zhu, Tomberlin, and Buongiorno, op. cit. note 3.

5. Per capita numbers from Miller Freeman, Inc., op. cit. note 3; industrial and developing country averages from FAO, op. cit. note 4; 30–40 kilograms from Mark P. Radka, "Policy and Institutional Aspects of the Sustainable Paper Cycle—An Asian Perspective" (Bangkok:

UNEP, Regional Office for Asia and the Pacific, 1994).

6. Proportions of different paper grades and consumption since 1980 from Miller Freeman, Inc., op. cit. note 3; Figure 6–2 from FAO, op. cit. note 4.

7. Paper.com, "Surprising Many, Paper Use Soars with Internet Growth," press release, <www.papercom.org/press6.htm>, viewed 22 April 1999.

8. Paper use in offices from Miller Freeman, Inc., *1999 North American Pulp and Paper Factbook* (San Francisco, CA: 1998), and from Raju Narisetti, "Pounded by Printers, Xerox Copiers Go Digital," *Wall Street Journal*, 12 May 1998; Bureau of Labor Statistics, *Employment and Earnings*, January 1999, <www.bls.gov/cpsaatab.htm>, viewed 15 September 1999; U.S. Postal Service, *United States Postal Service 1999 Annual Report* and *1997 Annual Report of the United States Postal Service*, <www.usps.gov/history>, viewed 21 April 1999.

9. See commentary in Miller Freeman, Inc., op. cit. note 8, and in Lauri Hetemäki, "Information Technology and Paper Demand Scenarios," in Matti Palo and Jussi Uusivuori, eds., *World Forests, Society and Environment* (Dordrecht, The Netherlands: Kluwer Academic Publishers, 1999).

10. Hou-Min Chang, "Economic Outlook for Asia's Pulp and Paper Industry," *TAPPI Journal*, January 1999.

11. Amount traded from FAO, op. cit. note 4; 45 percent from Bruce Michie, Cherukat Chandrasekharan, and Philip Wardle, "Production and Trade in Forest Goods," in Palo and Uusivuori, op. cit. note 9.

12. Growth calculated from Miller Freeman, Inc., op. cit. note 3; projections from Xiang-Ju Zhong, "Challenges and Opportunities in China," *Pulp and Paper International*, August 1998.

13. Zhang Yan, "Reliance on Foreign Lumber to Increase," *China Daily*, 7 September 1998; Bruce Gilley, "Sticker Shock," *Far Eastern Economic Review*, 14 January 1999; imports from FAO, op. cit. note 4.

14. Average annual natural forest cover loss between 1990 and 1995 was 13.7 million hectares according to FAO, *State of the World's Forests 1999* (Rome: 1999); causes of degradation from Dirk Bryant, Daniel Nielsen, and Laura Tangley, *The Last Frontier Forests, Ecosystems and Economies on the Edge* (Washington, DC: World Resources Institute, 1997).

15. Wood Resources International Ltd., *Fiber Sourcing Analysis for the Global Pulp and Paper Industry* (London: IIED, September 1996), reported that in 1993 wood pulp production required approximately 618 million cubic meters of wood—equal to 18.9 percent of the world's total wood harvest in 1993 and 41.8 percent of the total industrial wood harvest (wood harvest volumes as noted by FAO, op. cit. note 4); paper growing twice as fast as other major wood products from FAO, op. cit. note 14; 2050 from World Commission on Forests and Sustainable Development, *Our Forests, Our Future* (Cambridge, U.K.: Cambridge University Press, 1999).

16. Wood Resources International Ltd., op. cit. note 15, reported that 63 percent of the 618 million cubic meters was from the roundwood pulpwood supply and 37 percent was from manufacturing residues and off-site chipping operations.

17. Proportions of fiber sources and Figure 6–3 from FAO, op. cit. note 4.

18. Growth rates from Rita Pappens, "Chile Faces up to Financial and Forest Challenges," *Pulp and Paper International*, April 1999; cost from Nelson Noel, "Paper and Forest Products Industry Outlook" (New York: Moody's Investors Service, April 1999).

19. IIED, op. cit. note 2.

20. Trends in fiber supply from plantations from FAO, op. cit. note 14; Diana Propper de Callejon et al., "Sustainable Forestry within an Industry Context," in Sustainable Forestry Working Group, *The Business of Sustainable Forestry: Case Studies* (Chicago: John D. and Catherine T. MacArthur Foundation, 1998); subsidized plantation and foreign investment from Ricardo Carrere and Larry Lohmann, *Pulping the South, Industrial Tree Plantations and the World Paper Economy* (London: Zed Books, 1996); Japan Paper Association (JPA), *Pulp and Paper Statistics 1998* (Tokyo: 1999); JPA, *In Harmony with Nature* (Tokyo: October 1997).

21. Estimate of 13 million based on 10 percent of current plantation estate, from World Commission on Forests and Sustainable Development, op. cit. note 15; current plantation estate of approximately 130 million hectares calculated by adding the estimate of 60 million hectares in industrial countries to the 70 million in developing countries, from FAO, op. cit. note 14.

22. Benefits of plantations from FAO, op. cit. note 14; Roger Sedjo and Daniel Botkin, "Using Forest Plantations to Spare Natural Forests," *Environment*, December 1997; World Commission on Forests and Sustainable Development, op. cit. note 15.

23. IIED, op. cit. note 2; 80 percent from Nigel Dudley, "The Year the World Caught Fire" (Gland, Switzerland: World Wide Fund for Nature, December 1997); Fred Pearce, "Playing with Fire," *New Scientist*, 21 March 1998.

24. M.J. Mac et al., *Status and Trends of the Nation's Biological Resources, Vol. 1* (Reston, VA: U.S. Department of the Interior, U.S. Geological Survey, 1998); Sten Nilsson et al., "How Sustainable Are North American Wood Supplies?" Interim Report (Laxenburg, Austria: International Institute for Applied Systems Analysis, 1999); U.S. Department of Agriculture (USDA), U.S. Forest Service, *The South's Fourth Forest: Alternatives for the Future*, Forest Resource Report No. 24 (Washington, DC: June 1988).

25. Lesley Potter and Justin Lee, *Tree Planting in Indonesia: Trends, Impacts and Directions*, Occasional Paper No. 18 (Bogor, Indonesia: Center for International Forestry Research, December 1998); Tupinikim and Guarani from Carrere and Lohmann, op. cit. note 20; World Rainforest Movement, *Tree Plantations: Impacts and Struggles* (Montevideo, Uruguay: February 1999).

26. The Paper Task Force, *Paper Task Force Recommendations for Purchasing and Using Environmentally Preferable Paper* (New York: Environmental Defense Fund, 1995); IIED, op. cit. note 2.

27. World Energy Council, 1995, cited in IIED op. cit. note 2; Paper Task Force, op. cit. note 26.

28. Japan from JPA, *In Harmony with Nature*, op. cit. note 20, and from IIED op. cit. note 2; U.S. figure from Miller Freeman, Inc., op. cit. note 8.

29. Paper Task Force, op. cit. note 26; United States–Asia Environmental Partnership and the Civil Engineering Research Foundation, "Clean Technologies in U.S. Industries: Focus on Pulp and Paper" (Washington, DC: September 1997); JPA, *In Harmony with Nature*, op. cit. note 20.

30. Miller Freeman, Inc., op. cit. note 8; IIED, op. cit. note 2.

31. U.S. factories from IIED op. cit. note 2; Asia from Radka, op. cit. note 5.

32. Miller Freeman, Inc., op. cit. note 8.

33. IIED, op. cit. note 2.

34. Ibid.; Miller Freeman, Inc., op. cit. note 8.

35. IIED, op. cit. note 2; Miller Freeman, Inc., op. cit. note 8; U.S. Environmental Protection Agency (EPA), "Fact Sheet: EPA's Final Pulp, Paper, and Paperboard 'Cluster Rule'— Overview" (Washington, DC: November 1997);

Bill Nichols, "Four Years of Work, Debates Produce First Phase of EPA's Cluster Rules," *Pulp and Paper*, January 1998.

36. Ishiguro and Akiyama, 1994, cited in IIED, op. cit. note 2; Allen Hershkowitz, *Too Good To Throw Away: Recycling's Proven Record* (New York: Natural Resources Defense Council, February 1997); Dan Imhoff, *The Simple Life Guide to Tree-Free, Recycled and Certified Papers* (Philo, CA: SimpleLife, 1999). Note: the average conversion rate for recycled paper is 85–90 percent.

37. Douglas J. Burke, "1st Urban Fiber—A Pulp Mill in the Urban Forest," *TAPPI Journal*, January 1997; Weld F. Royal, "Paper Mill Project Plants Roots in the South Bronx," *Biocycle*, July 1994; Herbert Mushcamp, "Greening a South Bronx Brownfield," *New York Times*, 23 January 1998; John Holusha, "A New Life for a Polluted Old Bronx Rail Yard," *New York Times*, 6 July 1997.

38. U.S. figure from Franklin Associates, Ltd., "Characterization of Municipal Solid Waste in the United States: 1998 Update" (Washington, DC: Municipal and Industrial Solid Waste Division, Office of Solid Waste, EPA, July 1999).

39. Fiber supply and volumes from FAO, op. cit. note 4; rates from Miller Freeman, Inc., op. cit. note 3.

40. Rates from Miller Freeman, Inc., op. cit. note 3; Paper Recycling Promotion Center, "Information About Paper Recycling Promotion Center" (Tokyo: no date).

41. Miller Freeman, Inc., op. cit. note 8; "Divided EU Agrees on Packaging Directive, Joint Ratification of Climate Change Treaty," *International Environment Reporter*, 12 January 1994; "Council Agrees on FCP Phase-out, Packaging Recycling, Hazardous Waste List," *International Environment Reporter*, 11 January 1995; European Union Directive from "Report Highlights Benefits of Increasing Newspaper Recycling," *ENDS Report*, July 1998.

42. FAO, op. cit. note 4.

43. Fred D. Iannazzi, "A Decade of Progress in U.S. Paper Recovery," *Resource Recycling*, June 1999; Susan Kinsella, "Recycled Paper Buyers, Where Are You?" *Resource Recycling*, November 1998.

44. Miller Freeman, Inc., op. cit. note 8.

45. Lisa A. Skumatz and Robert Moylan, "How to Re-energize Recycling Progress," *Resource Recycling*, June 1999.

46. Process improvements from Jim Kenny, "More Recycled Fibers Generate Process Improvements," *Pulp and Paper Magazine*, June 1999, and from Marguerite Sykes et al., "Environmentally Sound Alternatives for Upgrading Mixed Office Wastes," *Proceedings of the 1995 International Environmental Conference*, Technical Association of the Pulp and Paper Industry (TAPPI), 7–10 May 1995, Atlanta, GA.

47. Maureen Smith, *The U.S. Paper Industry and Sustainable Production* (Cambridge, MA: The MIT Press, 1997).

48. Tsuen-Hsuin Tsien, *Written on Bamboo and Silk, The Beginnings of Chinese Books and Inscriptions* (Chicago: University of Chicago Press, 1962); 7 percent from FAO, op. cit. note 4.

49. FAO, op. cit. note 4.

50. Agricultural byproducts share from Leena Paavilainen, "European Prospects for Using Nonwood Fibers," *Pulp and Paper International*, June 1998; straw yields from Smith, op. cit. note 47; total available from George A. White and Charles G. Cook, "Inventory of Agro-Mass," in Roger M. Rowell, Raymond A. Young, and Judith K. Rowell, eds., *Paper and Composites from Agro-Based Resources* (Boca Raton, FL: CRC Press, 1997); Joseph E. Atchison, "Update on Global Use of Non-wood Plant Fibers and Some Prospects for Their Greater Use in the United States," in TAPPI, 1998 TAPPI Proceedings North American Nonwood Fiber Symposium, Atlanta, GA, 17–18 February 1998; burning of residues from Vaclav Smil,

"Crop Residues: Agriculture's Largest Harvest," *Bioscience*, April 1999, and from David O. Hall et al., "Biomass for Energy: Supply Prospects," in Thomas B. Johansson et al., eds., *Renewable Energy: Sources for Fuels and Electricity* (Washington, DC: Island Press, 1993).

51. Figure of 9 percent from Paavilainen, op. cit. note 50; lignin from James S. Han, "Properties of Nonwood Fibers," in *Proceedings of the Korean Society of Wood Science and Technology Annual Meeting 1998* (Seoul, Republic of Korea: Korean Society of Wood Science and Technology, 1998).

52. Sustainable Forestry Working Group, op. cit. note 20; Brazil from Jeremy Williams, *A Study of Plantation Timber Prices in Latin America and the Southern United States of America* (Rome: FAO, 1998).

53. "Company Watch," *Business Ethics*, May/June 1998; "McDonald's Calls for Disclosure of Forestry Practices," *Pulp and Paper Online*, <www.pponline.com>, viewed 9 March 1999.

54. World Wide Fund for Nature, "Certificatoin Hits More Than 15 Million Hectares Worldwide," press release (Gland, Switzerland: 21 January 1999); first approved pulp from The Wilderness Society, "New York Mill Produces North America's First 'Green Certified' Paper," press release (Washington, DC: 15 October 1998).

55. Lauren Blum, Richard A. Denison, and John F. Ruston, "A Life-cycle Approach to Purchasing and Using Environmentally Preferable Paper," *Journal of Industrial Ecology*, vol. 1, no. 3 (1997); Radka, op. cit. note 5; New Mexico mill in David J. Bentley, Jr., "McKinley Paper Goes from Zero to Success in Less than Four Years," *TAPPI Journal*, January 1999; "Green Bay Studying Plan to Build Series of Regional Mini-mills Based on Recycled Fiber," *Pulp and Paper Week*, 24 May 1993; "Case Closed," *Boxboard Containers*, October 1993.

56. IIED, op. cit. note 2; Paper Task Force, op. cit. note 26; David J. Senior et al., "Enzyme Use

Can Lower Bleaching Costs, Aid ECF Conversions," *Pulp and Paper*, July 1999; 80 percent in Jay Ritchlin and Paul Johnston, "Zero Discharge: Technological Progress Towards Eliminating Kraft Pulp Mill Liquid Effluent, Minimising Remaining Waste Streams and Advancing Worker Safety," prepared for Reach For Unbleached!, Zero Toxics Alliance Pulp Caucus and Greenpeace International (Whaletown, BC, Canada: Reach for Unbleached Foundation, no date).

57. USDA, Forest Service, Forest Products Lab (FPL), "Biopulping: Technology Learned from Nature that Gives Back to Nature" (Madison, WI: March 1997); Masood Akhtar et al., "Toward Commercialization of Biopulping," *Paper Age*, February 1997; Gary Myers, FPL, USDA, discussion with Janet Abramovitz, 30 July 1999; fiber loading from John H. Klungness et al., "Lightweight, High Opacity Paper: Process Costs and Energy Use Reduction," presented at the AIChE Symposium/TAPPI Pulping Conference, Montreal, 29 October 1998; Oliver Heise et al., "Industrial Scale-up of Fiber Loading on Deinked Wastepaper," TAPPI Pulping Conference Proceedings, Nashville, TN, 27–31 October 1996; Sykes et al., op. cit. note 46; payback period from John H. Klungness, chemical engineer, FPL, USDA, discussion with Janet Abramovitz, 30 July 1999.

58. Advantages of totally chlorine-free paper from Ritchlin and Johnston, op. cit. note 56, and from Paper Task Force, op. cit. note 26; production from Alliance for Environmental Technology (AET), "Trends in World Bleached Chemical Pulp Production: 1990–1998," October 1998, <www.aet.org>, viewed 16 July 1999; Greenpeace, "Chlorine Use Sectors— Pulp and Paper, <www.greenpeace.org/ toxics/ci>, viewed 16 July 1999.

59. AET, op. cit. note 58; Paper Task Force, op. cit. note 26.

60. Release amounts from Greenpeace, "Alternatives to Chlorine TCF vs ECF," <www. greenpeace.org/toxics/cp/cp-alts-pp3.html>, viewed 16 July 1999.

61. Mill study in Chad Nehrt, "Spend More to Show Rivals a Clean Pair of Heals," *Pulp and Paper International*, 1 June 1995.

62. Figure of 20 percent from Nick Robins and Sarah Roberts, "Rethinking Paper Consumption" (London: IIED, September 1996); EPA, WasteWise Program Updates (various), <www.epa.gov/wastewise>; Bruce Nordman, U.S. Department of Energy, Lawrence Berkeley National Laboratory, "Cutting Paper," <eetd. lbl.gov.paper>, viewed 2 August 1999; Xerox, "Xerox Business Guide to Waste Reduction and Recycling," <www.xerox.com>, viewed 4 November 1999.

63. Bank of America 1997 environment report at <www.bankamerica.com/environment>, viewed 17 August 1999.

64. Nordman, op. cit. note 62; size reductions from Kinsella, op. cit. note 43; Jerry Powell, "Seven Hot Trends in Paper Recycling," *Resource Recycling* (Recovered Paper Supplement), April 1998.

65. United Parcel Service (UPS) and The Alliance for Environmental Innovation (a project of the Environmental Defense Fund and The Pew Charitable Trusts), "Achieving Preferred Packaging: Report of the Express Packaging Project" (New York: November 1998); Elizabeth Sturcken, "Preferred Packaging: Accelerating Environmental Leadership in the Overnight Shipping Industry" (New York: Alliance for Environmental Innovation, December 1997); "Federal Express Joins with Recycled Paperboard Alliance," *Pulp and Paper Online*, <www.pponline.com>, viewed 25 October 1999.

66. Paper Recycling Promotion Center, "Manual for Used Office Paper Recycling" (Tokyo: March 1999).

67. Paper Task Force, op. cit. note 26; Susan Kinsella, "Environmentally Sound Paper Overview: The Essential Issues," *Conservatree's Greenline*, October 1996; Imhoff, op. cit. note 36; Coop America, *Woodwise Consumer Guide*

(Washington, DC: 1999).

68. Peter Ince, FPL, USDA, discussion with Janet Abramovitz, 30 July 1999.

69. Procter and Gamble from IIED, op. cit. note 2.

70. EPA, "WasteWise Fifth Year Progress Report," August 1999; EPA, "Wastewise Update," June 1996; Paul Hawken, Amory Lovins, and Hunter Lovins, *Natural Capitalism* (New York: Little, Brown and Company, 1999).

71. Nordman, op. cit. note 62; Nevin Cohen, "Greening the Internet: Ten Ways E-Commerce Could Affect the Environment," *Environmental Quality Management*, Autumn 1999.

72. UPS and Alliance for Environmental Innovation, op. cit. note 65; EPA, "Fifth Year Progress Report," op. cit. note 70.

73. Miller Freeman, Inc., op. cit. note 8.

74. World Wide Fund for Nature, Forests for Life Campaign, Buyers Groups information and contacts, <www.panda.org>, viewed 10 November 1999.

75. U.S. government from Todd Paglia, "The Big Deal Recycling Executive Order," *Multinational Monitor*, January/February 1998; Paper Recycling Promotion Center, op. cit. note 66.

76. Paper Task Force, op. cit. note 26; UPS and Alliance for Environmental Innovation, op. cit. note 65; Sturcken, op. cit. note 65; EPA, "Fifth Year Progress Report," op. cit. note 70.

77. Kinsella, op. cit. note 67; Imhoff, op. cit. note 36; Paper Task Force, op. cit. note 26; Coop America, op. cit. note 67.

78. Projection for 2010 from Zhu, Tomberlin, and Buongiorno, op. cit. note 3.

79. Smith, op. cit. note 47; International Energy Agency, *Indicators of Energy Use and Efficiency* (Paris: Organisation for Economic Co-operation and Development, 1997); Japan from Ministry of International Trade and Industry, *Pulp and Paper Yearbook* (various), analyzed by Nozaki Nao, Hokkaido University, June 1999, unpublished.

80. Recommended minimum from Radka, op. cit. note 5.

81. Stephan Schmidheiny, with the Business Council for Sustainable Development, *Changing Course* (Boston: The MIT Press, 1992); EPA, "Fifth Year Progress Report," op. cit. note 70.

82. Hawken, Lovins, and Lovins, op. cit. note 70.

83. Worldwatch Institute calculations based on FAO, op. cit. note 4 (using FAO categories of industrial and developing countries).

84. British Columbia and Indonesia from Nigel Sizer, David Downes, and David Kaimowitz, "Tree Trade: Liberalization of International Commerce in Forest Products: Risks and Opportunities" (Washington, DC: World Resources Institute and the Center for International Environmental Law, November 1999); U.S. subsidies from Susan Kinsella, "Welfare for Waste: How Federal Taxpayer Subsidies Waste Resources and Discourage Recycling," *Grassroots Recycling Network*, April 1999, and from Danna Smith, "Chipping Forests and Jobs: A Report on the Economic Impact of Chip Mills in the Southeast" (Brevard, NC: Dogwood Alliance and Native Forest Network, August 1997).

85. David Malin Roodman, *The Natural Wealth of Nations* (New York: W.W. Norton & Company, 1998); Duncan McLaren et al., *Tomorrow's World: Britain's Share in a Sustainable Future* (London: Earthscan, 1998).

86. Sizer, Downes, and Kaimowitz, op. cit. note 84; Christopher Bright, "Invasive Species: Pathogens of Globalization," *Foreign Policy*, fall 1999.

87. Ruth Walker, "Fee-Based Recycling of Trash is a Mixed Bag for Germans," *Christian*

Science Monitor, 22 May 1997; EPA, "Pay-As-You-Throw Introduction," <www.epa.gov/epaoswer/non-hw/payt>, viewed 13 April 1999; Smith, op. cit. note 47.

88. Radka, op. cit. note 5; British Columbia from "Zero AOX Law Faces Behind-the-Scenes Attack," press release (Whaletown, BC, Canada: 7 May 1999).

89. Current and historic R&D spending from Miller Freeman, Inc., op. cit. note 8; comparison to R&D in other industries from Smith, op. cit. note 47; research institutes in developing countries from Radka, op. cit. note 5.

Chapter 7. Harnessing Information Technologies for the Environment

1. Silicon Valley Toxics Coalition (SVTC), <www.svtc.org>, viewed 1 October 1999.

2. Claire L. Parkinson, *Earth From Above* (Sausalito, CA: University Science Books, 1997).

3. Richard Ehrlich, "Opening Up to the Outside World," *Bangkok Post*, 9 July 1999; Peter de Jonge, "Television's Final Frontier," *New York Times*, 22 August 1999; DrukNet, Bhutan's national Internet service provider, <www.druknet.net.bt>, viewed 11 August 1999.

4. Figure 7–1 from International Telecommunications Union (ITU), *World Telecommunications Indicators '98*, Socioeconomic Time-series Access and Retrieval System (STARS) database, downloaded 24 August 1999.

5. Phones from ibid; computers from Karen Petska, *Computer Industry Almanac* (forthcoming), e-mail to author, 31 July 1999; Internet hosts from Network Wizards, "Internet Domain Surveys, 1981–1999," <www.nw.com>, updated January 1999.

6. Earth observation missions from Draft Report of the Third United Nations Conference on the Exploration and Peaceful Uses of Outer Space, 16 April 1999; GIS software from Carmelle Côté, Environmental Systems Research Institute, Vienna, VA, discussion with author, 17 August 1999.

7. Information and communications technology industry from International Data Corporation (IDC), *Digital Planet* (Vienna, VA: World Information Technology and Services Alliance, 1999); Internet commerce from U.S. Department of Commerce, *The Emerging Digital Economy II* (Washington, DC: June 1999); net worth from Kerry Dolan, "200 Global Billionaires," *Forbes*, 5 July 1999.

8. Table 7–1 from the following sources: Tom Standage, *The Victorian Internet: The Remarkable Story of the Telegraph and the Nineteenth Century's Online Pioneers* (New York: Walker and Company, 1998); postal service from <www.ericsson.se/Connexion/connexion3-94/history.html>; telephone and telegraph from Ithiel de Sola Pool, ed., *The Social Impact of the Telephone* (Cambridge, MA: The MIT Press, 1977), and from John Bray, *The Communications Miracle* (New York: The Premium Press, 1995); Martin Campbell-Kelly and William Aspray, *Computer: A History of the Information Age* (New York: Basic Books, 1996); Heather E. Hudson, *Communication Satellites: Their Development and Impact* (New York: The Free Press, 1990); World Wide Web from Tim Berners-Lee, *Weaving the Web* (New York: HarperCollins, 1999); graphical browser from Katie Hafner and Matthew Lyon, *Where Wizards Stay Up Late: The Origin of the Internet* (New York: Simon and Schuster, 1996). Effects of printing press from Ithiel de Sola Pool and Eli M. Noam, eds., *Technologies Without Boundaries* (Cambridge, MA: Harvard University Press, 1990).

9. Standage, op. cit. note 8; Bray, op. cit. note 8.

10. Heather E. Hudson, *Global Connections: International Telecommunications Infrastructure and Policy* (New York: Van Nostrand Reinhold, 1997).

11. Industrial-country phone lines from ITU, *World Telecommunication Development Report*

1998 (Geneva: 1998). The ITU collects information on telephone "main lines." These can be either exclusive or shared, so the number of telephone subscribers exceeds the number of lines. They can be used for telephone sets, fax machines, or personal computers. Figure 7–2 from ITU, op. cit. note 4; Scandinavian countries from Organisation for Economic Co-operation and Development, *Communications Outlook 1999* (Paris: 1999).

12. Investment in developing countries from ITU, op. cit. note 11; gains in Latin America and Asia from ibid., and from ITU, *Asia-Pacific Telecommunication Indicators: New Telecommunication Operators* (Geneva: 1997); population of Buenos Aires and Eastern Africa from United Nations, *World Urbanization Prospects: 1996 Revision* (New York: United Nations, 1998); U.S. access from National Telecommunications and Information Administration, *Falling Through The Net* (Washington, DC: U.S. Department of Commerce, July 1999).

13. Computing power from Max Schulz, "The End of the Road for Silicon," *Nature*, 24 June 1999.

14. Fiber-optic capacity from Annabel Z. Dodd, *The Essential Guide to Telecommunications*, 2nd ed. (Upper Saddle River, NJ: Prentice Hall, 1999), and from David D. Clark, "High-Speed Data Races Home," *Scientific American*, October 1999; transmission of data surpassing voice from Alan Cane, "Operators Race to Surf 'Data Wave'," *Financial Times*, 10 June 1998; undersea cables from Charles W. Petit, "Spaghetti Under the Sea: A Network of Communications Cables Fill the Ocean Depths," *U.S. News and World Report*, 30 August 1999; satellites from Joseph N. Pelton, "Telecommunications for the 21st Century," Wireless Technologies: Special Report, *Scientific American*, April 1998.

15. Clark, op. cit. note 14.

16. Rachel Schwartz, *Wireless Communications in Development Countries: Cellular and Satellite Systems* (Boston, MA: Artech House, 1996);

Lisa Sykes, "Hanging on for the Phone," *New Scientist*, 14 June 1997; "Telecommunications Survey," *The Economist*, 9–15 October 1999; ITU, op. cit. note 4.

17. Heather E. Hudson, *Communication Satellites* (New York: The Free Press, 1990); Intelsat, "About Intelsat," 3 November 1997, <www.intelsat.int/cmc/info/intelsat.htm>, viewed 16 January 1998; Anthony Michaelis, *From Semaphore to Satellite* (Geneva: ITU, 1965); Bray, op. cit. note 8.

18. Low-orbiting satellites from Mike May, "High Hopes for Low Satellites," *Technology Review*, October 1997; Eric Schine et al., "The Satellite Biz Blasts Off," *Business Week*, 27 January 1997; Mike Mills, "Orbit Wars," *Washington Post Magazine*, 3 August 1997.

19. Kristi Coale, "Teledesic Mounts Lead in New Space Race," *Wired*, 14 October 1997; Teledesic Corporation, "Boeing to Build Teledesic's Internet-In-The-Sky," press release (Seattle, WA: 29 April 1997); "Iridium Kicks Off Satellite Service," *PC Week*, 9 November 1998; David S. Bennahum, "The United Nations of Iridium," *Wired*, October 1998; Iridium cost and 1-satellite systems from ITU, op. cit. note 11.

20. Communications satellite share of total from Jos Heyman, *Spacecraft Tables 1957–1997* (Riverton, Australia: Tiros Space Information, 1998), and from Jos Heyman, e-mail to author, 4 January 1999; private companies from Theresa Foley, "Commercial Spacefarers," *Air Force Magazine*, December 1998; commercial satellites eclipsing government satellites from Marco Caceres, "Commercial Satellites Surge Ahead," *Aerospace America*, November 1998; World-Space from "African Radio Satellite Launched," *Reuters*, 29 October 1998; Austin Bunn, "New Media: Information Affluence," *Wired*, February 1998.

21. J. Warren Stehman, *The Financial History of the American Telephone and Telegraph Company* (Boston, MA: Houghton Mifflin Co., The Riverside Press, 1925); World Trade Organization, "Ruggiero Congratulates Governments on Landmark Telecommunications Agreement," press release (Geneva: 17 February 1997); Alex Arena, "The WTO Telecommunications Agreement: Some Personal Reflections," in Gregory Staple, ed., *Telegeography 1997/98: Global Telecommunications Traffic Statistics and Commentary* (Washington, DC: Telegeography, Inc., 1997).

22. "FT Telecoms: Financial Times Review of the Telecommunications Industry," *Financial Times*, 10 June 1998, ITU, op. cit. note 11.

23. Calculation by Leslie Byster and Ted Smith, "High-Tech and Toxic," *Forum for Applied Research and Public Policy*, spring 1999, based on figures from Graydon Larabee, Texas Instruments, speech at the International Symposium on Semiconductor Manufacturing, September 1993.

24. U.S. Environmental Protection Agency (EPA), Office of Compliance Sector Notebook Project, *Profile of the Electronics and Computer Industry* (Washington, DC: September 1995); Microelectronics and Computer Technology Corporation, "Environmental Consciousness: A Strategic Competitiveness Issue for the Electronics and Computer Industry, Summary Report: International Analysis, Conclusions and Recommendations," March 1993, <www.mcc.com>, viewed 18 August 1999; calculation by John C. Ryan and Alan Thein Durning, *Stuff: The Secret Lives of Everyday Things* (Seattle, WA: Northwest Environment Watch, January 1997), based on Microelectronics and Computer Technology Corporation, op. cit. this note.

25. History of Silicon Valley from Aaron Sachs, "Virtual Ecology: A Brief Environmental History of Silicon Valley," *World Watch*, January/February 1999, and from Southwest Network for Environmental and Economic Justice and Campaign for Responsible Technology, *Sacred Waters: Life-Blood of Mother Earth* (Albuquerque, NM: 1997); health and environmental effects from Joseph LaDou and Timothy Rohm, "The International Electronics Industry," *International Journal of Occupational and Environmental Health*, January-March 1998;

Ricardo Alonso-Zaldivar, "Cancer Cases Cast Pall Over High-Tech Jobs," *Los Angeles Times*, 5 December 1998; David Bacon, "Toxic Technology," *In These Times*, 23 November 1997; Julie Schmit, "Dirty Secrets: Exposing the Dark Side of a 'Clean' Industry," *USA Today*, 12 January 1998.

26. Leah B. Jung, "The Conundrum of Computer Recycling," *Resource Recycling*, May 1999; National Safety Council, *Electronic Product Recovery and Recycling Baseline Report: Recycling of Selected Electronic Products in the United States,* (Washington, DC: 1999); "Computers: Disposal Becoming Global Concern," *UN Wire*, 13 August 1999; Ross Davies, "Dumping of Old PCs Becomes Problem in Britain," *Environmental News Network*, 29 October 1999.

27. Traci Watson, "USA Sitting on Mountain of Obsolete PCs," *USA Today*, 22 June 1999.

28. SVTC, op. cit. note 1; corporate reports from Motorola, *The Journey to a Sustainable World: Environmental, Health and Saftey Results for 1998*, <www.motorola.com/EHS>, viewed 21 September 1999, from IBM, *1999 Environment and Well-Being Report*, <www.ibm.com/environment>, viewed 21 September 1999, and from Intel, *Commitment Around the Globe: Designing for the Future*, <www.intel.com/other/ehs>, viewed 21 September 1999.

29. Carlo Vezzoli, "Life Cycle Design of a Telephone," *UNEP Industry and Environment*, January-June 1997.

30. European Commission, Directorate General XI, "Draft Directive on Waste from Electrical and Electronic Equipment," Brussels, 5 July 1999.

31. "Draft German Electronics Takeback Decree in Trouble," *CutterEdge Environment*, weekly e-mail service from Cutter Information Corp., 6 July 1999.

32. "Take Back Schemes Launched for Mobile Phones," *ENDS Report 264*, January 1997; Carl

Frankel, "Desperately Seeking Environment-ability," *Tomorrow*, September/October 1998.

33. Beverley Thorpe and Iza Kruszewska, "Strategies to Promote Clean Production: Extended Producer Responsibility," <www.igc. apc.org/svtc/cleancc/strat.htm>, viewed 2 August 1999; "Australia to Introduce Mobile Phone Recycling," *Reuters*, 7 August 1999

34. James R. Ghiles, "Casting a High-Tech Net for Space Trash," *Smithsonian*, January 1999; "The Long Arm of the Celestial Repairman," *The Economist*, 17 October 1998; Fred Guterl, "For Lease: Affordable Private Reusable Rockets," *Discover*, April 1999; Marcus Chown, "Reeling in Satellites," *New Scientist*, 21 February 1998; "Space Station Moved to Avoid 'Junk'," *Associated Press*, 26 October 1999.

35. Miller Freeman, Inc., *1999 North American Pulp and Paper FactBook* (San Francisco, CA: 1998); increase in consumption is a Worldwatch estimate based on U.N. Food and Agriculture Organization, *FAOSTAT Statistics Database*, <apps.fao.org>, viewed 23 August 1999; change in types of paper used from Matti Paolo and Jussi Uusivuori, eds., *World Forests, Society and Environment* (Dordrecht, The Netherlands: Kluwer Academic Publishers, 1999).

36. "Replacing Paper: Bad News for Trees," *The Economist*, 19 December 1998; Geoffrey E. Meredith, "The Demise of Writing," *The Futurist*, October 1999; Bennett Daviss, "Paper Goes Electric," *New Scientist*, 15 May 1999; "Electronic Paper Draws Beam on Pulp Industry," *Environmental News Service*, 29 June 1999.

37. "Energy Star Labeled Office Equipment," <www.epa.gov/appdstar/esoe/index.html>, viewed 23 July 1999; Johnathan Koomey et al., "Efficiency Improvements in U.S. Office Equipment: Expected Policy Impacts and Uncertainties" (Berkeley, CA: Lawrence Berkeley National Laboratory, December 1995); John B. Horrigan, Frances H. Irwin, and Elizabeth Cook, *Taking a Byte Out of Carbon* (Washington, DC: World Resources Institute (WRI), 1998).

38. Horrigan, Irwin, and Cook, op. cit. note 37; Chris Robertson, "Design for Energy Efficiency Adds Value to Semiconductor Company Shareholders," *Semiconductor Fabtech*, Edition 10, July 1999; U.S. Department of Energy, *Annual Energy Outlook 1999* (Washington, DC: U.S. Government Printing Office, 1999).

39. Representative of U.S. West quoted in Terrell J. Minger and Meredith Miller, "From Hydrocarbons to Bits and Bytes," *Our Planet*, June 1999.

40. Telecommunications stimulating travel from John S. Niles, "Beyond Telecommuting: A New Paradign for the Effect of Telecommunications on Travel," (Washington, DC: U.S. Department of Energy, Office of Energy Research, September 1994); history from Peter Hall, *Cities and Civilization* (New York: Pantheon, 1998).

41. Nevin Cohen, "Greening the Internet: Ten Ways E-Commerce Could Affect the Environment," *Environmental Quality Management*, Autumn 1999.

42. Advertising from Payal Sampat, "World Ad Spending Climbs," in Lester Brown, Michael Renner, and Brian Halweil, *Vital Signs 1999* (New York: W.W. Norton & Company, 1999); consumerism from Alan Thein Durning, *How Much is Enough?* (New York: W.W. Norton & Company, 1992).

43. "Astronauts Discuss Environment," *Associated Press*, 4 October 1999; quote from Mario Runco, astronaut on space shuttle missions in 1991 and 1993, cited in Daniel Glick, "Windows on the World," *National Wildlife*, February/March 1994.

44. Parkinson, op. cit. note 2; Landsat 6 did not reach orbit and Landsat 7 was launched in 1999.

45. Robert Bindschadler and Patricia Vornberger, "Changes in the West Antarctic Ice Sheet Since 1963 from Declassified Satellite Photography," *Science*, 30 January 1998; R. Monastersky,

"Spy Satellite Plumbs Secrets of Antarctica," *Science News*, 31 January 1998.

46. Parkinson, op. cit. note 2; R.B. Myneni et al., "Increase in Plant Growth in the Northern High Latitudes from 1981–1991," *Nature*, 17 April 1997.

47. Charles Kennel, Pierre Morel, and Gregory Williams, "Keeping Watch on the Earth: An Integrated Observing Strategy," *Consequences*, vol. 3, no. 2 (1997).

48. J.C. Farman et al., "Large Losses of Total Ozone in Antarctica Reveal Seasonal ClO_x/NO_x Interaction," *Nature*, 16 May 1985; Susan Solomon et al. "On the Depletion of Antarctic Ozone," *Nature*, 19 June 1986; R.S. Stolarski et al. "Nimbus 7 Satellite Measurements of the Springtime Antarctic Ozone Decrease," *Nature*, 28 August 1986.

49. Dana Mackenzie, "Ocean Floor Is Laid Bare by New Satellite Data," and Walter H.F. Smith and David T. Sandwell, "Global Sea Floor Topography from Satellite Altimetry and Ship Depth Soundings," *Science*, 26 September 1997.

50. "Sea WiFS Completes a Year of Remarkable Earth Observations," *Science Daily*, 18 September 1998; "EU, Russia Team Up with Companies on Creation of Global Satellite Data Source," *International Environment Reporter*, 5 August 1998.

51. Robert Wright, "Private Eyes," *New York Times*, 5 September 1999; Vernon Loeb, "Spy Satellite Will Take Photos for Public Sale," *Washington Post*, 25 September 1999.

52. Kevin Corbley, "Remote Sensing Skies Filling with Satellite Plans," *EOM* (Earth Observation Magazine), October 1996; "Spy-Quality Satellite Ready to Sell Images," *Reuters*, 25 December 1997; Tony Reichhardt, "Environmental GIS: The World in a Computer," *Environmental Science and Technology*, vol. 30, no. 8 (1996); Chuck Gilbert, "Evolution of GPS Data Collection for GIS," *EOM* (Earth Observation Magazine), February 1998.

53. B.D. Santer et al., "Detection of Climate Change and Attribution of Causes," in J.T. Houghton et al., eds., *Climate Change 1995: The Science of Climate Change*, Contribution of Working Group I to the Second Assessment Report of the Intergovernmental Panel on Climate Change (Cambridge, U.K.: Cambridge University Press, 1996); B.D. Santer et al., "A Search for Human Influences on the Thermal Structure of the Atmosphere," *Nature*, 4 July 1996; Robert Kaufmann and David Stern, "Evidence for Human Influence on Climate from Hemispheric Temperature Relations," *Nature*, 3 July 1997.

54. Jack M. Hollander and Duncan Brown, "Air Pollution" in Jack M. Hollander, ed., *The Energy-Environment Connection* (Washington, DC: Island Press, 1992); Patricia W. Birnie and Alan E. Boyle, *Basic Documents on International Law and the Environment* (Oxford, U.K.: Oxford University Press, 1995); Marc Levy, "European Acid Rain: The Power of Tote-Board Diplomacy," in Peter M. Haas et al., eds., *Institutions for the Earth* (Cambridge, MA: The MIT Press, 1993); International Institute for Applied Systems Analysis (IIASA), Laxenburg, Austria, "Regional Air Pollution Information and Simulation (RAINS)," <www.iiasa.ac.at/Research/TAP>, viewed 11 February 1998.

55. Francesca Lyman, "Fighting Toxins with New Technology," *MSNBC*, 31 March 1999, <www.msnbc.com>, viewed 2 September 1999.

56. Dennis Elliot and Marc Schwartz, "Wind Resource Mapping," Renewables for Sustainable Village Power Project Brief, July 1998, <www.rsvp.nrel.gov>, viewed 31 August 1999; Marc Schwartz, National Renewable Energy Laboratory, Golden, CO, e-mail to author, 1 October 1999.

57. Jeff Specht, "Mapping Endangered Diversity," *GIS World*, March 1996; Conservation International, <www.conservation.org>, viewed 29 September 1999.

58. Wesley Wettengel and Colby Loucks, World Wildlife Fund GIS Lab, discussion with author,

31 August 1999, and e-mail to author, 23 October 1999.

59. Dirk Bryant, Daniel Nielsen, and Laura Tangley, *The Last Frontier Forests, Ecosystems and Economies on the Edge* (Washington, DC: WRI, 1997); Dirk Bryant et al., *Reefs at Risk: A Map-Based Indicator of Threats to the World's Coral Reefs* (Washington, DC: WRI, ICLARM, World Conservation Monitoring Centre, and U.N. Environment Programme (UNEP), 1998).

60. Problems with data verification and funding led to the termination of the Global Environmental Monitoring System (GEMS) for Urban Air Pollution in 1996; see UNEP and World Health Organization (WHO), *City Air Quality Trends*, GEMS/AIR Data Vol. 3 (Nairobi: UNEP, 1995); WHO and UNEP, *Urban Air Pollution in Megacities of the World* (Oxford, U.K.: Blackwell, 1992); and UNEP and WHO, *Assessment of Urban Air Quality* (Nairobi: GEMS, 1998). Data problems from Dieter Schwela, WHO, Geneva, e-mail to author, 3 March 1999; indicators from Eric Rodenburg, "Eyeless in Gaia: The State of Global Environmental Monitoring" (Washington, DC: WRI, March 1991).

61. The two international research programs are the International Geosphere-Biosphere Programme and the World Climate Research Programme; John Townshend, University of Maryland, discussion with author, 22 September 1999.

62. Population Services International, <www.psi.org>, viewed 13 September 1999.

63. WRI, *GFWatcher: The Global Forest Watch Newsletter*, October 1998.

64. Tim Thwaites, "Bioninformatics: Spinning a Worldwide Web of Life," *Ecos 97*, October-December 1998.

65. Information technologies as tools for nongovernmental organizations from Jessica Mathews, "Power Shift," *Foreign Affairs*, January/February 1997; Union of International Organizations, *Yearbook of International Organizations 1996–1997* (Munich: K.G. Saur Verlag, 1997); P.J. Simmons, "Learning to Live with NGOs," *Foreign Affairs*, fall 1998.

66. Rainforest Action Network from "How CSOs are Using the Internet," *CIVICUS World* (CIVICUS: World Alliance for Citizen Participation, Washington, DC), May-June 1999.

67. Richard Civille, "New Media Strategies for Internet Organizing," <www.rffund.org/techproj/newmedia.html>, viewed 16 July 1999; Bill Lambrecht, "Biotechnology Companies Face New Foe: The Internet," *St. Louis Post-Dispatch*, 19 September 1999.

68. Civille, op. cit. note 67; "150,000 Americans Use the Power of the Web to Help Save America's Last Unprotected Forests," *E-Wire*, 30 June 1999, <ens.lycos.com/e-wire/>; Michael Gilbert, "Interview with Rob Stuart of the Rockefeller Family Fund," *NonProfit Online News*, 24 August 1999, <www.gilbert.org/news>.

69. "SDNP Pakistan: Success in Networking for Development," <www.sdnp.undp.org/>, viewed 24 August 1999.

70. Richard Civille, Center for Civic Networking, discussion with author, 12 August 1999.

71. Raul Zambrano, Project Manager, Sustainable Development Networking Programme, U.N. Development Programme, New York, e-mail to author, 1 October 1999.

72. Ann Florini, "The End of Secrecy," *Foreign Policy*, Summer 1998; EPA, *1997 Toxics Release Inventory Public Data Release Report* (Washington, DC: 1999), <www.epa.gov/tri/>, viewed 12 August 1999.

73. Environmental Defense Fund, <www.scorecard.org>, viewed 12 August 1999.

74. World Bank, *World Development Report 1998/99* (Washington, DC: 1999); Julia Royall, "SatelLife—Linking Information and People:

The Last Ten Centimetres," *Development in Practice*, February 1998.

75. HealthNet, <www.healthnet.org/>, viewed 7 July 1999; Mary Jander, "Power to the People," *Data Communications*, 21 October 1998.

76. Civille, op. cit. note 70.

77. Center for Tele-Information, Technical University of Denmark, "Telecottages in Estonia," at ITU, <www.itu.int/ITU-D-UniversalAccess/casestudies/estonia.html>, viewed 16 August 1998; Terefe Ras-Work, ed., *Tam-Tam to Internet: Telecoms in Africa* (Johannesburg, South Africa: Mafube Publishing, 1998); Richard Fuchs, "Little Engines That Did: Case Histories from the Global Telecentre Movement," IDRC Study/Acacia Initiative, June 1998, <www.idrc.ca/acacia/>, viewed 16 December 1998.

78. Geoffrey Naim, "European Telecentres: Showcases for Remote Working," *Financial Times*, 1 April 1998; Miriam Jordan, "It Takes A Cell Phone," *Wall Street Journal*, 25 June 1999; S. Kamaluddin, "Calling Countryfolk: Cellular-Phone Operator Targets Rural Bangaldesh," *Far Eastern Economic Review*, 24 April 1997; ITU, "Phones for All the World?" <www.itu.int/newsarchive/press/WTDC98/Feature2.html>, viewed 16 December 1998; Pernille Tranberg, "From Chickens to Cellphones, the New Face of Microcredit in Dhaka," *The Earth Times*, 1–15 August 1998.

79. Robert Freling, Solar Electric Light Fund, discussion with author, 25 October 1999; Palestine from Greenstar Foundation, <www.greenstar.org>; west India from Rashmi Mayur and Bennett Daviss, "The Technology of Hope: Tools to Empower the World's Poorest Peoples," *The Futurist*, October 1998; Uganda from Margot Higgins, "Village Power 2000 Energizes Uganda," *Environmental News Network*, 18 August 1999.

80. Sally Bowen, "Peru Introduces the Science of Fishing—By Satellite," *Financial Times*, 27 August 1999; "Italians to Spot Toxic Waste from Space," *Associated Press*, 25 October 1999.

81. David Rejeski, "Electronic Impact," *The Environmental Forum*, July/August 1999.

82. Cohen, op. cit. note 41; Rejeski, op. cit. note 81.

83. Illiterate population from U.N. Development Programme, *Human Development Report 1999* (Oxford, U.K.: Oxford University Press, 1999); English on the Internet from Global Reach, "Global Internet Statistics by Language," <www.euromktg.com/globstats/>, viewed 20 July 1999.

84. Data glut from Emma Brockes, "Information Overload," *The Guardian*, 15 February 1999; growth of the World Wide Web from Members of the Clever Project, "Hypersearching the Web," *Scientific American*, June 1999.

85. "Internet Hazardous to Social Health? Study on Social and Psychological Effects of Internet," *Science*, 11 September 1998; Bill McKibben, *The Age of Missing Information* (New York: Random House, 1992).

Chapter 8. Sizing Up Micropower

1. Paul Bennett, "Big Ideas Down Under," *Landscape Architecture*, February 1999; Andrew Darby, "Countdown to the Green Games," *Environment News Service*, 25 May 1999; K.W. Riley, Commonwealth Scientific and Industrial Research Organization, Bangor, NSW, Australia, e-mail to Seth Dunn, 12 November 1999.

2. Thomas B. Ackerman, Goran Andersson, and Lennart Soder, "What Is Distributed Generation?" (Stockholm, Sweden: Department of Electric Power Engineering, Royal Institute of Technology, 1999); Amory B. Lovins and Andre Lehmann, *Small is Profitable: The Hidden Economic Benefits of Making Electrical Resources the Right Size* (Boulder, CO: Rocky Mountain Institute, forthcoming).

3. E.F. Schumacher, *Small is Beautiful: Eco-

nomics as if People Mattered (San Bernardino, CA: Borgo Press, 1973); Gerard L. Cler and Michael Shepard, *Distributed Generation: Good Things Are Coming in Small Packages* (Boulder, CO: E Source, November 1996); Worldwatch estimates based on Gerard Cler and Nicholas Lenssen, *Distributed Generation: Markets and Technologies in Transition* (Boulder, CO: E Source, December 1997), and on Thomas R. Casten, *Turning Off the Heat: Why America Must Double Energy Efficiency to Save Money and Reduce Global Warming* (Amherst, NY: Prometheus Books, 1998).

4. Lovins and Lehmann, op. cit. note 2.

5. Larry Flowers, "Renewables for Sustainable Village Power," Presented at International Conference of Village Electrification through Renewable Energy, New Delhi, India, 3–5 March 1997 (Golden, CO: National Renewable Energy Laboratory (NREL), March 1997).

6. Larry Armstrong, "I Am Your Local Power Plant," *Business Week*, 30 August 1999; Joseph F. Shuler, Jr., "Distributed Generation: A 'Hot Corner' for Venture Capital?" *Public Utilities Fortnightly*, 15 October 1998; Robert W. Shaw, Jr., "Distributed Generation: The Emerging Option," Presentation to Board on Energy and Environmental Systems, National Research Council, Washington, DC, 6 May 1999.

7. Daniel Yergin and Joseph Stanislaw, *The Commanding Heights: The Battle Between Government and Marketplace That Is Remaking the Modern World* (New York: Simon and Schuster, 1998); Walt Patterson, *Transforming Electricity* (London: Royal Institute of International Affairs and Earthscan Publications, 1999).

8. Gas Research Institute (GRI), *The Role of Distributed Generation in Competitive Energy Markets* (Chicago: March 1999); Matthew Josephson, *Edison: A Biography* (New York: McGraw-Hill Book Company, 1959); Lovins and Lehmann, op. cit. note 2.

9. Thomas P. Hughes, *Networks of Power: Electrification in Western Society, 1880–1930*

(Baltimore: Johns Hopkins University Press, 1983); Stephen Heiser, "Distributed Generation: So Old It's New," *Power Online*, <www.poweronline.com>, 11 December 1998; GRI, op. cit. note 8.

10. Richard F. Hirsh, *Technology and Transformation in the American Public Utility Industry* (Cambridge, U.K.: Cambridge University Press, 1989); Cler and Lenssen, op. cit. note 3; U.S. Department of Energy (DOE), Energy Information Administration (EIA), *Annual Electric Generator Report*, electronic database, Washington, DC, 1999.

11. Cler and Lenssen, op. cit. note 3; KPMG, Bureau voor Economische Argumentatie, *Solar Energy: From Perennial Promise to Competitive Alternative*, Final Report, written for Greenpeace Netherlands (Hoofddorp, the Netherlands: August 1999).

12. Cler and Lenssen, op. cit. note 3; Table 8–1 contains Worldwatch estimates based on Lovins and Lehmann, op. cit. note 2, on Casten, op. cit. note 3, and on Cler and Lenssen, op. cit. note 3.

13. Cler and Lenssen, op. cit. note 3; Jason Makansi, "Venerable Engine/Generator Repositioned for On-Site, Distributed Power," *Power*, January-February 1999; Sarah McKinley, "Not All Small-Scale Generation is Emerging," *Public Utilities Fortnightly*, 1 February 1999; Table 8–2 based on Cler and Lenssen, op. cit. note 3, on GRI, op. cit. note 8, and on Gale Morrison, "Stirling Renewal," *Mechanical Engineering*, May 1999.

14. Bruce Wadman, "Advances in Smaller Gas Engine Gen-Sets," *Diesel Progress, North American Edition*, September 1998; Cler and Lenssen, op. cit. note 3.

15. Cler and Lenssen, op. cit. note 3; R. Neal Elliott and Mark Spurr, *Combined Heat and Power* (Washington, DC: American Council for an Energy-Efficient Economy, May 1999); Cler and Lenssen, op. cit. note 3.

16. Cler and Lenssen, op. cit. note 3.

17. Ibid.

18. Ibid.; "Microturbine Poses Competition for Utilities," *National Journal Daily Energy Briefing,* 2 December 1997; Ivan Amato, "May the Microforce Be With You," *Technology Review,* September/October 1999.

19. Cler and Lenssen, op. cit. note 3; Ake Almgren, Capstone Turbine Corporation, "Micro-Turbines, An Enabling Technology," Presentation to Board on Energy and Environmental Systems, National Research Council, Washington, DC, 6 May 1999.

20. Ann Keeton, "Future Generations," *Wall Street Journal,* 13 September 1999; Pat Maio, "New England Heat Wave Sparks Interest in Microturbines," *Wall Street Journal,* 23 June 1999; Almgren, op. cit. note 19.

21. Morrison, op. cit. note 13; Maurice A. White et al., "Generators That Won't Wear Out," *Mechanical Engineering,* February 1996; Cler and Lenssen, op. cit. note 3.

22. Cler and Lenssen, op. cit. note 3; White et al., op. cit. note 21; Jeremy Harrison, "Domestic Stirling Engine-Based Combined Heat & Power," *CADDET Energy Efficiency,* Newsletter No. 2, 1998.

23. "The Future of Fuel Cells," Special Issue of *Scientific American,* July 1999.

24. Ibid.; H. Frank Gibbard, H Power Corporation, "Fuel Cells," Presentation to Board on Energy and Environmental Systems, National Research Council, Washington, DC, 6 May 1999.

25. Brian C.H. Steele, "Running on Natural Gas," *Nature,* 12 August 1999; E. Perry Murray, T. Tsai, and S.A. Barnett, "A Direct-Methane Fuel Cell With a Ceria-Based Anode," *Nature,* 12 August 1999; Christopher K. Dyer, "Replacing the Battery in Portable Electronics," *Scientific American,* July 1999; Arthur Kaufman, H Power Corporation, "Compact Fuel Cells for Portable Applications," Presentation to Conference on Small Fuel Cells and Battery Technologies for Use in Portable Applications, Bethesda, MD, 29 April 1999.

26. Alan C. Lloyd, "The Power Plant in Your Basement," *Scientific American,* July 1999; Table 8–3 based on Cler and Lenssen, op. cit. note 3, on GRI, op. cit. note 8, on Gibbard, op. cit. note 24, on Henry R. Linden, "Distributed Power Generation—The Logical Response to Restructuring and Convergence," Regulated Industries Dinner/Discussion Series, Putnam, Hayes, & Bartlett, Inc., 8 April 1998, on KPMG, op. cit. note 11, and on BTM Consult, *Wind Force 10: A Blueprint to Achieve 10% of the World's Electricity from Wind Power by 2020,* Report Commissioned by European Wind Energy Association, Forum for Energy and Development, and Greenpeace International (London: 1999); Lloyd, op. cit. this note. Costs and efficiencies vary in part with engine size; reciprocating engine figures based on natural-gas-fired, not diesel-fired, engines.

27. A. John Appleby, "The Electrochemical Engine for Vehicles," *Scientific American,* July 1999; "Fuel Cells Meet Big Business," *The Economist,* 24 July 1999.

28. Appleby, op. cit. note 27; Lloyd, op. cit. note 26.

29. Lloyd, op. cit. note 26.

30. Ibid.; Seth Borenstein, "You Can Have Your Own Power Plant," *Knight Ridder Newspapers,* 23 August 1999.

31. Laurent Belsie, "Bringing Home the Sun," *Christian Science Monitor,* 23 August 1999; James McVeigh et al., *Winner, Loser or Innocent Victim: Has Renewable Energy Performed as Expected?* Research Report No. 7 (Washington, DC: Renewable Energy Policy Project (REPP), April 1999); Michael Rieke, "Old Vs. New," *Wall Street Journal,* 13 September 1999; Kathryn S. Brown, "Bright Future—Or Brief Flare—For Renewable Energy?" *Science,* 30 July 1999.

32. Paul Gipe, *Wind Energy Basics: A Guide to Small and Micro Wind Systems* (White River Junction, VT: Chelsea Green Publishing Company, 1999).

33. Andreas Wagner, German Wind Energy Association, "The Growth of Wind Energy in Europe—An Example of Successful Regulatory and Financial Incentives," Presentation to Windpower '99 Conference, American Wind Energy Association, Burlington, VT, 21 June 1999; Larry Goldstein, John Mortensen, and David Trickett, *Grid-Connected Renewable-Electric Policies in the European Union* (Boulder, CO: NREL, May 1999); Joe Cohen, Princeton Economic Research, Inc., "Draft Distributed Wind Power Assessment for the U.S.," Presentation to Windpower '99 Conference, Burlington, VT, 23 June 1999; John Byrne, Bo Shen, and William Wallace, "The Economics of Sustainable Energy for Rural Development: A Study of Renewable Energy in Rural China," *Energy Policy*, January 1998.

34. BTM Consult, *International Wind Energy Development: World Market Update 1998* (Copenhagen: March 1999); William Clairborne, "Towers Reach Across Midwest for Ethereal Source of Power," *Washington Post*, 19 September 1999; President's Committee of Advisors on Science and Technology (PCAST), *Powerful Partnerships: The Federal Role in International Cooperation on Energy Innovation* (Washington, DC: June 1999); Liang Chao, "Wind to Bring Light to Poor Rural Areas," *China Daily*, 6 January 1998.

35. John Perlin, *From Space to Earth: The Story of Solar Electricity* (Ann Arbor, MI: AATEC Publications, 1999); Mark A. Farber, Evergreen Solar, "Photovoltaics Industry Survey," Presentation to Board on Energy and Environmental Systems, National Research Council, Washington, DC, 6 May 1999.

36. Rieke, op. cit. note 31; Paul Maycock, "Photovoltaic Technology: Performance, Manufacturing Costs, and Markets," *Renewable Energy World*, July 1999; Paul D. Maycock, *Photovoltaic Technology Performance, Cost, and*

Markets: 1975–2010 (version 7) (Warrenton, VA: PV Energy Systems, August 1998); Paul D. Maycock, PV Energy Systems, Warrenton, VA, e-mail to Seth Dunn, 29 October 1999.

37. C. Wu, "The Secret to a Solar Cell's Stability?" *Science News*, 14 August 1999; A. Shah et al., "Photovoltaic Technology: The Case for Thin-Film Solar Cells," *Science*, 30 July 1999; KPMG, op. cit. note 11; Daniel McQuillen, "Harnessing the Sun: Building-Integrated Photovoltaics are Turning Ordinary Roofs Into Producers of Clean, Green Energy," *Environmental Design and Construction*, July/August 1998; Richard Duke and Daniel M. Kammen, "The Economics of Energy Transformation Programs," *The Energy Journal*, October 1999; Matthew Wald, "Where Some See Rusting Factories, Government Sees a Source of Solar Energy," *New York Times*, 4 August 1999.

38. Maio, op. cit. note 20; Pam Belluck and David Barboza, "After Summer's Power Failures, Concerns About Large Utilities," *New York Times*, 13 September 1999; "Lessons From the Blackout," *New York Times*, 13 July 1999.

39. Helwig quoted in Allanna Sullivan, "Electric Utilities Act to Update Distribution Networks," *Wall Street Journal*, 30 September 1999.

40. Alexandra von Meier, "Integrating Supple Technologies into Utility Power Systems: Possibilities for Reconfiguration," in Jane Summerton, ed., *Changing Large Technical Systems* (Boulder, CO: Westview Press, 1994); Thomas E. Hoff, "Using Distributed Resources to Manage Risks Caused by Demand Uncertainty," in Yves Smeers and Adonis Yatchew, eds., *Distributed Resources: Toward a New Paradigm of the Electricity Business*, Special Issue of *The Energy Journal*, 1997; Lovins and Lehmann, op. cit. note 2.

41. Borenstein, op. cit. note 30; Joseph Romm, "With Energy, We're Simply Too Demanding," *Washington Post*, 1 August 1999; Maio, op. cit. note 20.

42. Lovins and Lehmann, op. cit. note 2; Ann Deering and John P. Thornton, "Applications of Solar Technology for Catastrophe Reponse, Claims Management, and Loss Prevention," (Golden, CO: NREL, February 1999); Fred Gordon, Joe Chaisson, and Dave Andrus, *Helping Distributed Resources Happen: A Blueprint for Regulators, Advocates, and Distribution Companies*, Final Report for The Energy Foundation, 21 December 1998, as submitted to Harvard Electricity Policy Group, Cambridge, MA.

43. Roberta Stauffer, "Nature's Power on Demand: Renewable Energy Systems as Emergency Power Sources" (Washington, DC: DOE, October 1995); Edward Vine, Evan Mills, and Allen Chen, "Energy Efficiency and Renewable Energy Options for Risk Management and Insurance Loss Reduction: An Inventory of Technologies, Research Capabilities and Research Facilities at the U.S. Department of Energy's National Laboratories," (Berkeley, CA: Lawrence Berkeley National Laboratory, August 1998); Deering and Thornton, op. cit. note 42.

44. Lewis Milford, "The Lesson Hidden in the Blackout," *New York Times*, 13 July 1999; Wald, op. cit. note 37.

45. Milford, op. cit. note 44; Wald, op. cit. note 37; Borenstein, op. cit. note 30.

46. "Anchorage Mail Processing Center to be Powered by World's Largest Commercial Fuel Cell System," E-Wire Press Release, *Environment News Service*, 18 August 1999; Milford, op. cit. note 44.

47. Richard F. Hirsh and Adam H. Serchuk, "Power Switch: Will the Restructured Electric Utility System Help the Environment?" *Environment*, September 1999; World Bank, *1999 World Bank Development Indicators* (Washington, DC: April 1999); Natural Resources Defense Council (NRDC) and Public Service Electric and Gas Company (PSE&G), *Benchmarking Air Emissions of Electric Utility Generators in the United States* (New York: NRDC, June 1998); International Energy Agency, *Energy and Climate Change: An IEA Source-Book for Kyoto and Beyond* (Paris: 1997).

48. Worldwatch estimate based on Cler and Lenssen, op. cit. note 3, and on Casten, op. cit. note 3; Worldwatch estimate based on NRDC and PSE&G, op. cit. note 47, and on DOE, EIA, *Emissions of Greenhouse Gases in the United States 1997* (Washington, DC: October 1998).

49. Lovins and Lehmann, op. cit. note 2.

50. Ibid.; Patterson, op. cit. note 7.

51. "Technologies for Tomorrow," *EPRI Journal*, January/February 1998; Figures 8–1 and 8–2 based on Joseph Iannucci, Distributed Utility Associates, "Distributed Generation: Barriers to Market Entry," Presentation to Board on Energy and Environmental Systems, National Research Council, Washington, DC, 6 May 1999, and on Distributed Power Coalition of America, *What Is Distributed Generation*, information brochure (Washington, DC: 1999).

52. Taylor Moore, "Beyond Silicon: Advanced Power Electronics," *EPRI Journal*, November/December 1997.

53. Iannucci, op. cit. note 51; Gerald P. Ceasar, NIST Advanced Technology Program, "Overview of the ATP Premium Power Program: Power for the Digital Information Age," Distributed at Meeting of Board on Energy and Environmental Systems, National Research Council, Washington, DC, 6 May 1999; Dan W. Reicher, Assistant Secretary for Energy Efficiency and Renewable Energy, DOE, Testimony Before the Committee on Energy and Natural Resources, U.S. Senate Hearing on Distributed Power Generation, 22 June 1999; Hirsh and Serchuk, op. cit. note 47.

54. Iannucci, op. cit. note 51; Amory B. Lovins and Brett D. Williams, "A Strategy for the Hydrogen Transition," Presentation to the National Hydrogen Association, Vienna, VA, 7–9 April 1999; Amory B. Lovins, L. Hunter Lovins, and Paul Hawken, "A Road Map for Natural Capitalism," *Harvard Business Review*, May-June 1999; Walter Schroeder, PROTON

Energy Systems, Inc., "Hydrogen Supply for Distributed Power," Presentation to Board on Energy and Environmental Systems, National Research Council, Washington, DC, 6 May 1999; John A. Turner, "A Realizable Renewable Energy Future," *Science*, 30 July 1999; "Five Years in the Making, $18 Million Hydrogen Production/Fueling Station Opens in Munich," *Hydrogen & Fuel Cell Letter*, June 1999.

55. David E. Nye, *Electrifying America* (Cambridge, MA: The MIT Press, 1990); Patterson, op. cit. note 7.

56. European Commission (EC), Non-Nuclear Energy Programme, Joule 3, *The Value of Renewable Electricity*, Final Report, Coordinated by the Science Policy Research Unit, University of Sussex, U.K., June 1998.

57. Thomas J. Starrs and Howard Wenger, "Policies to Support a Distributed Energy System," in Adam Serchuk and Virinder Singh, eds., *Expanding Markets for Photovoltaics: What To Do Next*, Special Report (Washington, DC: REPP, December 1998); Goldstein, Mortensen, and Trickett, op. cit. note 33; "Summary of State Net Metering Programs," *Home Power*, <www.homepower.com/netmeter.htm>, viewed 20 October 1999; Sadayuki Niikumi, New Energy Foundation, Tokyo, e-mail to Seth Dunn, 9 November 1999; Worldwatch estimate based on BTM Consult, op. cit. note 26.

58. Mark Bernstein et al., RAND Corporation, *Developing Countries and Global Climate Change: Electric Power Options for Growth*, Prepared for the Pew Center on Global Climate Change (Washington, DC: June 1999); Keith Kozloff, *Electricity Sector Reform in Developing Countries*, Research Report No. 2 (Washington, DC: REPP, April 1998).

59. "Promising Distributed Technologies Face Barriers to Widespread Use," *Inside Energy*, 16 November 1998; Richard Stavros, "Distributed Generation: Last Big Battle for State Regulators?" *Public Utilities Fortnightly*, 15 October 1999; Starrs and Wenger, op. cit. note 57.

60. Starrs and Wenger, op. cit. note 57; Francis H. Cummings and Philip M. Marston, "Paradigm Buster: Why Distributed Power Will Rewrite the Open-Access Rules," *Public Utilities Fortnightly*, 15 October 1999; Reicher, op. cit. note 53.

61. Starrs and Wenger, op. cit. note 57; Reicher, op. cit. note 53.

62. Starrs and Wenger, op. cit. note 57.

63. Casten, op. cit. note 3; Tina Kaarsberg et al., "The Outlook for Small-Scale CHP in the USA," *CADDET Energy Efficiency*, Newsletter No. 2, 1998.

64. Reicher, op. cit. note 53; Iannucci, op. cit. note 51.

65. Iannucci, op. cit. note 51; Joseph F. Schuler, "Distributed Generation: Regulators' Next Challenge?" *Public Utilities Fortnightly*, 15 May 1999.

66. Perlin, op. cit. note 35; Lloyd, op. cit. note 26; Joseph J. Romm, *Cool Companies: How the Best Businesses Boost Profits and Productivity by Cutting Greenhouse Gas Emissions* (Washington, DC: Island Press, 1999).

67. Patterson, op. cit. note 7; Lovins and Lehmann, op. cit. note 2.

68. Richard Munson and Tina Kaarsberg, "Unleashing Innovation in Electricity Generation," *Issues in Science and Technology*, spring 1998; Worldwatch estimates based on ibid., on American Automobile Manufacturers Association (AAMA), *World Motor Vehicle Data, 1998 edition* (Washington, DC: 1998), on AAMA, *Motor Vehicle Facts and Figures* (Washington, DC: 1998), and on DOE, EIA, *Electric Power Annual 1998, Volume I* (Washington, DC: 1998).

69. Cler and Lenssen, op. cit. note 3; "More Power From Off the Grid," *Industries in Transition*, June 1999; Business Communications Company, *Small-Scale Power Generation: How Much, What Kind* (Norwalk, CT: July 1999).

70. Thomas Ackermann, Royal Institute of Technology, "Distributed Power Generation in a Deregulated Market Environment," Working Paper, First Draft, Royal Institute of Technology, Department of Electric Power Engineering, Stockholm, Sweden, June 1999; Iannucci, op. cit. note 51; "Allied Sees Enormous Opportunities for Renewables-Assisted Fuel Cells," *Solar & Renewable Energy Outlook*, 27 August 1999.

71. Office of Economic, Electricity, and Natural Gas Analysis, DOE, *Supporting Analysis for the Comprehensive Electricity Competition Act* (Washington, DC: May 1999); Patterson, op. cit. note 7; EC, op. cit. note 56.

72. Flowers, op. cit. note 5; Table 8–7 based on Byrne, Shen, and Wallace, op. cit. note 33, on Daniel M. Kammen, "Bringing Power to the People: Promoting Appropriate Energy Technologies in the Developing World," *Environment*, June 1999, on Mridula Chhetri, "Gone With the Wind," *Down to Earth*, 30 June 1999, on Anil Cabraal, Mac Cosgrove-Davies, and Loretta Schaeffer, *Best Practices for Photovoltaic Household Electrification Programs*, World Bank Technical Paper No. 324 (Washington, DC: 1996), on Richard H. Acker and Daniel M. Kammen, "The Quiet (Energy) Revolution: Analysing the Dissemination of Photovoltaic Power Systems in Kenya," *Energy Policy*, January/February 1996, on Daniel M. Kammen, "Household Power in a New Light: Policy Lessons, and Questions, for Photovoltaic Technology in Africa," *Tiempo: Global Warming and the Third World*, August 1996, on "Mandela Launches Record Solar Project," *BBC News Online*, 24 February 1999, and on Thomas Johansson, U.N. Development Programme, "Photovoltaics for Household and Community Use," HORIZON Solutions Site Case Study, <www.solutions-site.org>, viewed 25 August 1999.

73. Flowers, op. cit. note 5; Antonio C. Jimenez and Ken Olson, "Renewable Energy for Rural Health Clinics, (Boulder, CO: NREL, September 1998); Thomas Lynge Jensen, *Renewable Energy on Small Islands* (Copenhagen: Forum for Energy and Development,

April 1998).

74. Kammen, "Bringing Power to the People," op. cit. note 72.

75. "Public Acceptance Obstacle to Distributed Technology, Rather than Technology," *The Energy Report*, 23 November 1998; Daniel L. Berman and John T. O'Connor, *Who Owns the Sun? People, Politics, and the Struggle for a Solar Economy* (White River Junction, VT: Chelsea Green Publishing Company, 1996); a distributed generation Internet discussion group has been established by Thomas Ackermann of the Royal Institute of Technology at <www.egroups.com/list/distributed-generation>; NREL's Renewables for Sustainable Village Power Web site is <www.rsvp.nrel.gov>; Barbara C. Farhar, *Willingness to Pay for Electricity from Renewable Resources: A Review of Utility Market Research* (Boulder, CO: NREL, July 1999); Ackermann, op. cit. note 70; Ed Smeloff and Peter Asmus, *Reinventing Electric Utilities* (Washington, DC: Island Press, 1997); Donald E. Osborn, "Commercialization and Business Development of Grid-Connected PVs at SMUD," Presented at Solar 98: Renewable Energy for the Americas, Albuquerque, NM, June 1998; "Selling a Solar-Cell Future," *Christian Science Monitor* (editorial), 18 August 1999.

76. Joel N. Gordes and Jeremy Leggett, *Electrofinance: A New Insurance Product for a Restructured Electric Market*, Issue Brief No. 13 (Washington, DC: REPP, August 1999).

77. Daniel M. Kammen and Michael R. Dove, "The Virtues of Mundane Science," *Environment*, July/August 1997; Matthew S. Mendis, *Financing Renewable Energy Projects: Constraints and Opportunities* (Silver Spring, MD: Alternative Energy Development, July 1998); Nancy Wimmer, "Micro Credit in Bangladesh Introducing Renewable Energies," *Sustainable Energy News*, February 1999; "Grameen Shakti: Development of Renewable Energy Resources for Poverty Alleviation," <www.grameen-info.org>, viewed 25 August 1999; PCAST, op. cit. note 34; Jenniy Gregory et al., *Financing Renewable Energy Projects: A Guide for Develop-*

ment Workers (London: Intermediate Technology Publications, 1997); NREL, *A Consumer's Guide to Buying a Solar Electric System* (Washington, DC: September 1999).

78. Ceasar, op. cit. note 53; Acker and Kammen, op. cit. note 72.

79. Kammen, "Bringing Power to the People," op. cit. note 72.

80. Tim Forsyth, "Technology Transfer and the Climate Change Debate," *Environment*, November 1998; White House report is PCAST, op. cit. note 34; Robert M. Margolis and Daniel M. Kammen, "Underinvestment: The Energy Technology and R&D Policy Challenge," *Science*, 30 July 1999; Jim Williams et al., "The Wind Farm in the Cabbage Patch," *Bulletin of the Atomic Scientists*, May/June 1999.

81. Patterson, op. cit. note 7.

82. Joseph L. Bower and Clayton M. Christensen, "Disruptive Technologies: Catching the Wave," *Harvard Business Review*, January/February 1995; Clayton M. Christensen, *The Innovator's Dilemma: When New Technologies Cause Great Firms to Fail* (Boston: Harvard Business School Press, 1997).

83. Bower and Christensen, op. cit. note 82.

Chapter 9. Creating Jobs, Preserving the Environment

1. Kirkpatrick Sale, *Rebels Against the Future: The Luddites and Their War on the Industrial Revolution* (Reading, MA: Addison-Wesley Publishing Co., 1995).

2. Ibid.

3. Ibid.

4. Pace of job destruction and creation in Europe from European Commission (EC), *Living and Working in the Information Society: Peo-*

ple First (Brussels: Employment, Industrial Relations and Social Affairs, 1996).

5. Economic growth has been de-coupled from energy consumption, but not yet from materials consumption. Opportunities for boosting energy and materials efficiency are explored, and many practical examples offered, in Paul Hawken, Amory B. Lovins, and L. Hunter Lovins, *Natural Capitalism* (Boston: Little, Brown and Company, 1999).

6. International Labour Organization (ILO), *World Employment Report 1998–99* (Geneva, 1998); Brian Halweil and Lester R. Brown, "Unemployment Climbing as World Approaches 6 Billion," press release (Washington, DC: Worldwatch Institute, 2 September 1999).

7. Ibid.; long-term unemployment from United Nations Development Programme (UNDP), *Human Development Report 1999* (New York: Oxford University Press, 1999).

8. ILO, op. cit. note 6; EC, op. cit. note 4; overtime from Peter Merry, "Green Works," as posted on the British Green Party Web site, <www.gn.apc.org/www.greenparty.org>, viewed 5 August 1999.

9. Joseph A. Schumpeter, *Capitalism, Socialism and Democracy* (New York: Harper & Brothers, 1943), as cited in Robert Kuttner, *Everything for Sale: The Virtues and Limits of Markets* (Chicago: University of Chicago Press, 1996).

10. In assessing the impact of new technologies and automation on factory and office workplaces, Fred Block observes a "tension between the impulse to reduce workers' skill and the need for improved quality and flexibility." Fred Block, *Postindustrial Possibilities: A Critique of Economic Discourse* (Berkeley: University of California Press, 1990).

11. Richard J. Barnet, "Lords of the Global Economy," *The Nation*, 19 December 1994; U.N. Research Institute for Social Development, *States of Disarray: The Social Effects of Globaliza-*

tion (Geneva: 1995); Keith Bradsher, "Skilled Workers Watch Their Jobs Migrate Overseas," *New York Times*, 28 August 1995.

12. Skilled and unskilled manufacturing employment from ILO, op. cit. note 6; Britain from Merry, op. cit. note 8; Germany from Wolgang Bonß, "Das Ende der Normalität," *Politische Ökologie*, May/June 1998.

13. Bonß, op. cit. note 12; Martine Bulard, "What Price the 35-Hour Week?" *Le Monde Diplomatique*, September 1999, on-line English language version at <www.monde-diplomatique. fr/en/1999/09>; Jeremy Rifkin, *The End of Work* (New York: G.P. Putnam's Sons, 1995); Diana Bronson and Stéphanie Rousseau, "Working Paper on Globalization and Workers' Human Rights in the APEC Region," International Centre for Human Rights and Democratic Development, Expert Meeting on Globalization and Workers' Rights in the APEC Region, Kyoto, Japan, 12 November 1995.

14. International Monetary Fund (IMF), *World Economic Outlook* (Washington, DC: May 1999).

15. U.S. real hourly earnings and poverty-level wages from Economic Policy Institute, "Real Average Weekly and Hourly Earnings of Production and Non-Supervisory Workers, 1967–97," and "Share of Employment for All Workers by Wage Multiple of Poverty Wage, 1973–97," <epinet.org/datazone>, viewed 2 February 1999; comparison of manufacturing compensation from U.S. Department of Labor, Bureau of Labor Statistics (BLS), "International Comparisons of Hourly Compensation Costs for Production Workers in Manufacturing, 1997," released 16 September 1998; IMF, op. cit. note 14; Organisation for Economic Co-Operation and Development (OECD), *Making Work Pay: Taxation, Benefits, Employment and Unemployment*, The OECD Jobs Strategy Series (Paris: 1997).

16. ILO, op. cit. note 6; United Nations, *Monthly Bulletin of Statistics* (New York: various editions); lagging wage payments from Michael

Specter, "Protesting Privation, Millions of Russian Workers Strike," *New York Times*, 28 March 1997.

17. ILO, "Asian Labour Market Woes Deepening," press release (Geneva: 2 December 1998); ILO, op. cit. note 6; Seth Mydans, "Bad News, Silver Lining for Indonesian Laborers," *New York Times*, 6 February 1998; Sheryl WuDunn, "South Korea's Mood Swings from Bleak to Bullish," *New York Times*, 24 January 1999; Erik Eckholm, "Joblessness: A Perilous Curve on China's Capitalist Road," *New York Times*, 20 January 1998.

18. ILO, "Despite Decade-Long Reforms, Social Progress Risks Stalling in Latin America, Caribbean, Warns ILO in New Report," press release (Geneva: 23 August 1999); ILO, op. cit. note 6.

19. Sub-Saharan Africa from ILO, op. cit. note 6, and from ILO, *World Labour Report 1993* (Geneva: 1993); Latin America from UNDP, op. cit. note 7; Michael P. Todaro, *Urbanization, Unemployment and Migration in Africa: Theory and Policy* (New York: Population Council, 1997).

20. Current world youth unemployment from ILO, op. cit. note 6; share of population under 15 from Population Reference Bureau, "1999 World Population Data Sheet," wall chart (Washington, DC: June 1999).

21. Calculated from U.S. Department of Labor, BLS, "Multifactor Productivity in U.S. Manufacturing and in 20 Manufacturing Industries, 1949–1996" (Washington, DC: January 1999), and from additional data provided by Larry Rosenblum, BLS, e-mail to author, 1 July 1999.

22. Figure 9–1 based on BLS, op. cit. note 21, and on Rosenblum, op. cit. note 21.

23. Hawken, Lovins, and Lovins, op. cit. note 5.

24. Table 9–2 calculated from U.S. Department

of Commerce, Census Bureau, *1996 Annual Survey of Manufactures. Statistics for Industry Groups and Industries* (Washington, DC: U.S. Government Printing Office (GPO), February 1998), from U.S. Department of Energy (DOE), Energy Information Administration (EIA), *Manufacturing Consumption of Energy 1994* (Washington, DC: GPO, December 1997), and from U.S. Environmental Protection Agency, *The 1997 Toxics Release Inventory*, <www.epa.gov/opptintr/tri/tri97/drhome.htm>.

25. Manufacturing output and job trend calculated from U.S. Department of Labor, BLS, "Comparative Civilian Labor Force Statistics, Ten Countries, 1959–1998" (Washington, DC: 13 April 1999).

26. Manufacturing to service employment ratio calculated from U.S. Department of Labor, op. cit. note 25; Figure 9–2 from U.S. Department of Labor, BLS, "National Employment, Hours, and Earnings," data extracted from BLS database, <146.142.4.24/cgi-bin/surveymost>, viewed 12 July 1999.

27. U.S. hourly earnings from U.S. Department of Labor, op. cit. note 26.

28. University of Würzburg from Bonß, op. cit. note 12.

29. Bruce Guile and Jared Cohon, "Sorting Out a Service-Based Economy," in Marian R. Chertow and Daniel C. Esty, eds., *Thinking Ecologically: The Next Generation of Environmental Policy* (New Haven, CT: Yale University Press, 1997); T. Gameson et al., *Environment and Employment. Report for the Committee on Environment, Public Health and Consumer Protection of the European Parliament* (Seville, Spain: Institute for Prospective Technological Studies, April 1997).

30. Richard Kazis and Richard H. Grossman, *Fear at Work: Job Blackmail, Labor and the Environment* (Philadelphia: New Society Publishers, 1991); OECD, *Environmental Policies and Employment* (Paris: 1997); number of pollution

control jobs worldwide is a Worldwatch estimate based on Environmental Business International, "The Global Environmental Market and United States Environmental Industry Competitiveness" (San Diego, CA: undated), on David R. Berg and Grant Ferrier, *The U.S. Environment Industry* (Washington, DC: U.S. Department of Commerce, Office of Technology Policy, September 1998), and on Roger H. Bezdek, "Jobs and the Economic Opportunities During the 1990s in the U.S. Created by Environmental Protection" (Oakton, VA: Management Information Services, Inc., June 1997).

31. Michael Renner, *Jobs in a Sustainable Economy*, Worldwatch Paper 104 (Washington, DC: Worldwatch Institute, September 1991); innovation effects from Rainer Walz et al., *A Review of Employment Effects of European Union Policies and Measures for CO_2-Emission Reductions* (Karlsruhe, Germany: Fraunhofer Institute for Systems and Innovation Research, May 1999).

32. David Malin Roodman, *The Natural Wealth of Nations* (New York: W.W. Norton & Company, 1998).

33. AFL-CIO, "U.S. Energy Policy," Executive Council Statement (Washington, DC: 17 February 1999); for the argument in favor of a proactive policy in the U.S. context, see Judith M. Greenwald, *Labor and Climate Change* (Washington, DC: Progressive Policy Institute, October 1998).

34. World coal mining employment from Seth Dunn, "King Coal's Weakening Grip on Power," *World Watch*, September/October 1999; Figure 9–3 and U.S. trends from DOE, EIA, *Annual Energy Review 1998* (Washington, DC: GPO, 1999), and from U.S. Department of Labor, op. cit. note 26, viewed 17 August 1999.

35. Germany from Uwe Fritsche et al., *Das Energiewende-Szenario 2020* (Berlin, Germany: Oko-Institut, 1996); British miners from Peter Colley; *Reforming Energy: Sustainable Futures and Global Labor* (Chicago: Pluto Press, 1997); coal mined from British Petroleum, *BP Statistical Review of World Energy* (London: Group

Media & Publications, various years); China from James Kynge, "China Plans to Close Down 25,800 Coal Mines This Year," *Financial Times*, 11 January 1999.

36. European Wind Energy Association from European Commission (EC), "Energy for the Future: Renewable Sources of Energy," White Paper for a Community Strategy and Action Plan (Brussels: 26 November 1997); Danish data from EC, Directorate-General for Energy, *Wind Energy—The Facts, Vol. 3* (Brussels: 1997).

37. European firms' market share from U.S. General Accounting Office, "Renewable Energy: DOE's Funding and Markets for Wind Energy and Solar Cell Technologies" (Washington, DC: May 1999); India turbine manufacturers from Christopher Flavin, "Wind Power Blows to New Record," in Lester R. Brown, Michael Renner, and Brian Halweil, *Vital Signs 1999* (New York: W.W. Norton & Company, 1999).

38. Job per megawatt formula, 1996 capacity, and 2001 capacity projection from EC, *Wind Energy*, op. cit. note 36; job calculation from Worldwatch Institute, based on EC formula.

39. Bundesverband Wind Energie e.V., "Windenergie—25 Fakten," Osnabrück, Germany, 1999, as posted at <www.wind-energie.de>, viewed 10 July 1999; trend toward larger wind turbines from European Wind Energy Association, <www.ewea.org/summary.htm>, viewed 8 July 1999.

40. Solar Energy Industries Association (SEIA), "Solar Facts: Solar Jobs for Today & Tomorrow," <www.seia.org/sf/sfjobs.htm>, viewed 8 July 1999; Eurosolar, "Zukunftsmarkt Solartechnologie: Die Herausforderung Europas durch Japan und die USA," <www.eurosolar.org/mitteilungen/USA-JapanI.html>, viewed 8 July 1999.

41. Current jobs from European Solar Industry Federation (ESIF), <erg.ued.ie/esif/welcome_to_esif.html>, viewed 3 August 1999; projection from ESIF, "Solar Thermal Systems in Europe," booklet produced with support from the Euro-

pean Commission, Directorate-General for Energy, as posted at <erg.ued.ie/esif/welcome_to_esif.html>, viewed 3 August 1999, and from EC, "Energy for the Future," op. cit. note 36.

42. SEIA, op. cit. note 40; EC, "Energy for the Future," op. cit. note 36.

43. Howard Geller, John DeCicco and Skip Laitner, *Energy Efficiency and Job Creation* (Washington, DC: American Council for an Energy-Efficient Economy, 1992).

44. Table 9–4 from the following: Christine Lottje, "Climate Change and Employment in the European Union," Climate Network Europe, Brussels, May 1998, <www.climatenetwork.org/cne/joblink.htm>; Stephen Bernow et al., *America's Global Warming Solutions* (Washington, DC: WWF Global Climate Campaign, August 1999); Friends of the Earth UK, *Cutting CO₂—Creating Jobs* (London: Friends of the Earth, 1998); Fritsche et al., op.cit. note 35.

45. Gerd U. Scholl, "Employment Effects of a Changing Structure of Economic Activity," Institut für ökologische Wirtschaftsforschung, Heidelberg, Germany, unpublished paper.

46. Hawken, Lovins, and Lovins, op. cit. note 5; Amory B. Lovins, L. Hunter Lovins, and Paul Hawken, "A Road Map for Natural Capitalism," *Harvard Business Review*, May-June 1999.

47. OECD, op. cit. note 30.

48. U.S. re-manufacturing from Remanufacturing Industries Council International, "Frequently Asked Questions," <www.remanufacturing.org/frfaqust.htm>, viewed 28 October 1999.

49. Impact of discount retailers, in the U.S. context, from Stephen A. Herzenberg, John A. Alic, and Howard Wial, *New Rules for a New Economy* (Ithaca, NY: Cornell University Press, 1998).

50. Gerhard Scherhorn, "Das Ende des fordistischen Gesellschaftsvertrags," *Politische Ökologie*, March/April 1997.

51. OECD, op. cit. note 15; Lorenz Jarass, "More Jobs, Less Tax Evasion, Better Environment—Towards a Rational European Tax Policy," Contribution to the Hearing at the European Parliament, Brussels, 17 October 1996. In European Union countries, labor's tax burden is far higher than in the United States; see Gameson et al., op. cit. note 29, and Theodore Panayotou, "Market Instruments and Consumption and Production Patterns," *Consumption for Human Development* (New York: UNDP, Human Development Report Office, 1998).

52. Lottje, op. cit. note 44; Carsten Krebs and Danyel Reiche, "Vier Typen, drei Optionen," and Frank Steffe, "Die Evolution der Konzepte," *Politische Ökologie*, September/October 1998; Friends of the Earth UK, op. cit. note 44; Stefan Bach et al., "Ökologische Steuerreform: Umwelt- und steuerpolitische Ziele zusammenführen," *DIW-Wochenbericht*, No. 36/99 (Berlin: Deutsches Institut für Wirtschaftsforschung (DIW)); 1994 German study is Stefan Bach, Michael Kohlhaas, and Barbara Praetorius, "Ecological Tax Reform Even if Germany Has to Go it Alone," *Economic Bulletin* (DIW, Berlin) July 1994.

53. Martin Jänicke and Lutz Mez, "Einsam im Alleingang?" *Politische Ökologie*, September/October 1998; Roodman, op. cit. note 32; Stefan Bach and Michael Kohlhaas, "Nur zaghafter Einstieg in die ökologische Steuerreform," *DIW-Wochenbericht* (Berlin: DIW), No. 36/99; Reinhard Loske and Kristin Heyne, "Ökologische Steuerreform: Die Stufen 2-5," press release (Berlin: German Green Party parliamentary group, 29 June 1999).

54. Jänicke and Mez, op. cit. note 53; German provisions from Bach and Kohlhaas, op. cit. note 53, and from Loske and Heyne, op. cit. note 53.

55. See, for example, Alan Durning, *How Much Is Enough?* (New York: W.W. Norton & Company, 1992); Kuttner, op. cit. note 9; Juliet Schor, "The New Politics of Consumption," *Boston Review*, <bostonreview.mit.edu/BR24.3/schor.html>, viewed 24 August 1999.

56. Work hours in 1900 from Dieter Seifried, "Wer Später Kommt, Darf Früher Gehn," *Politische Ökologie*, March/April 1997; Block, op. cit. note 10; Marcus Stewen, "Die Basis für Integrierende Strategien Schaffen," *Politische Ökologie*, March/April 1997; Hans Diefenbacher and Volker Teichert, "Arbeit Statt Arbeitslosigkeit Fördern," *Politische Ökologie*, May/June 1998; Table 9–6 from U.S. Department of Labor, BLS, Office of Productivity and Technology, "Underlying Data for Indexes of Output per Hour, Hourly Compensation, and Unit Labor Costs in Manufacturing, Twelve Industrial Countries, 1950–1998, and Unit Labor Costs in Korea and Taiwan, 1970–1998" (Washington, DC: August 1999).

57. France from Bulard, op. cit. note 13; Germany from Volker Hielscher and Eckart Hildebrandt, "Weniger Arbeiten, Besser Leben?" *Politische Ökologie*, March/April 1997; Denmark from Diefenbacher and Teichert, op. cit. note 56.

58. "Jobmaschine Teilzeit," *Spiegel Online*, <www.spiegel.de/politik/deutschland/0,1588,36697,00.html>, 24 August 1999; IG Metall (German Metal Workers Union), *Die Ökonomischen Hintergründe des "Niederländischen Modells"* (Frankfurt, Germany: July 1998); United States from Vicki Robin, "Is Simplicity Necessary in this Era of Abundance?" (op-ed), *Christian Science Monitor*, 1 November 1999.

59. Block, op. cit. note 10.

60. Ibid.

61. Ibid.; André Gorz, "Jenseits der Erwerbsarbeit," *Politische Ökologie*, May/June 1998; Diefenbacher and Teichert, op. cit. note 56.

62. Alan Thein Durning, *Green-Collar Jobs: Working in the New Northwest*, NEW Report 8 (Seattle, WA: Northwest Environment Watch, 1999).

63. Colley, op. cit. note 35; David Malin Roodman, "Reforming Subsidies," in Lester R. Brown et al., *State of the World 1997* (New York: W.W. Norton & Company, 1997).

Chapter 10. Coping with Ecological Globalization

1. For different takes on the meaning of globalization, see, for example, Wolfgang H. Reinicke, *Global Public Policy* (Washington, DC: Brookings Institution Press, 1998); David C. Korten, *When Corporations Rule the World* (West Hartford, CT: Kumarian Press, Inc., 1995); George Soros, *The Crisis of Global Capitalism* (New York: Public Affairs, 1998); and Thomas L. Friedman, *The Lexus and the Olive Tree* (New York: Farrar, Straus and Giroux, 1999).

2. For a warning of instability on the horizon, see Eugene Linden, *The Future in Plain Sight: Nine Clues to the Coming Instability* (New York: Simon & Schuster, 1998).

3. Norman Myers, "The World's Forests and Their Ecosystem Services," in Gretchen C. Daily, ed., *Nature's Services* (Washington, DC: Island Press, 1997); overall forest loss from Dirk Bryant, Daniel Nielsen, and Laura Tangley, *The Last Frontier Forests* (Washington, DC: World Resources Institute (WRI), 1997); 14 million hectares based on estimates in U.N. Food and Agriculture Organization (FAO), *State of the World's Forests 1999* (Rome: 1999).

4. Janet N. Abramovitz, *Taking a Stand: Cultivating a New Relationship with the World's Forests*, Worldwatch Paper 140 (Washington, DC: Worldwatch Institute, April 1998); growth of plywood exports in Indonesia and Malaysia are volume figures from FAO, *FAOSTAT Statistics Database*, electronic database, <apps.fao.org>, viewed 22 October 1999.

5. Trends in forest products trade and Figure 10–1 based on FAO, op. cit. note 4; 1998 figures are preliminary estimates. Figures deflated using the U.S. GNP Implicit Price Deflator, pro-

vided in U.S. Department of Commerce, *Survey of Current Business*, July 1999.

6 . John E. Young, *Mining the Earth*, Worldwatch Paper 109 (Washington, DC: Worldwatch Institute, July 1992); gold-to-waste ratio based on U.S. Bureau of Mines data provided in John E. Young, "Gold Production at Record High," in Lester R. Brown, Hal Kane, and David Malin Roodman, *Vital Signs 1994* (New York: W.W. Norton & Company, 1994); frontier forests from Bryant, Nielsen, and Tangley, op. cit. note 3; indigenous peoples figure from Roger Moody, "The Lure of Gold—How Golden Is the Future?" Panos Media Briefing No. 19 (London: Panos Institute, May 1996).

7. Minerals imports and exports are volume figures from U.N. Conference on Trade and Development, *Handbook of World Mineral Trade Statistics, 1992–1997* (Geneva: 1995).

8. Figure 10–2 based on FAO, op. cit. note 4, with figures deflated as described in note 5; 11 percent and continental breakdown from World Trade Organization (WTO), *Annual Report 1998* (Geneva: 1998); grain trade statistics from U.S. Department of Agriculture, *Production, Supply, and Distribution*, electronic database, Washington, DC, updated November 1999.

9. Developing countries' share of exports from FAO, op. cit. note 4; nontraditional exports and Colombian flower workers from Lori Ann Thrupp, *Bittersweet Harvests for Global Supermarkets* (Washington, DC: WRI, 1995).

10. Figure 10–3 based on FAO, *Yearbook of Fishery Statistics*, vol. 45 (Rome: 1978), and on Sara Montanaro, Statistical Clerk, Fishery Information, Data, and Statistics Unit, FAO Fisheries Department, e-mail to Lisa Mastny, Worldwatch Institute, 6 September 1999, with figures deflated as described in note 5; share of fish traded today (1997 data) from FAO, *Yearbook of Fishery Statistics*, vol. 84 (Rome: 1999), and from FAO, *Yearbook of Fishery Statistics*, vol. 85 (Rome: 1999); share traded in 1970 from FAO, *Yearbook of Fishery Statistics*, vol. 45, op. cit. this note; fishing grounds from Maurizio

Perotti, Fishery Information, Data and Statistics Unit, FAO, Rome, e-mail to Anne Platt McGinn, Worldwatch Institute, 14 October 1997; overexploitation from FAO, *The State of World Fisheries and Aquaculture, 1996* (Rome: 1997).

11. Share of industrial and developing countries in 1997 based on data supplied by Montanaro, op. cit. note 10; share of developing countries in 1970 from FAO, *Yearbook of Fishery Statistics*, vol. 45, op. cit. note 10; four developing countries from FAO, *Yearbook of Fishery Statistics*, vol. 53 (Rome: 1983) and from FAO, *Yearbook of Fishery Statistics*, vol. 85, op. cit. note 10.

12. Deprivation of small-scale fishers from George Kent, "Fisheries, Food Security, and the Poor," *Food Policy*, vol. 22, no. 5 (1997), and from Mahfuzuddin Ahmed, "Fish for the Poor Under a Rising Global Demand and Changing Fishery Regime," *NAGA, The ICLARM Quarterly*, July-December 1997; Stephen Buckley, "Senegalese Fish for a Living in Sea Teeming with Industrial Rivals," *International Herald Tribune*, 4 November 1997; 1 billion and Asian dependence from WRI, *World Resources 1996–97* (Washington, DC: Oxford University Press, 1996).

13. Estimate of $10 million does not include timber and fisheries products; this and species trade figures from TRAFFIC North America, "World Trade in Wildlife," information sheet (Washington, DC: July 1994). Higher estimate of $20 million, illegal trade, and primate numbers from Statement by Shafqat Kakakhel, Deputy Executive Director, U.N. Environment Programme (UNEP), on behalf of the Executive Director, at UNEP Workshop on Enforcement and Compliance with Multilateral Environmental Agreements, Geneva, 12 July 1999.

14. Chris Bright, *Life Out of Bounds* (New York: W.W. Norton & Company, 1998); Christopher Bright, "Invasive Species: Pathogens of Globalization," *Foreign Policy*, Fall 1999.

15. Role of air travel and new diseases from World Health Organization (WHO), *Report on Infectious Diseases* (Geneva: 1999); quote and environmental changes from WHO, *Health and Environment in Sustainable Development* (Geneva: 1998).

16. On expectations for the Seattle meeting, see, for example, Anne Swardson, "Trade Body Summit Targeted for Protests," *Washington Post*, 2 November 1999. On transparency issues, see Lori Wallach and Michelle Sforza, *Whose Trade Organization?* (Washington, DC: Public Citizen, 1999); Daniel C. Esty, "Non-Governmental Organizations at the World Trade Organization: Cooperation, Competition, or Exclusion," *Journal of International Economic Law*, March 1998; and Gary P. Sampson, "Trade, Environment, and the WTO: A Framework for Moving Forward," ODC Policy Paper (Washington, DC: Overseas Development Council, February 1999).

17. Jeffrey S. Thomas and Michael A. Meyer, *The New Rules of Global Trade: A Guide to the World Trade Organization* (Scarborough, ON, Canada: Carswell, 1997).

18. "Text of Uruguay Round Ministerial Decision on Trade and Environment," 14 April 1994, provided in National Resources Defense Council and Foundation for International Environmental Law and Development, *Environmental Priorities for the World Trading System* (Washington, DC: January 1995); for a useful discussion of the deliberations of the Committee on Trade and Environment, see Hector Rogelio Torres, "Environmental Rent: Cooperation and Competition in the Multilateral Trading System" (Winnipeg, MN, Canada: International Institute for Sustainable Development (IISD), 1998).

19. General Agreement on Trade and Tariffs (GATT), *United States—Restrictions on Imports of Tuna: Report of the Panel* (Geneva: 3 September 1991).

20. Ibid.

21. For a useful discussion of the clash between the trade and environmental systems on the question of process standards, see the paper by Konrad Von Moltke included in Sadruddin Aga Khan, ed., *Policing the Global Economy: Why, How, and For Whom?* Proceedings of an International Conference held in Geneva, March 1998 (London: Cameron May Ltd., 1998).

22. For a discussion of the tuna-dolphin ruling and its implications, see Eric Christensen and Samantha Geffin, "GATT Sets Its Net on Environmental Regulation: The GATT Panel Ruling on Mexican Yellowfin Tuna Imports and the Need for Reform of the International Trading System," *Inter-American Law Review*, Winter 1991–1992; on the issue of conflicts between international environmental and trade agreements, see Steve Charnovitz, "Restraining the Use of Trade Measures in Multilateral Agreements: An Outline of the Issues," presentation at the Conference on the Relationship Between the Multilateral Trading System and the Use of Trade Measures in Multilateral Agreements: Synergy or Friction? The Hague, 22–23 January 1996.

23. Urgency and effectiveness of trade measures from Peter Fugazzotto and Todd Steiner, *Slain by Trade: The Attack of the World Trade Organization on Sea Turtles and the US Endangered Species Act* (San Francisco: Sea Turtle Restoration Project, July 1998); 16 countries from David Hogan, Office of Marine Conservation, Bureau of Oceans and International Environmental and Scientific Affairs, U.S. Department of State, letter to author, 24 September 1999.

24. Adam Entous, "WTO Rules Against U.S. on Sea Turtle Protection Law," *Reuters*, 6 April 1998; Anne Swardson, "Turtle-Protection Law Overturned By WTO," *Washington Post*, 13 October 1998; WTO, *United States—Import Prohibition of Certain Shrimp and Shrimp Products* (Geneva: 12 October 1998). For a discussion of the new dispute resolution rules, see Thomas and Meyer, op. cit. note 17.

25. New guidelines from U.S. Department of State, "Revised Guidelines for the Implementa-tion of Section 609 of Public Law 101-162 Relating to the Protection of Sea Turtles in Shrimp Trawl Fishing Operations," Public Notice 3086, *Federal Register*, 8 July 1999; on the status of the shrimp-turtle case at the WTO, see "Implementation Status of Shrimp-Turtle Ruling," *BRIDGES Between Trade and Sustainable Development*, July-August 1999.

26. "Trade Court Backs Protecting Turtles From Shrimpers' Nets," *Miami Herald*, 9 April 1999; Nancy Dunne, "Legal Wrangle Engulfs US Shrimp Dispute," *Financial Times*, 14 April 1999.

27. WTO, "Agreement on the Application of Sanitary and Phytosanitary Measures," in *Final Act: Agreement Establishing the World Trade Organization*, <www.wto.org/wto/eol/e/pdf/15-sps.pdf>, viewed 1 November 1999; see also Thomas and Meyer, op. cit. note 17.

28. For the historical backdrop to the beef hormone case, see U.S. Office of Technology Assessment, *Trade and Environment: Conflicts and Opportunities* (Washington, DC: U.S. Government Printing Office, May 1992). For more recent events, see Paul Jacobs, "U.S., Europe Lock Horns in Beef Hormone Debate," *Los Angeles Times*, 9 April 1999; Mark Suzman, "American Farmers Baffled as Europe Steers Clear of Beef Treated By Hormones," *Financial Times*, 22 July 1999; and Michael Balter, "Scientific Cross-Claims Fly in Continuing Beef War," *Science*, 28 May 1999.

29. WTO, "EC Measures Concerning Meat and Meat Products (Hormones)," Report of the Appellate Body (Geneva: 16 January 1998); "U.S. Imposes Sanctions in Beef Fight," *New York Times*, 19 July 1999; on the European response to the ruling, see "Big Mac Targeted by French Farmers," *BRIDGES Weekly Trade News Digest*, 30 August 1999, and Anne Swardson, "Something Is Rotten in Roquefort," *Washington Post*, 21 August 1999.

30. WTO, op. cit. note 29; on the precautionary principle, see James Cameron and Julie Abouchar, "The Status of the Precautionary

Principle in International Law," in David Free-stone and Ellen Hey, eds., *The Precautionary Principle and International Law* (The Hague: Kluwer Law International, 1996); Rio Declaration from Lakshman Guruswamy, Geoffrey Palmer, and Burns Weston, *International Environmental Law and World Order* (St. Paul, MN: West Publishing, 1994).

31. WTO, op. cit. note 27; Thomas and Meyer, op. cit. note 17.

32. Guy de Jonquières, "Genetically Modified Trade Wars," *Financial Times*, 18 February 1999; "EU Finalizes Labeling Rules for Genetically Modified Foods," *Business and the Environment*, July 1998; "Japan Risks U.S. Ire With GMO Label Plan," *Reuters*, 5 August 1999; "Australia, NZ Require Mandatory GM Labels on Food," *Reuters*, 4 August 1999; use of GMOs in U.S. food products from Rick Weiss, "In Europe, Cuisine du Gene Gets a Vehement Thumbs Down," *Washington Post*, 24 April 1999; on the move of European companies away from GMOs, see John Willman, "Consumer Power Forces Food Industry to Modify Approach," *Financial Times*, 10 June 1999, and Deborah Hargreaves, "Consumers' Unease Leads to Rethink on Modified Food Supplies," *Financial Times*, 29–30 May 1999; on the trade conflicts over labeling, see "Europe and US in Confrontation Over GM Food Labelling Criteria," *Nature*, 22 April 1999, and "U.S., Canada Concerned About Increase in Labeling Measures Affecting GMOs," *International Environment Reporter*, 23 June 1999.

33. On the breakdown of the biosafety talks, see Adam Thomson, "Efforts to Adopt UN Biosafety Protocol Fail," *Financial Times*, 25 February 1999, and "World Trade: Trade Tensions: The Biosafety Protocol Has Been Undermined by a Clash Between the Interests of US Multinationals and European Consumers," *Financial Times*, 26 February 1999; on continuing negotiations, see "Governments to Discuss Restarting Stalemated Talks on Biosafety Protocol," *International Environment Reporter*, 23 June 1999.

34. Requirements of proposed agreement from "U.S. Says Trade Plan Won't Hurt Forests," *Reuters*, 4 November 1999; Office of the U.S. Trade Representative and Council on Environmental Quality, *Accelerated Tariff Liberalization in the Forest Products Sector: A Study of the Economic and Environmental Effects* (Washington, DC: November 1999). For a general discussion of the expected impact of trade liberalization on forests, see David Kaimowitz, "The Potential Environmental Impacts of Trade Liberalisation in Forest Products," *BRIDGES Between Trade and Sustainable Development*, July-August 1999.

35. Earthjustice Legal Defense Fund and Northwest Ecosystem Alliance, *Our Forests at Risk: The World Trade Organization's Threat to Forest Protection* (Seattle, WA: Earthjustice Legal Defense Fund, 1999).

36. For acknowledgement of the need for WTO reforms, see William Jefferson Clinton, "Remarks at the Commemoration of the 50th Anniversary of the World Trade Organization," Geneva, 18 May 1999, and Group of Eight, "G-8 Communiqué Köln 1999," press release (Cologne: 20 June 1999); proposals to reduce subsidies from "World News: Appeal to End Fishing Subsidies," *Financial Times*, 2 August 1999, and from "Agriculture and the Environment: The Case of Export Subsidies," submission by Argentina et al. to the WTO's Committee on Trade and Environment (Geneva: 11 February 1999); $14–20 billion from Matteo Milazzo, *Subsidies in World Fisheries: A Reexamination*, World Bank Technical Paper No. 406, Fisheries Series (Washington, DC: World Bank, April 1998).

37. For ideas on needed steps to reform the rules of world trade, see "The World Trade Organization and the Environment," Technical Statement by U.S. Environmental Organizations, 16 July 1999; National Wildlife Federation, *What's TRADE Got To Do With It?* (Washington, DC: 1999); World Wide Fund for Nature (WWF-International), *Sustainable Trade for a Living Planet: Reforming the World Trade Organization* (Gland, Switzerland: September 1999); and Wallach and Sforza, op. cit. note 16.

38. On the reasons for the surge of private capital flows into the developing world in the early 1990s, see World Bank, *Private Capital Flows to Developing Countries* (New York: Oxford University Press, 1997), and Jacques de Larosière, "Financing Development in a World of Private Capital Flows: The Challenge for Multilateral Development Banks in Working with the Private Sector," the Per Jacobsson Lecture, Washington, DC, 29 September 1996.

39. Private capital flows and Figure 10–4 from World Bank, *Global Development Finance 1999* (Washington, DC: 1999), with figures deflated as described in note 5; $1.5 trillion from Bank for International Settlements, *Central Bank Survey of Foreign Exchange and Derivatives Market Activity in April 1998* (Basel, Switzerland: May 1999).

40. History of crisis from World Bank, *East Asia: The Road to Recovery* (Washington, DC: October 1998); capital outflows from International Monetary Fund (IMF), *World Economic Outlook, October 1999* (Washington, DC: 1999); $40 billion from Brian Duffy, "Market Chaos Goes Global," *U.S. News and World Report*, 14 September 1998; Brazil from Carlos Lozado, "Brazilian Domino Effect?" *Christian Science Monitor*, 9 November 1998; $42-billion bailout from "The Real Thing," *The Economist*, 21 November 1998.

41. Poverty and social fallout from World Bank, op. cit. note 40, from James D. Wolfensohn, President, World Bank Group, "The Other Crisis," Annual Meetings Address, Washington, DC, 6 October 1998, and from Kevin Sullivan, "A Generation's Future Goes Begging: Asia's Children Losing to Destitution," *Washington Post*, 7 September 1998; Indonesian poaching from Peter Waldman, "Desperate Indonesians Devour Country's Trove of Endangered Species," *Wall Street Journal*, 26 October 1998; on the environmental impacts of the crisis, see generally World Bank, East Asia Environment and Social Development Unit, *Environmental Implications of the Economic Crisis and Adjustment in East Asia* (Washington, DC: January 1999), and Peter Dauvergne, *The Environment*

in Times of Crisis, report commissioned by the Australian Agency for International Development, April 1999.

42. Signs of economic recovery from IMF, op. cit. note 40; poverty rates from World Bank, *Poverty Trends and Voices of the Poor* (Washington, DC: 1999), available at <wb.forumone.com/poverty/data/trends>, and from Jean Michel Severino, Vice President, East Asia and Pacific Region, World Bank, "East Asia Regional Overview" (Washington, DC: 23 September 1999); future instability from Jeffrey E. Garten, "A Crisis Without a Reform," *New York Times*, 18 August 1999.

43. Dani Rodrik, "The Global Fix," *New Republic*, 2 November 1998.

44. See, for example, Jeffrey Sachs, "IMF Is a Power Unto Itself," *Financial Times*, 11 December 1997.

45. Total lending and adjustment lending from World Bank, *Annual Report 1999* (Washington, DC: 1999); policy reforms from Joseph Stiglitz, "More Instruments and Broader Goals: Moving Toward the Post-Washington Consensus," The 1988 WIDER Annual Lecture, Helsinki, Finland, 7 January 1998, <www.worldbank.org/html/extdr/extme/js-010798/wider.htm>, viewed 19 October 1999.

46. David Reed, ed., *Structural Adjustment, the Environment, and Sustainable Development* (London: Earthscan Publications, Ltd., 1996).

47. William D. Sunderlin, Consultative Group on International Agricultural Research, "Between Danger and Opportunity: Indonesia's Forests in an Era of Economic Crisis and Political Challenge," 11 September 1998, <www.cgiar.org/cifor/>, viewed 8 October 1998; Indonesian palm exports from FAO, op. cit. note 4; recent fires from "Indonesian Fires Blamed on Plantations," *Reuters*, 10 August 1999, from "With Nothing Left to Burn, Fires Mostly Out," *Washington Post*, 25 April 1998, and from "Asian Nations Reach Accord On Fighting Haze," *Washington Post*, 24 December

1997.

48. Asian cuts in environmental spending from World Bank, *Environment Matters, Annual Review* (Washington, DC: 1999); Russia from Friends of the Earth–US, "Environmental Consequences of the IMF's Lending Policies," information sheet (Washington, DC: undated); Brazil from Diana Jean Schemo, "Brazil Slashes Money for Project Aimed at Protecting Amazon," *New York Times*, 1 January 1999; high rates of deforestation from Daniel C. Nepstad et al., "Large-scale Impoverishment of Amazonian Forests by Logging and Fire," *Nature*, 8 April 1999.

49. "IMF Decides Once Again to Halt Disbursement of Loans Because of Illegal Timber Practices," *International Environment Reporter*, 15 October 1997; Government of Indonesia, "Supplementary Memorandum of Economic and Financial Policies: Fourth Review Under the Extended Agreement," document submitted to Michel Camdessus, Managing Director of the IMF, 16 March 1999, <www.imf.org/external/np/loi/1999/031699.htm>, viewed 19 October 1999; Sunderlin, op. cit. note 47.

50. Government of Indonesia, op. cit. note 49; Sunderlin, op. cit. note 47; abuses in reforestation fund from World Bank, op. cit. note 41.

51. Andrea Durbin and Carol Welch, "Greening the Bretton Woods Institutions: Sustainable Development Recommendations for the World Bank and the International Monetary Fund" (Washington, DC: Friends of the Earth–US, December 1998); Ved P. Gandhi, *The IMF and the Environment* (Washington, DC: IMF, 1998). I am indebted to Frances Seymour of the World Resources Institute for suggesting that the IMF's surveillance function might be expanded to cover environmental issues.

52. Evolving role of IMF from "Time For a Redesign?" *The Economist*, 30 January 1999, and from Carol Welch, "In Focus: The IMF and Good Governance," *U.S. Foreign Policy in Focus*, October 1988; see also Gandhi, op. cit. note 51.

53. World Bank policy on adjustment lending from "OD 8.60, Adjustment Lending Policy," in World Bank, *The World Bank Operational Manual*, <www.wbln0018.worldbank.org/Institutional/Manuals/OpManual.nsf>; 60 percent from World Bank, "The Evolution of Environmental Concerns in Adjustment Lending: A Review," prepared for the CIDIE Workshop on Environmental Impacts of Economywide Policies in Developing Countries, Washington, DC, 23–25 February 1993; less than 20 percent from Nancy Dunne, "World Bank: Projects 'Fail' the Poor," *Financial Times*, 24 September 1999; World Bank environmental assessment policy as it applies to adjustment lending from Kathryn McPhail, World Bank, comment at meeting of NGOs with Ian Johnson, Vice President, Environmentally & Socially Sustainable Development, World Bank, 26 May 1999; failure of IMF to require environmental impact assessments from Welch, op. cit. note 52.

54. Description of World Bank private-sector operations from World Bank, op. cit. note 45; 10 percent from World Bank, *Annual Report 1995* (Washington, DC: 1995). The World Bank has not updated this estimate in more recent annual reports.

55. For a description of the World Bank Group's environmental policies and guidelines, see World Bank, op. cit. note 48; International Finance Corporation, "Procedure for Environmental and Social Review of Projects," information sheet (Washington, DC: December 1998); Multilateral Investment Guarantee Agency, "Environmental and Social Review Procedures," <www.miga.org/disclose/soc_rev.htm>, viewed 4 November 1999; on the Bank's shortcomings in implementing policies, see Leyla Boulton, "World Bank Admits to Weakness on Environment," *Financial Times*, 4 October 1996, and Mark Suzman, "World Bank Accuses Itself of 'Serious Violations'," *Financial Times*, 7 January 1998; on importance of World Bank standards as a point of reference for private investors, see "Attorney Says Environmental Trends Will Have Far-Reaching Impact on U.S. Firms," *International Environment Reporter*, 1 November 1995, and World Bank, op. cit. note 48.

56. Danielle Knight, "Finance-Environment: Increased Lending for Destructive Projects," *Inter Press Service*, 22 February 1999; Berne Declaration et al., *A Race to the Bottom* (Washington, DC: March 1999); export credit support in 1988 from Anthony Boote, *Official Financing for Developing Countries* (Washington, DC: IMF, 1995); support in 1996 from Anthony Boote and Doris Ross, *Official Financing for Developing Countries* (Washington, DC: IMF, 1998); 10 percent is a Berne Union estimate cited in Nancy Dunne, "Environmentalists Damn Export Credit Agencies' Policies," *Financial Times*, 31 July 1998.

57. Overseas Private Investment Corporation, *OPIC Environmental Handbook* (Washington, DC: April 1999); U.S. Export-Import Bank, "Ex-Im Bank and the Environment," information sheet (Washington, DC: 8 June 1998); Three Gorges from Berne Declaration et al., op. cit. note 56, and from "Breaking the Wall: China and the Three Gorges Dam," *Harvard International Review*, summer 1998; countries working on standards from Pam Foster, Halifax Initiative, Ottawa, ON, Canada, discussion with Lisa Mastny, Worldwatch Institute, 4 November 1999; international discussions on common standards from Group of Eight, op. cit. note 36.

58. Vanessa Houlder, "Greens Gun for Finance," *Financial Times*, 9 February 1999; Ann Monroe, "The Looming EcoWar," *Investment Dealers' Digest*, 24 May 1999; "Three Gorges Dam Sneaking Its Way into Capital Markets," *The Bull and Bear Newsletter* (National Wildlife Federation and Friends of the Earth–US) 20 January 1999.

59. UNEP, "Bankers to Link Environment and Financial Performance," press release (New York: 16 May 1997); signatories and countries from "19 New Signatories to the Financial Services Initiatives," *The Bottom Line* (UNEP Financial Services Initiative newsletter), spring 1999, and from "New Signatories to the Initiatives," *The Bottom Line* (UNEP Financial Services Initiative newsletter), summer 1999; UNEP, "UNEP Statement by Financial Institutions on the Environment and Sustainable Development," revised version, May 1997, <www.unep.ch/eteu/finserv/english.html>, viewed 3 November 1999; Abid Aslam, "Finance-Environment: Chinese Dam Tests 'Green' Banking Club," *Inter Press Service*, 13 September 1999; Julie Hill, Doreen Fedrigo, and Ingrid Marshall, *Banking on the Future: A Survey of Implementation of the UNEP Statement by Banks on Environment and Sustainable Development* (London: Green Alliance, March 1997).

60. Stephan Schmidheiny and Federico J. L. Zorraquín, with the World Business Council for Sustainable Development, *Financing Change* (Cambridge, MA: The MIT Press, 1996); Patareeya Benjapolchai, Senior Vice President, Stock Exchange of Thailand, "Stock Exchange Policies for Protecting the Environment," address to The Environment and Financial Performance, UNEP's Third International Roundtable Meeting on Finance and the Environment, Columbia University, New York, 22–23 May 1997.

61. "Global Reporting Initiative," background information available at <www.ceres.org/reporting/globalreporting.html>; "Ceres Green Reporting Guidelines Launched," *Environmental News Service*, 3 March 1999.

62. On the need for a global society to complement the global economy, see George Soros, "Toward a Global Open Society," *The Atlantic Monthly*, January 1998, and Rodrik, op. cit. note 43.

63. Treaties and Figure 10–5 are compiled from UNEP, *Register of International Treaties and Other Agreements in the Field of the Environment 1996* (Nairobi: 1996), from United Nations, *The United Nations Treaty Collection*, electronic database, <www/un/org/Depts/Treaty>, viewed 9 March 1999, and from UNEP, "International Conventions and Protocols in the Field of the Environment: Report of the Executive Director," prepared for UNEP Governing Council Twentieth Session, 13 November 1998.

64. UNEP, *Data Report on Production and Consumption of Ozone-Depleting Substances,*

1986–1996 (Nairobi: November 1998); Caroline Taylor, "The Challenge of African Elephant Conservation," *Conservation Issues* (World Wildlife Fund), April 1997; whale takes in 1961 from Elizabeth Kemf and Cassandra Phillips, *Wanted Alive! Whales in the Wild* (Gland, Switzerland: WWF-International, October 1995); current whale takes is a Worldwatch estimate based on WWF-International, "Japanese Fleet Returns from Whale Sanctuary With 389 Minkes," press release (Gland, Switzerland: 23 April 1999), and on Kieran Mulvaney, "The Whaling Effect," news feature (Gland, Switzerland: WWF-International, October 1999); decline in oil spills from Anne Platt McGinn, *Safeguarding the Health of Oceans*, Worldwatch Paper 145 (Washington, DC: Worldwatch Institute, March 1999); The Antarctic Project, "The Protocol on Environmental Protection to the Antarctic Treaty," 14 June 1999, <www.asoc. org/currentpress/protocol.htm>, viewed 3 November 1999.

65. For a prominent proposal to create a Global Environmental Organization, see Daniel C. Esty, "GATTing the Greens," *Foreign Affairs*, November/December 1993, and Daniel C. Esty, "The Case for a Global Environmental Organization," in Peter B. Kenen, ed., *Managing the World Economy: Fifty Years After Bretton Woods* (Washington, DC: Institute for International Economics, 1994). Both James Gustave Speth, former Administrator of the U.N. Development Programme, and Renato Ruggiero, former Director-General of the WTO, have voiced support for creating a World Environment Organization over the last few years; see James Gustave Speth, "Note Regarding Questions on UN Reform Proposals," unpublished document dated 4 October 1996, and Renato Ruggiero, "Opening Remarks to the High Level Symposium on Trade and the Environment," 15 March 1999, <www.wto.org/wto/hlms/dgenv.htm>, viewed 14 April 1999.

66. On the distinction between governance and governments, see Commission on Global Governance, *Our Global Neighborhood* (New York: Oxford University Press, 1995), and Reinicke, op. cit. note 1.

67. IISD, *Global Green Standards: ISO 14000 and Sustainable Development* (Winnipeg, MN, Canada: 1996); Riva Krut and Harris Gleckman, *ISO 14001: A Missed Opportunity for Sustainable Global Industrial Development* (London: Earthscan Publications Ltd., 1998).

68. International Federation of Organic Agriculture Movements from J. Patrick Madden and Scott G. Chaplowe, *For ALL Generations: Making World Agriculture More Sustainable* (Glendale, CA: OM Publishing, 1997); Forest Stewardship Council, "Who We Are," information sheet, <www.fscoax.org>, viewed 29 June 1999; Marine Stewardship Council, "Our Empty Seas: A Global Problem, A Global Solution," information brochure (London: April 1999).

69. For a general discussion of the growing strength of NGOs, see Curtis Runyan, "The Third Force: NGOs," *World Watch*, November/December 1999; on NGO activism in the developing world, see Helmut K. Anheier and Lester M. Salamon, eds., *The Nonprofit Sector in the Developing World* (New York: Manchester University Press, 1998), and Jonathan Friedland, "Across Latin America, New Environmentalists Extend Their Reach," *Wall Street Journal*, 26 March 1997; growth of international NGOs is a Worldwatch Institute estimate based on Union of International Associations, *Yearbook of International Organizations 1998–1999* (Munich: K.G. Saur Verlag, 1999); environmental NGOs as share of total from Margaret E. Keck and Kathryn Sikkink, *Activists Beyond Borders* (Ithaca, NY: Cornell University Press, 1998), based on data from the Union of International Organizations.

70. R.C. Longworth, "Activists on Internet Reshaping Rules of Global Economy," *Chicago Tribune*, 5 July 1999; Climate Action Network, "Climate Action Network: A Force for Change," <www.climatenetwork.org>, viewed 6 July 1999; Pesticide Action Network North America, "What is PANNA?" <www.panna. org/panna/whatis.html>, viewed 1 July 1999; World Forum of Fish Workers & Fish Harvesters," <www.south-asian-initiative.org/

wff/intro.htm>, viewed 25 June 1999; International POPs Elimination Network, "Organizational Structure," <www.psr.org/ipen/ipen_structure.htm>, viewed 1 July 1999; Women's Environment & Development Organization, "About WEDO," <www.wedo.org/about/about.htm>, viewed 1 July 1999.

71. On the role of activists in overturning the Multilateral Agreement on Investment, see Madelaine Drohan, "How the Net Killed the MAI," *Toronto Globe and Mail*, 29 April 1998; for activists' role in opposing biotechnology, see Bill Lambrecht, "Biotechnology Companies Face New Foe: The Internet," *St. Louis Post Dispatch*, 20 September 1999.

Index

Aarhus Protocol, 93–94
acid rain, 28, 33–36
Afghanistan, 62, 65
AFL-CIO, 172
Africa
   Internet access, 125–26
   malnutrition, 61–64, 69–72
   paper consumption, 102
   small-scale irrigation, 53–54
   see also AIDS; specific countries
Agarwal, Ravi, 85
agriculture
   acid rain effects, 28, 33–36
   bioinvasion, 24, 36, 188
   biotechnology, 184, 192–93
   climate change effects, 23, 29, 36, 132–33
   crop diversity, 38
   cropland loss, 7, 26
   cropland productivity, 7–8, 48–57
   erosion, 23
   farm subsidies, 56
   fertilizer use, 30–31, 35, 55–56
   grain, 14, 42, 65, 187
   livestock, 192
   organic farming, 99
   soil, 23, 28, 33–36
   water scarcity, 14, 41–45, 61–62
   world exports, 186–87
   see also food; forest products; forests;
      irrigation; pesticides; technology

AIDS
   epidemic spread, 4–5, 20, 188
   mortality rates, 4–5, 15–16
   prevention, 20, 136
air pollution, 32–37, 93, 108
Alacatel, 128
Almgren, Ake, 146
Alsthom, 148
Amazon basin, 26–28, 196
American Council for an Energy-Efficient
    Economy, 174
American Institute for Cancer Research, 71, 74
American Public Health Association, 97
Approtech, 54
aquifer depletion, 6–7, 13, 41–44, 49
Aracruz Celulose, 107
ARCO, 12
Argentina, 127
Association for Progressive Communications,
    137–38
Australia, 46, 129, 187
Austria, 67, 84
automobiles, 26, 33, 35
Avianto, Teten, 45

Baby-Friendly Hospital Initiative, 75
Ballard, 148
Bangladesh
   cellular telephones, 140
   micro-loans, 160

Bangladesh *(continued)*
   treadle pump irrigation, 52–53
   water scarcity, 42, 61–62
Bank of America, 115
Barker, David, 70
Belgium, 67, 79, 84
Bhutan, 122
biodiversity
   agricultural crops, 38
   bioinvasion, 24, 36, 188
   convention, 193
   extinctions, 8, 24–26, 80, 85–88
   fisheries, 29–32, 135, 189–91
   genetically modified organisms, 184,
      192–93
   indigenous people, 107
   management, 37–38
   see also endangered species
bioinvasion, 24, 36, 188
biotechnology, 184, 192–93
Birnbaum, Linda, 85
Block, Fred, 181–82
Bower, Joseph, 161
Bowlin, Mike, 12
BP Amoco, 143
Brazil
   deforestation, 26–28, 196
   ecological spending, 196
   forest fires, 26–28
   indigenous people, 107
   integrated pest management, 113
   obesity, 63
British Petroleum, 12
Broadview Water District, 57
Brownell, Kelly, 68, 78
Bruns, Bryon, 45
Business Communications Company,
      157

California Alliance for Distributed Energy
      Resources, 157
California Irrigation Management
      Information Service, 50
Cambodia, 196

Canada
   forest subsidies, 119
   International Development Resource
      Centre, 140
   paper production, 107
   persistent organic pollutants, 86–87, 120
cancer, 69, 71–74, 89–90
Capstone Turbine, 146
carbon dioxide, 5, 29, 32
carbon emissions
   climate change, 28, 36
   global rise, 29, 32
   policy, 132, 136, 172, 176
   stabilization, 16–19
   see also climate change; fossil fuels;
      renewable energy
Carnegie Endowment for International
      Peace, 138
Casten, Thomas, 156
Caterpillar, 144–45
CDH District Heating (Canada), 145
cellular telephones, 122–27, 140
Center for the Biology of Natural Systems,
      97
Center for Civic Networking, 137
Chaturvedi, Bharati, 85
Child and Adolescent Trial for
      Cardiovascular Health, 76
Chile, 105
China
   acid rain effects, 34–35
   aquifer depletion, 6–7, 42
   DDT use, 83
   flooding, 12, 24
   grain production, 42
   obesity, 62–63, 66, 71
   paper production, 102, 104, 111
   population growth, 19, 167
   Three Gorges dam project, 145, 199
   unemployment, 166–67
   water scarcity, 42, 45
   wind power, 17, 149, 173
chlorofluorocarbons, 200
Christensen, Clayton, 161

chronic disease
    cancer, 69, 71–74, 89–90
    diabetes, 69, 72–73
    dietary contributions, 59–64, 69–72
    economic burden, 73–74
    heart disease, 69, 71, 76
    persistent organic pollutant effects, 87–90
    smoking effects, 11
    Yucheng, 89
Civille, Richard, 137
Clay, Charlie, 95
Climate Action Network, 202
climate change
    agricultural effects, 23, 29, 36, 132–33
    carbon emissions, 28, 36
    convention, 172, 176
    ecological effects, 22, 28–29, 36, 132
    El Niño effects, 28, 132
    fisheries effects, 29–30, 32
    forest effects, 28, 36
    jobs impact, 175
    policy, 132, 136, 172, 175–76
    recorded rise, 5–6, 29
    satellite monitoring, 132–33
    sea level rise, 6, 32
    stabilization strategies, 16–19
    tropical storms, 23
    see also carbon emissions
coal, 17, 173, 182
Coalition for Environmentally Responsible
    Economies, 199
Coca-Cola, 67
Codex Alimentarius Commission, 91
Cohen, Nevin, 130
Coimbatore, 50
Colborn, Theo, 88
Colditz, Graham, 71, 73
Colombia, 63
Commonwealth Edison, 150
communications technology, *see* information
    technologies; *specific types*
computers
    energy use, 129–30
    environmental monitoring, 131–36, 140–41

historical perspectives, 3–4
    Internet, 121–27, 138–41
    production, 127–30
    recycling, 127–28
Congo, 16, 29
conservation
    electronics take-back plans, 128
    energy reform, 149–53
    forest management, 12, 106, 113, 137
    integrated pest management, 99, 113
    natural areas, 97, 127–31
    water management, 6, 50, 56–58, 145,
        199
    see also biodiversity; endangered species;
        recycling; waste management
Convention on Biological Diversity, 193
Convention on Long-Range Transboundary
    Air Pollution, 93
Costa Rica, 96
Côte d'Ivoire, 4, 28, 127, 136
cropland loss, 7, 26
cropland productivity, 7–8, 48–57
Czech Republic, 95

DaimlerChrysler, 12, 147–48, 153
dams, 55, 145, 199
DDT, 80–83, 86–91, 98–99
deforestation, 8, 22–29, 185, 196
Democratic Republic of Congo, 16, 29
Denmark, 17, 20, 149, 154, 173, 181
Detroit Diesel, 144
developing countries
    agricultural dependence, 187
    fisheries, 187–88
    foreign aid, 194
    forest management, 12, 107, 119, 185,
        196
    international investment, 194–200
    irrigation projects, 52–56
    labor force, 162–66, 170, 173
    malnutrition, 61–64, 69–74, 76
    microgenerators, 10, 142–46
    micro-loans, 160
    mining, 186

developing countries *(continued)*
  paper consumption, 102–03
  population growth, 5, 7, 15–16, 19, 167
  poverty, 64
  unemployment, 164–67
  water conflicts, 44–48
  wind energy, 173–74
  see also specific countries
diet, *see* food
dioxins, 81–86, 97–98, 109, 120
disease, *see* chronic disease; infectious disease
Distributed Utility Associates, 158
Dombeck, Michael, 12
Dominican Republic, 135
Durning, Alan, 182

Earth Summit, 92, 95, 138, 192
ecological globalization, 184–202
  definition, 184–85
  policy reform, 37–38, 189–202
  trade, 185–89
economy
  agricultural exports, 186–87
  aquifer depletion effects, 6–7, 13,
      41–44, 49
  chronic disease burden, 73–74
  cropland productivity, 7–8, 48–57
  ecological impact, 185–89, 194–200
  e-commerce, 130–31, 139
  environment policy, 37–38, 189–202
  Export-Import Bank, 198
  fisheries trade, 187–91
  forest products trade, 184–86, 193,
      196–97
  forest subsidies, 119
  free trade zones, 166
  global economic crisis, 194–97
  globalization, 189–200
  grain trade, 187
  green investment, 8–10, 194–200
  industrialization, 162–66
  international investment, 194–200
  irrigation subsidies, 56
  malnutrition costs, 72–74

micro-loans, 160
paper costs, 104–09
productivity, 167–71
smoking costs, 11
stock market, 4, 194–95, 199
sustainable development, 10–16, 113,
    136–40
tariffs, 189–90, 192–93
telecommunications markets, 126–27
unemployment, 164–67
water markets, 57–58
wildlife trade, 188, 195
  see also agriculture; jobs; recycling; subsidies;
      taxes; technology; World Bank
ecosystems
  coral reefs, 6, 29–32, 135
  disaster management, 37–38
  fertilizer runoff, 30–31, 55–56
  persistent organic pollutant effects, 80,
      85–86
  sustainable forest management, 113
  wetlands, 24, 38, 54
  see also biodiversity; endangered species;
      freshwater ecosystems
Egypt, 95
El Niño, 28, 132
employment, *see* jobs
endangered species
  Act (U.S.), 191
  bioinvasion, 24, 36, 188
  Convention on Biological Diversity, 193
  extinction rates, 8, 24–26, 80, 85–88
  fisheries, 29–32, 135, 189–91
  illegal hunting, 28–29, 200
  pollution effects, 80, 85–86, 88
  sustainable forest management, 113
  wildlife trade, 188, 195
  see also biodiversity
energy
  conservation, 149–53
  consumption, 144, 149–51, 157
  distribution weaknesses, 149–52, 155
  fuel cells, 12, 147–51, 153, 157
  global trends, 16–21

information technology use, 129–30
microgenerators, 10, 142–46
paper production, 107–08
pollution, 151
reform, 17, 153–57
solar, 140, 149, 154, 160, 174
subsidies, 149, 160
tax reform, 179–80
technology production, 127
wind, 135, 148–49
see also fossil fuels; micropower; pollution;
    renewable energy; *specific energy sources*
environment
acid rain effects, 28, 33–36
carbon dioxide effects, 5, 29, 32
climate change effects, 23, 29, 36, 132–33
ecological globalization, 184–202
erosion, 23
forest degradation, 8, 22–29, 185, 196
green investment, 8–10, 194–200
job creation, 4, 171–76, 183
monitoring through technology, 131–36,
    140–41
ozone depletion, 36, 38, 97, 133
policy, 37–38, 189–202
taxes, 18–19, 178–80
see also conservation; ecosystems; pollution;
    sustainable development; water
Environmental Protection Agency (U.S.),
    85, 90, 95, 109, 129, 138
environmental surprise, 22–38
air pollution, 32–37, 93, 108
coral reef destruction, 29–32
deforestation, 8, 22–29, 185, 196
discontinuity of stable states, 24
management, 37–38
synergism, 24
unnoticed trends, 24
Ericsson, 128
erosion, 23
Estonia, 139
Estrada, Joseph, 96
Ethiopia, 7, 61–62
European Commission, 174

European Photovoltaic Industry Association,
    174
European Solar Industry Federation, 174
European Union, 128, 140, 192
    see also specific countries
exotic organisms, 24, 36, 188
extinction, 8, 24–26, 80, 85–88

farmland, *see* agriculture; cropland loss;
    cropland productivity
fertilizer
nutrient flow, 35
runoff, 30–31, 55–56
sewage sludge, 99
Finland, 77
First National Bank of Omaha, 151
fisheries
agricultural runoff effects, 30–31, 55–56
biodiversity, 29–32, 135, 189–91
catch, 8
climate change effects, 28–32, 132
coral reef destruction, 30–31, 55–56
decline threshold, 13
endangered species, 29–32, 135, 189–91
fishing gear effects, 189–91
global trade, 187–88
habitat destruction, 6, 29–32, 135
infectious disease epidemic, 31–32
invasive species, 188
irrigation impact, 46
overfishing, 24, 187
persistent organic pollutant effects, 88–89,
    92
regulation, 189–91
satellite monitoring, 135, 140
floods, 12, 24, 38
    see also dams
Florini, Ann, 138
food, 59–78
agricultural crop diversity, 38
biosafety, 192–93
cropland productivity, 7–8, 48–57
dietary health, 59–78
economy effects, 65

food *(continued)*
    fat consumption, 61, 63, 65–69, 71
    global management, 64–65, 74–78, 192
    grain, 14, 42, 65, 187
    junk, 68, 77–78
    livestock, 192
    malnutrition, 61–64, 69–74, 76
    organic farming, 99
    overeating, 59–74, 76
    see also agriculture; cropland productivity;
        fisheries
Food and Agriculture Organization (U.N.)
    cultivated wetlands, 54
    fisheries report, 187
    nutrition surveys, 61, 64
    paper consumption projections, 117–18
    pesticide evaluation, 91
Ford Motor Company, 147
foreign investment, 194
forest products
    international trade, 185–86, 196
    new products, 105
    paper consumption, 102–03, 105, 115–18
    production management, 112
Forest Products Lab, 117
Forest Stewardship Council, 113, 201
forests
    acid rain effects, 28, 33–36
    carbon dioxide effects, 28, 36
    climate change effects, 28, 36
    coercion, 196–97
    conversion taxes, 119
    deforestation, 8, 22–29, 185, 196
    fires, 26–28
    illegal hunting, 28–29
    management, 12, 106, 113, 137
    nitrogen imbalance, 35–36
    paper fiber production, 105–06, 112
    plantations, 106–07, 112
    subsidies, 119
    sustainable forest management, 113
fossil fuels
    coal, 17, 173, 182
    nitrogen oxide emissions, 33, 35

sulfur dioxide emissions, 33–35, 134
    see also carbon emissions; renewable energy
Framingham Heart Study, 71
France, 67, 131, 181
freshwater ecosystems
    Amazon River basin, 26–28, 196
    dam effects, 55, 145, 199
    floods, 12, 24, 38
    Murray-Darling River basin, 46
    wetlands, 24, 38, 54
    Yangtze River, 12–13, 24
Friends of the Earth, 97
fuel cells, 12, 147–51, 153, 157
furans, 81–82, 84–85

General Agreement on Tariffs and Trade,
    189–90
General Electric, 143, 148
General Motors, 3, 96
genetically modified organisms, 184, 192–93
Germany
    dioxin emissions, 84
    employment, 181
    environmental taxes, 19, 179
    Federal Environment Agency, 89
    jobs, 165
    packaging laws, 128
    solar power, 18
    wind energy, 149, 154
Ghana, 127, 139
GIS software company, 123, 131, 134–35
Giuliani, Rudy, 150
Global Biological Information Facility, 136
Global Coral Reef Alliance, 32
Global Environmental Service, 133
Global Environment Monitoring System, 135
Global Forest Watch program, 136
Global Water Policy Project, 6
Gore, Albert Jr., 137
Goreau, Thomas, 32
grain, 14, 42, 65, 187
    see also agriculture; cropland loss; cropland
        productivity
Grameen Shakti (Bangladesh), 160

Great Man-Made River Project (Libya), 43
Green Alliance (U.K.), 199
green investment, 8–10, 194–200
Green Revolution, 38
greenhouse gases, *see* carbon emissions;
    climate change
Greenpeace, 95

Haiti, 62
Harvard School of Public Health, 61, 71
Hawken, Paul, 118, 169
hazardous waste, 93–95
health
    cancer, 69, 71–74, 90
    diabetes, 69, 72–73
    dietary contributions, 59–78
    economic burden, 73–74
    fisheries epidemic, 31–32
    heart disease, 69, 71, 76
    malaria, 22–23, 83, 98–99
    new diseases, 188–89
    persistent organic pollutant effects, 87–90
    smoking effects, 11
    Yucheng, 89
    see also AIDS; World Health Organization
heart disease, 69, 71, 76
Helwig, David, 150
Heritage Forest Campaign, 137
High Plains Underground Water
    Conservation District, 50
Honduras, 22–23
Horowitz, Michael, 55
Horton, Susan, 73, 75
Hubbard Brook Experimental Forest, 34
hunger, 61–64, 69–74, 76
hydrogen fuel, 12, 17, 147–53, 157

Iannucci, Joseph, 158
IBM, 3, 128
Iceland, 12
IKEA, 96
Imperial Irrigation District, 46
India
    aquifer depletion, 6–7, 41–42

Central Pollution Control Board, 97
    check dams, 55
    DDT use, 83
    drip irrigation, 50
    fisheries, 191
    grain production, 41
    malnutrition, 61, 70, 76
    persistent organic pollutants, 85–86, 97
    water scarcity, 45
    wind energy, 173
indigenous people, 107
Indonesia
    economic crisis, 195–97
    foreign investment, 194
    foreign trade, 185
    forest plantations, 107
    forest subsidies, 119
    integrated pest management, 99
    malnutrition, 61, 72
    water scarcity, 45
industrialization, 162–66
infectious disease
    fisheries epidemic, 31–32
    malaria, 22–23, 83, 98–99
    new diseases, 188–89
    see also AIDS
information technologies, 121–41
    cellular telephones, 122–27, 140
    digitization, 124–25
    e-commerce, 130–31, 139
    environmental monitoring, 131–36,
        140–41
    fixed-line telephones, 124–25
    global network, 122–27
    historical perspectives, 123–24
    Internet, 121–27, 138–41
    jobs, 163–64
    natural resources role, 127–31
    networking, 136–40
    radio, 122, 124
    satellites, 123–26, 129, 131–36
    telecommunications markets, 126–27
    television, 122–24, 130–31, 141
Institute for Global Communications, 138

Institute of Ocean Sciences, Fisheries, and
    Oceans (Canada), 86
Intel, 3, 128, 130
Intelsat, 126
Inter-American Development Bank, 160
International Campaign for Responsible
    Technology, 128
International Conference on Population and
    Development (Cairo), 20
International Council of Chemical
    Associations, 93
International Development Enterprises, 53
International Diabetes Federation, 72
International Federation of Organic
    Agriculture Movements, 201
International Finance Corporation, 198
International Food Policy Research Institute,
    44, 75
International Forum on Chemical Safety, 92
International Institute for Communications
    and Development (Netherlands), 140
International Joint Commission on the
    Great Lakes, 92
International Labour Organisation, 167
International Legally Binding Instrument
    for Implementing International Action
    on Certain Persistent Organic
    Pollutants, 92
International Monetary Fund, 64, 185,
    194–98, 200
International Organization for
    Standardization, 201
International Water Management Institute,
    41, 50
International Whaling Commission, 200
Internet, 121–27, 138–41
Iridium, 126
irrigation, 39–58
    agriculture productivity, 40–41
    aquifer depletion, 6–7, 13, 41–44, 49
    Broadview Water District, 57
    costs, 57
    cropland productivity enhancement, 48–57
    dams, 55, 145, 199
    drip irrigation, 48–51
    efficiency strategies, 48–52
    fisheries impact, 46
    global irrigated area, 40–41
    historical perspectives, 39–41
    low-energy precision application sprinklers,
        51
    non-farm diversion, 46
    runoff systems, 30–31, 55–56
    small-scale irrigation, 52–56
    subsidies, 56
    treadle pumps, 52–53
    water conflicts, 44–48
    water management, 56–58
    water markets, 57–58

Japan
    acid rain effects, 34
    dioxin emissions, 84
    grain imports, 187
    organic farming, 99
    paper production, 107–08, 116–18
    solar power, 18, 149, 154
    unemployment, 166
jobs, 162–83
    creation, 172–76, 183
    durability enhancement, 176–78
    environment policy, 171–72, 183
    industrialization, 162–66
    information technologies, 163–64
    productivity, 167–71
    tax shifts, 178–80
    unemployment, 164–67
    wind generation, 173–74
    work force, 164–67
    work reform, 180–83

Kammen, Daniel, 158–59
Kazakhstan, 4
Kellogg's, 68
Kenya, 15, 54, 160
Kuhn, Thomas, 9, 13
Kurnia, Ganjar, 45
Kyoto Protocol, 172, 176

labor, *see* jobs
land, *see* cropland loss; cropland productivity
Lego, 96
Lehman Brothers, 199
Lehmann, Andre, 151
Libya, 43
livestock, 192
logging, *see* forest products; forests
Lopez, Alan, 70
Lovins, Amory, 118, 151, 169
Lovins, Hunter, 118, 169

Macdonald, Robie, 86
Malaysia, 28, 185, 191, 194
Mali, 55
malnutrition, 61–64, 69–74, 76
MAN, 144
Mankiw, N. Gregory, 19
manufacturing
    durability enhancement, 176–78
    productivity enhancement, 167–71
    recycle economy, 127–28
    see also technology
Marine Mammal Protection Act (U.S.),
        189–90
Marine Stewardship Council, 201–02
Mason, John, 70
McDonald's, 67–68, 113, 117, 192
McKibben, Bill, 141
Metropolitan Water District, 46
Mexico
    fisheries, 189–90
    food scarcity, 65
    free trade zones, 166
    irrigation reforms, 56–57
    telecommunications networks, 127
Microelectronics and Computer Technology
        Corporation, 127
micropower, 142–61
    barriers to use, 155–57
    benefits, 152–53, 160
    distribution, 149–53, 155, 160
    energy reform, 153–57
    fuel cells, 147–49, 151, 153, 157

microgenerators, 142–46
miniaturized machines, 143–47
Microsoft, 3
mining, 186
Montreal Protocol on Substances That
        Deplete the Ozone Layer, 38, 97, 200
Morgan Stanley, 199
Motorola, 128
Müller, Paul, 81, 83
Multilateral Investment Guarantee Agency,
        198
Myers, Norman, 24
Myneni, Ranga, 132

Naomi Berrie Diabetes Center, 72
National Environmental Engineering
        Research Institute, 41
natural disaster, *see* environmental surprise;
        specific types
natural resources
    government subsidies, 56, 119, 149, 160
    mining, 186
    sustainable use, 113
    see also biodiversity; endangered species;
        fisheries; forests; fossil fuels; recycling;
        renewable energy; water
Nautilus Institute, 160
Netherlands
    electronics take-back plans, 128
    employment, 181
    environmental taxes, 119
    overeating, 73
    persistent organic pollutants, 84, 89
New York Public Interest Research Group,
        134
Niger, 62
Nigeria, 7
Nike, 96
nitrogen cycle, 35–36
nitrogen oxide emissions, 33, 35
Nokia, 128
nongovernmental organizations
    global governance role, 185, 199, 202
    networking, 136–39, 202

nongovernmental organizations *(continued)*
  persistent organic pollutants policy, 92
North American Free Trade Agreement, 65
North Korea, 160
Northwest Environment Watch, 182
nuclear energy, 17
nutrient cycling, 35–36
nutrition, *see* food

oceans
  coral reef destruction, 6, 29–32, 135
  persistent organic pollutant effects, 88–89,
    92
  sea level rise, 6, 32
  sea surface temperatures, 6, 29–30, 32
  see also fisheries
Oil, Chemical, and Atomic Workers'
    International Union, 96
Oldways Preservation & Exchange Trust,
    61, 78
OrbImage, 133
Organisation for Economic Co-operation
    and Development, 136
Overseas Private Investment Corporation,
    198
ozone depletion
  environmental effects, 36
  monitoring, 133
  Montreal Protocol, 38, 97, 200

Pakistan, 7, 137–38, 191
Pan American Health Organization, 138
Panasonic, 128
paper, 101–20
  consumption, 102–05, 115–18, 129
  costs, 104–09, 115–17
  efficient use, 115–17
  fiber sources, 105–07, 110–12
  historical perspectives, 101–04
  pollution, 108, 114
  production, 101–04, 107–15, 169
  recycling, 105, 109–19
Papua New Guinea, 75
Patterson, Walt, 154

PCBs, 79–91, 94, 98
persistent organic pollutants, 79–100
  chemical nature, 80–85
  environmental fate, 85–87
  exposure routes, 85–87
  health effects, 87–90
  policy, 90–95
  production, 80–82
  regulation strategies, 95–100, 120
Peru, 140
Pesticide Action Network, 202
pesticides
  banned chemicals, 87, 91–93, 98
  consumer exposure, 22–23, 87–90
  DDT, 80–83, 86–91, 98–99
  exports, 87
  integrated pest management, 99, 113
  pest resistance, 38, 83
  taxes, 97
  world use, 83, 91
Pew Center on Global Climate Change, 155
Philippines, 4, 72, 96, 135, 194
Physicians Committee for Responsible
    Medicine, 73
Pilkington Solar International, 18
Plug Power, 148
Polak, Paul, 53
Polaroid, 161
policy
  climate, 132, 136, 172, 175–76
  ecological globalization, 37–38, 189–202
  economic reform, 194–200
  energy reform, 153–57
  environment versus jobs, 171–72
  fisheries management, 189–91
  global food management, 64–65, 74–78,
    192
  nongovernmental organization role, 185,
    199, 202
  persistent organic pollutants, 90–95
  water management, 6, 56–58
pollution, 79–100
  acid rain, 28, 33–36
  agricultural runoff, 30–31, 55–56

air, 32–37, 93, 108
chlorine, 81, 109, 114
dioxins, 81–86, 97–98, 109, 120
energy production, 151
environmental taxes, 18–19, 178–80
furans, 81–82, 84–85
hazardous waste, 93–95
incineration, 84–85, 96–97
manufacturing, 108, 114, 121–22
mining, 186
persistent organic pollutants, 79–100
polychlorinated biphenyls, 79–91, 94, 98
polyvinyl chloride, 81–82, 97
species extinction, 80, 85–88
toxic waste, 95, 121–22, 127–28
water, 30–31, 88–89, 108–09
see also carbon emissions; pesticides
Pollutant Release and Transfer Registers,
    95
Pollution Prevention and Abatement
    Handbook, 198
population growth
  job demand, 167
  projections, 5, 7, 15–16
  stabilization, 19–20
Population Services International, 136
Postel, Sandra, 6, 39
poverty, 64
Procter and Gamble, 116
Program for Monitoring Emerging Diseases,
    138
Punyarachun, Anand, 20
PVCs, 81, 97

radio, 122, 124
Rainforest Action Network, 137
RAND Corporation, 155
recycling
  computers, 127–28
  landfills, 119
  paper, 105, 109–19
  persistent organic pollutants, 98
  sewage sludge, 99
Rejeski, David, 140

renewable energy
  hydrogen fuel, 12, 147–51, 153, 157
  photovoltaic cells, 149, 159, 174
  tax incentives, 149, 160
  wind power, 135, 148–49
  see also solar energy
rice, *see* grain
Ringler, Claudia, 44–45
river systems, *see* dams; floods; freshwater
    ecosystems; water
Rocky Mountain Institute, 151–52, 169
Rodrik, Dani, 195
Rongji, Zhu, 12
Rosegrant, Mark, 44–45
Royal Dutch Shell, 12, 18, 153
Royal Institute for International Affairs, 154
Russia, 36, 83, 166

Sale, Kirkpatrick, 162
satellites, 123–26, 129, 131–36
Saudi Arabia, 43
Schumacher, E.F., 142
Schumpeter, Joseph, 164
Seckler, David, 41
Sen, Amartya, 64
Senegal, 188
sewage treatment, *see* waste management
Sharp, 130
Shell Oil Company, 12
Siemens, 128, 148
Silicon Valley Toxics Coalition, 121–22, 128
Singapore, 76
Sivanappan, R. K., 50
Slow Food Movement, 78
Smith Barney, 199
smoking, 11
soil
  acidification, 28, 33–36
  erosion, 23
  nitrogen cycle, 35–36
  see also cropland loss; cropland productivity
solar energy
  annual growth rate, 17
  job creation, 174

solar energy *(continued)*
  photovoltaic cells, 149, 159, 174
  sell-back programs, 154
  small scale power generation, 140, 149,
      154, 159–60
  Solar Electric Light Fund, 140
  Solar Energy Industries Association, 174
  Solar Turbines, 146
  solar/hydrogen economy, 9, 17
  subsidies, 149, 160
Somalia, 62
Sony, 128
South Africa, 127
South Korea, 76, 194
Soviet Union (former), 166
Space Imaging, 133
Spain, 72, 149, 154
Ssewankambo, Godfrey, 16
Starrs, Thomas, 156
Stuart, Rob, 137
subsidies
  agriculture, 56
  forest use, 119
  irrigation, 56
  solar energy, 149, 160
  water use, 56
Sudan, 59
sulfur dioxide emissions, 33–35, 134
Superfund sites, 127
sustainable development
  decline threshold, 13–16
  forests, 113
  natural resources, 113
  networking, 136–40
  world commitment, 10–13
Sustainable Development Networking
      Programme, 138
Sweden, 84, 109, 128, 181

Taiwan, 76, 89, 128
taxes
  environmental, 18–19, 178–80
  forest conversion, 119
  landfills, 119

pesticides, 97
technology
  biotechnology, 184, 192–93
  cellular telephones, 122–27, 140
  digitization, 124–25
  drip irrigation, 48–51
  energy use, 129–30
  environmental monitoring, 131–36,
      140–41
  fixed-line telephones, 124–25
  fuel cells, 147–49, 151, 153, 157
  historical perspectives, 3–4, 101, 123–24
  Internet, 121–27, 138–41
  microgenerators, 142–46
  paper production, 114
  pedal pump, 54
  radio, 122, 124
  recycling, 127–29
  satellites, 123–26, 129, 131–36
  small-scale irrigation, 52–56
  Stirling engine, 146–47, 158
  television, 122–24, 130–31, 141
  treadle pumps, 52–53
  weather monitoring, 132–33, 136
  see also computers; information technologies
Technology Project, 137
Teledesic, 126
telephones, 122–27, 140
television, 122–24, 130–31, 141
Thailand, 20, 85–86, 191, 199
Third World, *see* developing countries
3M, 129
tobacco, 11
Töpfer, Klaus, 92
toxic waste, 95, 121–22, 127–28
Toxics Release Inventory (U.S.), 95, 138
Toyota Motor Company, 147
trade, *see* economy
Trigen Energy Corporation, 156
tropical storms, 23

Uganda, 20
unemployment, *see* jobs
UNESCO, 140

UNICEF, 63, 75
Union of International Organizations, 137
United Kingdom, 62, 97, 128, 179, 181
United Nations
  Conference on Environment and
    Development (Rio), 92, 95, 138, 192
  Conference on Population and
    Development (Cairo), 20
  Convention on Biological Diversity, 193
  Development Programme, 138
  Environment Programme, 84, 130, 199,
    201
  Framework Convention on Climate
    Change, 172, 176
  Global Environment Monitoring System,
    135
  International Maritime Organization, 200
  Sustainable Development Networking
    Programme, 138
  World Environment Program, 201
  World Meteorological Organization, 132
  see also Food and Agriculture Organization
United Parcel Service, 115
United States
  Agency for International Development,
    140
  air pollution effects, 34, 36
  aquifer depletion, 6, 42–43, 49
  beef hormone dispute, 192
  biosafety, 192–93
  Broadview Water District, 57
  carbon emissions, 34, 36
  Department of Agriculture, 62, 68, 114,
    117, 137
  dioxin emissions, 84
  Emergency Planning and Community
    Right-to-Know Act, 95, 138
  employment, 166, 181
  Endangered Species Act, 191
  energy consumption, 144, 149–51, 157
  environmental policy, 38, 189–94, 198
  Environmental Protection Agency, 85, 90,
    95, 109, 129, 138
  Export-Import Bank, 198

forest management, 12, 107
Forest Service, 12, 107
global trade, 189–94
Internet access, 125
invasive species, 24
irrigation, 44–46, 49–51
job creation, 4, 172–73
junk foods, 68, 77–78
landfills, 119
malnutrition, 62
National Aeronautics and Space
  Administration, 132–33
National Oceanographic and Atmospheric
  Administration, 132
National Renewable Energy Laboratory,
  134–35, 159
National Safety Council, 127
obesity, 59, 62–63, 67, 72, 76
Overseas Private Investment Corporation,
  198
paper consumption, 102–03
paper production, 107–08, 110–11
persistent organic pollutants, 80–92, 96
population growth, 19
Postal Service, 115, 117
recycling, 110–11
Securities and Exchange Commission,
  199
solar power, 149, 154, 174
space command, 129
Superfund sites, 127
unemployment, 166
water conflicts, 44–46
water markets, 57–58
wind energy, 17–18

Viet Nam, 61, 85–86
Volkswagen, 147, 181

waste management
  hazardous materials, 93–95
  landfills, 119
  toxic waste, 95, 121–22, 127–28
  see also recycling

water
    aquifer depletion, 6–7, 13, 41–44, 49
    conflicts, 44–48
    conservation, 50, 56–58
    dams, 55, 145, 199
    floods, 12, 24, 38
    irrigation, 39–58
    policy, 56–58
    pollution, 30–31, 88–89, 108–09
    runoff levels, 30–31, 55–56
    scarcity, 14, 41–45, 61–62
    subsidies, 56
    wetlands, 24, 38, 54
    see also freshwater ecosystems
Waukesha, 145
weather, *see* climate change
Wenger, Howard, 156
wetlands
    cultivation, 54
    flood control, 24, 38
    see also floods; freshwater ecosystems
wheat, *see* grain
wind energy, 17, 135, 148–49, 173–74
Women's Environment and Development
    Organization, 202
work force, *see* jobs
World Bank
    crisis prevention, 195–96
    disease study, 70
    energy projects, 159–60
    environmental lending, 195–98
    medical waste incinerator promotion, 84

micro-loans, 160
nutrition study, 72
Pollution Prevention and Abatement
    Handbook, 198
World Cancer Research Fund, 71, 74
World Commission on Forests and
    Sustainable Development, 106
World Conservation Union–IUCN, 137
World Forum of Fish Workers & Fish
    Harvesters, 202
World Health Organization
    beef-hormone ban, 192
    DDT study, 83, 88, 98–99
    environmental monitoring, 135, 189–94
    fisheries policy, 190–91
    infectious disease study, 188–89
    nutrition study, 59, 62, 70, 75
    pesticide toxicity effects, 88, 92
World Meteorological Organization, 132
World Resources Institute, 129–30, 136,
    186
World Trade Organization, 65, 119, 126,
    200–01
World Weather Watch, 132, 136
World Wide Web, 121–27, 138–41
World Wildlife Fund, 135
World-Space, 126

Xerox Corporation, 129

Zaire, 16, 29
Zimbabwe, 15–16, 53–54, 86

Now you can import all the tables and graphs from *State of the World 2000* and all other Worldwatch publications into your spreadsheet program, presentation software, and word processor with the ...

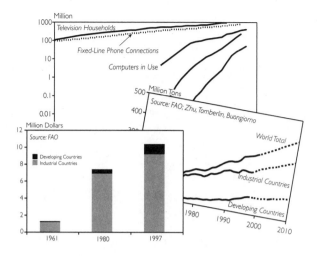

2000 Worldwatch Database Disk

The Worldwatch Database Disk Subscription gives you current data from all Worldwatch publications, including the *State of the World* and *Vital Signs* annual book series, *World Watch* magazine, Worldwatch Papers, and other Worldwatch books.

The disk covers trends from mid-century onward . . . much not readily available from other sources. All data are sourced, and are accurate, comprehensive, and up-to-date. Researchers, students, professors, reporters, and policy analysts use the disk to—

- ◆ *Design graphs to illustrate newspaper stories and policy reports*
- ◆ *Prepare overhead projections on trends for policy briefings, board meetings, and corporate presentations*
- ◆ *Create specific "what if?" scenarios for energy, population, or grain supply*
- ◆ *Overlay one trend on another, to see how they relate*
- ◆ *Track long-term trends and discern new ones*

Order the 2000 Worldwatch Database Disk for just $89 plus $4 shipping and handling. To order by credit card (Mastercard, Visa, or American Express), call (800) 555-2028 or fax to (202) 296-7365. Our e-mail address is <wwpub@worldwatch.org>. You can also order from our Web site at <www.worldwatch.org>, or by sending your check or credit card information to: Worldwatch Institute, 1776 Massachusetts Ave., NW, Washington, DC 20036.

WORLDWATCH INSTITUTE
1776 Massachusetts Ave., NW
Washington, DC 20036
www.worldwatch.org